Constantin Paul

Diagnosis and treatment of diseases of the heart

Constantin Paul
Diagnosis and treatment of diseases of the heart
ISBN/EAN: 9783744737180
Printed in Europe, USA, Canada, Australia, Japan
Cover: Foto ©berggeist007 / pixelio.de

More available books at **www.hansebooks.com**

DIAGNOSIS AND TREATMENT

OF

DISEASES OF THE HEART

BY

CONSTANTIN PAUL,

MEMBER OF THE ACADEMY OF MEDICINE; PHYSICIAN TO THE LARIBOISIÈRE HOSPITAL

TRANSLATED FROM THE FRENCH.

NEW YORK
WILLIAM WOOD & COMPANY
56 & 58 LAFAYETTE PLACE
1884

TABLE OF CONTENTS.

GENERAL CONSIDERATIONS ON THE TOPOGRAPHY OF THE HEART.

CHAPTER I.

	PAGE
GENERAL CARDIAC TOPOGRAPHY OF THE HEART...........................	1

CHAPTER II.

CLINICAL EXAMINATION OF THE NORMAL HEART	8
Diagnosis of the shape and position of the heart................	8
Clinical measurement of the heart .:......................................	8

CHAPTER III.

DIAGNOSIS OF DISPLACEMENTS OF THE HEART	16

CHAPTER IV.

CLINICAL EXPLORATION OF THE MOVEMENTS OF THE HEART, AND THE CARDIAC SOUNDS	22

CHAPTER V.

EXAMINATION OF THE MOVEMENTS OF THE HEART BY AUTOMATIC APPARATUS.	26

CHAPTER VI.

THE MUSCLES AND NERVES OF THE HEART—THE INFLUENCE OF THE NERVOUS SYSTEM ON THE CARDIAC RHYTHM.............................	30

CHAPTER VII.

VARIATIONS IN THE PULSE—PALPITATIONS—LIPOTHYMIA—SYNCOPE..........	41

CHAPTER VIII.

ANÆMIA	47

DISEASES OF THE HEART AND ITS MEMBRANES.

CHAPTER IX.
The Pericardium and Pericarditis.. 59

CHAPTER X.
Complications of Pericarditis—Various Forms of the Disease—Pericardial Adhesions—Course—Termination............................ 68

CHAPTER XI.
Hydropericardium.. 74

CHAPTER XII.
The Endocardium and Endocarditis... 77

CHAPTER XIII.
Ulcerative Endocarditis... 87

CHAPTER XIV.
Chronic or Sclerotic Endocarditis... 91

CHAPTER XV.
Lesions of the Mitral Valve and Orifice—Stenosis—Insufficiency.... 98

CHAPTER XVI.
Résumé of the Signs of Mitral Induration—Course and Prognosis of this Affection.. 112

CHAPTER XVII.
Aortic Insufficiency, or Corrigan's Disease................................ 118

CHAPTER XVIII.
Sclerotic Endocarditis of the Aortic Semilunar Valves............... 129
   Stenosis, or narrowing of the aortic orifice............................ 129

CHAPTER XIX.
Diseases of the Origin of the Aorta—Aortitis—Atheroma.......... 134
   Aneurismal dilatation, or true aneurism—Hodgson's disease...... 134

CHAPTER XX.
Symptoms and Diagnosis of Hodgson's Disease, or Fatty Degeneration Producing True Aneurism of the Origin of the Aorta........ 140

CHAPTER XXI.

Continuation of the Study of Hodgson's Disease 147
    Compression of adjacent organs—course—termination—etiology 147

CHAPTER XXII.

Lesion of the Descending Portion of the Thoracic Aorta........... 151

CHAPTER XXIII.

Affections of the Right Heart 154

CHAPTER XXIV.

Obliteration of the Superior Vena Cava............................. 167

CHAPTER XXV.

Diseases of the Pulmonary Artery 172

CHAPTER XXVI.

Diseases of the Pulmonary Artery (continued)...................... 180
    Insufficiency of the valves of the pulmonary orifice 180

CHAPTER XXVII.

Diseases of the Myocardium.. 187

CHAPTER XXVIII.

Partial Aneurisms of the Heart...................................... 196

CHAPTER XXIX.

Evolution of Diseases of the Heart.................................. 200

CHAPTER XXX.

Hypertrophy of the Heart... 210

CHAPTER XXXI.

Hypertrophy of the Heart (continued) 217

CHAPTER XXXII.

The Causes of Cardiac Hypertrophy 222
    Cardiac lesions... 222
    Changes in the aorta and arteries.. 225
    Pregnancy... 228

CONTENTS.

CHAPTER XXXIII.

                                                                                                                      PAGE

HYPERTROPHY AND DILATATION SECONDARY TO DISEASES OF OTHER ORGANS, 233
    Lesions of the kidneys and urinary passages... 233
    Diseases of the respiratory organs 237
    Hypertrophy of gastro-hepatic origin 241

CHAPTER XXXIV.

THE SEQUELÆ OF HYPERTROPHY... 244
    Effect on menstruation, gestation, and delivery........................... 244

CHAPTER XXXV.

SECONDARY AFFECTIONS OF THE RESPIRATORY PASSAGES.................. 246

CHAPTER XXXVI.

SECONDARY AFFECTIONS OF THE LIVER AND DIGESTIVE TRACT 251

CHAPTER XXXVII.

SECONDARY AFFECTIONS OF THE KIDNEYS................................. 254

CHAPTER XXXVIII.

SECONDARY AFFECTIONS OF THE NERVOUS SYSTEM 256

CHAPTER XXXIX.

BASEDOW'S DISEASE (ANEURISMAL BRONCHOCELE)......................... 260

CHAPTER XL.

NEW GROWTHS AND PARASITES.. 263
    Clots and polyps 265

CHAPTER XLI.

ANEURISMAL TUMORS OF THE THORACIC AORTA (FALSE ANEURISMS)........ 268
    Aneurisms of the pulmonary artery 280
    Aneurisms of the coronary artery 280
    DISEASES OF THE HEART IN THE FŒTUS 281
    Congenital diseases of the heart.. 281
    Isolated persistence of the foramen ovale 282
    Persistence of the ductus arteriosus 283
    Aneurism of the ductus arteriosus...................................... 285
    Congenital affections of the right heart 285
    Simple stenosis of the pulmonary artery, with closure of the foramen ovale.. 285
    Stenosis of the pulmonary artery, with persistence of the foramen ovale..... 286
    Stenosis of the infundibulum of the pulmonary artery, with communication
        between both ventricles.. 288

CONTENTS.   vii

	PAGE
Stenosis or obliteration of the pulmonary artery, with communication between both ventricles	288
Endocarditis of the right heart, with insufficiency of the tricuspid valve	292
Stenosis of the aortic orifice	293
Stenosis of the mitral orifice	294
Transposition of the arterial trunks	294
Transposition of the venous trunks	294

TREATMENT.

CHAPTER XLII.

THE CHIEF DRUGS WHICH ACT UPON THE CARDIAC MOVEMENTS 295
Moderators of the Heart 295

CHAPTER XLIII.

TREATMENT OF PERICARDITIS AND ENDOCARDITIS 302

CHAPTER XLIV.

HYGIENE OF PATIENTS SUFFERING FROM HEART DISEASE 307

CHAPTER XLV.

TREATMENT OF THE CLOSE OF THE PERIOD OF TOLERANCE 311

CHAPTER XLVI.

TREATMENT OF CARDIAC DISTRESS AND ASYSTOLE 314

CHAPTER XLVII.

TREATMENT OF HYPERTROPHY 319

CHAPTER XLVIII.

TREATMENT OF SECONDARY AFFECTIONS OF THE DIGESTIVE ORGANS, AND OF RENAL AFFECTIONS AND DROPSY 323

CHAPTER XLIX.

TREATMENT OF SECONDARY AFFECTIONS OF THE NERVOUS SYSTEM AND EXOPHTHALMIC GOITRE 327
Treatment of aortic aneurisms 328

INDEX 331

DIAGNOSIS AND TREATMENT

OF

DISEASES OF THE HEART.

GENERAL CONSIDERATIONS ON THE TOPOGRAPHY OF THE HEART.

CHAPTER I.

GENERAL CARDIAC TOPOGRAPHY.

THE cardiac region, including the heart, the origin of the large vessels, and the serous pericardium, is enclosed completely by the fibrous sac of the pericardium. The latter forms a pyramidal box with three surfaces, and possesses the most complete fixation possible. Next to the pelvis the pericardial sac is the least movable region in the human body.

This triangular pyramid is fixed by its base to the aponeurotic centre of the diaphragm and indirectly to the vertebral column by the intervention of tendinous bands.

The apex of this pyramid is also fixed to the vertebral column, the hyoid bone, and the sternum, by very firm aponeurotic fibres. As all these ligaments have very slight extensibility, the fibrous sac forms as fixed a region as possible. This stability and protection are necessary for so important an organ as the heart, since it is the only one which is never at rest.

Situation.—Let us first determine the exact position of the pericardial sac. This is situated in the median line of the thorax (the whole extent of which it occupies) between the two lungs and pleuræ.

Posteriorly it is bounded by the posterior mediastinum, *i.e.*, it is separated from the spine by the connective tissue enclosing the trachea, descending aorta, intercostal arteries and veins, vena azygos major, thoracic duct, lymphatic glands, œsophagus, the two pneumogastrics and both sympathetic nerves.

Anteriorly the pericardial sac is separated from the sternum by connective tissue and the anterior border of the pleuræ.

Shape.—The pericardial sac has the exact shape of a triangular pyramid, the apex of which is situated upon the median line, and corresponds to the first piece of the sternum, at a distance of 15 to 20 mm. from the sternal notch. The base of the pericardial sac is triangular, and is attached very firmly to the fibrous centre of the diaphragm by the intercrossing of its fibres with those of the diaphragmatic aponeurosis. The apex of this triangle, situated posteriorly, corresponds to the inferior vena cava, behind which it is situated. I may here remark that the inferior vena cava is absolutely rectilinear and vertical, and does not form any angle in entering the right auricle. The base of the triangle is situated anteriorly behind the sternum, extending on the right 3 ctm. beyond the median line, and on the left 8 to 10 ctm., according to the individual. The right border of the base of the pericardium passes directly from before backward, almost parallel to the median plane of the body; the left border is oblique and passes from the right side of the spinal column behind the vena cava to the apex of the heart. Finally, the base of the pericardium forms a flat, almost horizontal surface, slightly oblique from behind forward and from right to left. On the average the right apex of this triangle is more elevated than the left by 2 ctm.

Means of Fixation of the Aponeurotic Sac of the Pericardium.—The apex of the pericardial sac is fixed to the skeleton by three important ligaments:

1. It is fixed to the vertebral column by Beraud's [1] ligament, or the suspensory ligament of the pericardium. This first superior ligament is situated a little in front of and to the left of the aorta. It arises from the pericardium at the level of the arches, and is inserted into the middle and left lateral portion of the third cervical vertebra, coalescing with the intervertebral ligament. It passes forward and to the left, is 3 ctm. in height and 2 ctm. in width. The inner or right border corresponds to the origin of the left subclavian artery, its outer or left border is covered by the left pleura.

2. The second superior ligament is the cervico-pericardial pseudo-aponeurosis of Richet, which coalesces with the middle aponeurosis of the neck and is inserted into the hyoid bone.[2]

The third superior ligament passes anteriorly, and was first described by Luschka under the name sterno-pericardial ligament,[3] then by Lannelongue and Le Dentu, under the term costo-pericardial ligament.[4]

It is thus evident that the pericardium is fixed solidly to the skeleton,

---

[1] Gaz. Médicale, 1862, p. 162. [2] Richet: Traité d'anatomie medico-chirurgicale.
[3] Luschka: Anatomie des Menschen, i. Tuebingen, 1863.
[4] Lannelongue et Le Dentu: Archives de Physiologie, i., p. 448. 1868.

and that it is upon these ligaments that the vessels passing to and from the heart are supported.

The base of the aponeurotic sac is fixed to the centre of the diaphragm, beyond which it extends about 3 ctm. anteriorly toward the left in the normal condition, and often much more in cases of hypertrophy. Through the medium of the diaphragm the aponeurotic sac is fixed posteriorly to the spinal column, to the first three lumbar vertebræ by the right crus, to the second and third lumbar vertebræ by the left crus. Anteriorly the fibres are adherent to the posterior part of the xiphoid appendix.

It is important here to correct a physiological error. It is generally stated that the diaphragm is lowered during inspiration. This is a mistake. If the diaphragm were lowered during each inspiration, the heart would also descend, just as the thyroid body follows the movements of the trachea; but this is not so. I will go even further: not alone is the diaphragm not lowered at the level of the central tendon, but it is not even lowered at the ribs, otherwise the apex of the heart, especially when the organ is hypertrophied, would be depressed during each inspiration.

This is also proved by other phenomena. In inspiration the abdominal pressure produced by the contraction of the diaphragm is not produced vertically toward the hypogastric region, but on the contrary, in the epigastric region. If the resultant of the diaphragmatic pressure is directed anteriorly toward the epigastrium, it is because the fibres of the diaphragm in shortening draw the costal insertions toward the centre by elevating them slightly. As Sappey[1] has shown (vide Fig. 1), the external intercostal muscles enlarge the thoracic cavity by raising the ribs.

If we wish to assure ourselves of this fact, it is sufficient to observe a rachitic child whose upper ribs are dilated and the lower ones driven in by the enlargement of the stomach, so that the chest appears strangulated by the diaphragmatic insertions, and the trunk assumes the shape of a violin.

Finally, it must be remembered that the external intercostal muscles are inspiratory organs, and the internal intercostal muscles expiratory organs. The obliquity of the fibres of the former is such that they act as if fixed to the cervical spine, and that of the fibres of the latter (which is in the opposite direction) as if fixed to the lumbar spine.

It follows from all these anatomical facts, as I have stated before, that the aponeurotic sac of the pericardium is fixed firmly and directly to the skeleton. It is not lowered during inspiration, and therefore, in the normal condition, the patient may be examined either lying down or sitting. Slight displacements may be observed in pathological cases in which the weight of the heart is considerably increased and the resistance of the ligaments has been overcome gradually. On the other hand, the heart is displaced transversely during left decubitus, and even during right decubitus.

---

[1] Sappey : Traité d'anatomie descriptive, 3d édit., t. ii., p. 257. 1876.

General Relations of the Heart to the Anterior Wall of the Thorax.—
In order to obtain an exact idea of the shape and relations of the heart, it should be examined *in situ* in the thoracic cavity. An opening must be made in the anterior wall of the chest, care being taken not to enter the abdominal cavity. A transverse section of the sternum is made at the insertion of the right and left fifth cartilages, then the costal cartilages divided at their outer extremity at the level of their union with the ribs. The sternum is then divided a second time below the attachment of the

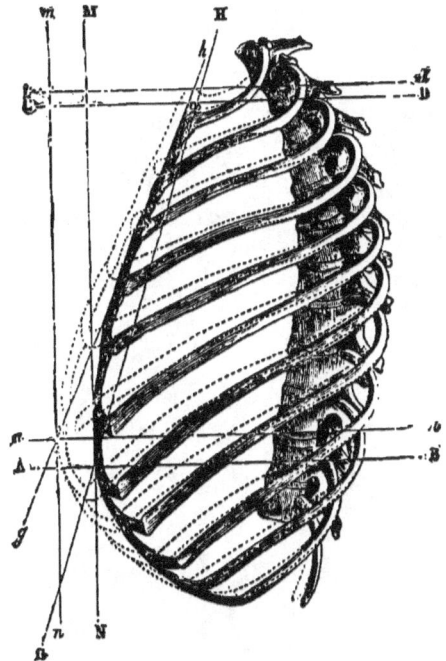

FIG. 1.—Reprinted from Béclard's Physiology.

cartilages of the first ribs. When this is raised we are confronted by the pericardium and the two pleuræ, separated by connective tissue.

When the pericardium is opened the heart is found lying transversely upon the diaphragm, while almost all anatomical plates since the time of Bourgery and Jacob represent it with the apex below and the base above. Sénac had already made this observation in the following terms : "According to the description of these anatomists, or rather, according to that which may be observed by the most careless eyes, the heart rests upon the diaphragm. This transverse muscle forms for it a sort of floor."[1]

---

[1] Sénac: Traité de la structure du cœur, de son action et de ses maladies, t. 1er, p. 185. 1783.

For the past ten years I have also followed the example of Sénac in describing the heart as shaped like a triangular pyramid. This description,

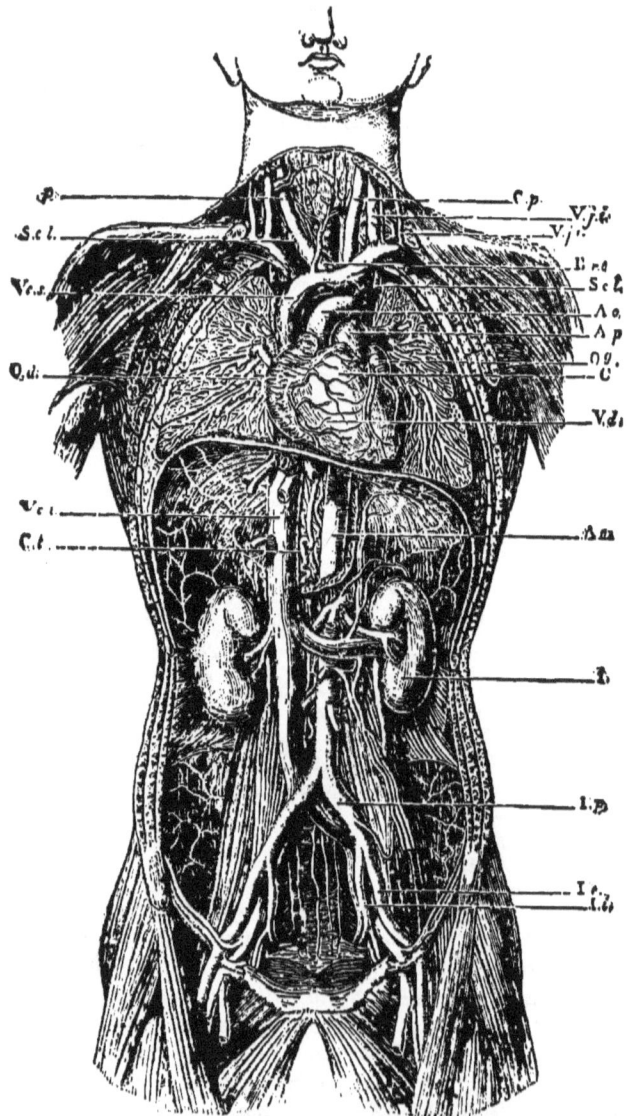

FIG. 2.—Reprinted from Cruveilhier's Anatomy.

which approaches the truth much more closely than others, furnishes the greatest facilities in giving an exact account of the relations of the organ.[1]

---

[1] Vide my paper entitled, Du rétrécissement des orifices de l'artère pulmonaire, read before the Société Médicale des Hôpitaux, August 11, 1871.

The heart, then, regarded as a pyramid, presents three surfaces and three borders:

1. An anterior vertical surface corresponding to the posterior wall of the sternum and the costal cartilages; this surface is triangular. The apex of the triangle is formed by the apex of the heart, *i.e.*, by the tip of the left ventricle. The base is formed by the vertical border (convex toward the right) of the right auricle, which is elongated vertically like a spindle, receiving directly the superior and inferior venæ cavæ.

The superior vena cava descends vertically and enters the right auricle at the level of the second intercostal space, at 1½ ctm. from the right border of the sternum, or better still at 3 ctm. from the median line. The inferior angle is right or nearly right; it is formed by the inferior vena cava, which is situated along the right border of the spinal column, at a height corresponding to the insertion of the right fifth cartilage into the sternum.

We must here correct an error made by almost all anatomists, who regard the heart as presenting only two surfaces and describe it as suspended by its base from the large vessels. They teach that the vena cava, after its entrance into the diaphragm, makes a curve, passing horizontally forward before entering the right auricle through the Eustachian orifice. On the contrary, the vena cava, after having traversed the diaphragm, continues its vertical ascending course until its entrance into the right auricle, and then passes through the inferior or diaphragmatic surface of the heart, which is horizontal, or almost so. As the diaphragm is slightly higher behind than in front (about 1 ctm.), the anterior surface of the vena cava is 1¼ ctm. long, its posterior surface ½ ctm. at the most. The two venæ cavæ thus prolong the two angles of the base like the hinges of a door.

The inferior border of the triangle passes from the apex of the heart to the insertion of the inferior vena cava. This border is very irregular, and is marked particularly by a fatty band. The left border is oblique; it starts from the insertion of the superior vena cava, passes in front of the aorta and the origin of the pulmonary artery and follows very accurately the anterior border of the interventricular notch; it corresponds to the groove followed by the cardiac vessels and nerves, except at the apex, at which the left ventricle makes a slight projection.

2. The inferior surface of the heart is flat, approximately horizontal, slightly inclined from right to left and from behind forward. It rests upon the diaphragm and is formed by the base of the right auricle and the inferior surface of both ventricles. The interventricular groove divides this surface into two equal parts.

3. The left surface of the cardiac pyramid is convex; it is formed by the pulmonary surface of the left ventricle and is slightly smaller than the two others. It is oblique from right to left, from above downward, and from behind forward.

The base of the cardiac pyramid is formed by the two auricles. The right auricle forms the base of the anterior surface; it has the shape of a vertical spindle, a sort of enlargement of the venæ cavæ.

The left auricle, on the contrary, is situated more posteriorly and is stretched out transversely to the left to receive the arterial blood of both series of pulmonary veins. It is situated immediately beneath the bifurcation of the trachea, which separates it from the ramifications of the pulmonary artery. It corresponds to the level of the sixth dorsal vertebra. Posteriorly the left auricle is separated from the vertebral column by the posterior mediastinum, containing the cardiac plexus derived from both pneumogastrics, the cardiac plexus derived from the branches of the sympathetic, the œsophagus, aorta, thoracic duct, and vena azygos.

CHAPTER II.

CLINICAL EXAMINATION OF THE NORMAL HEART.

DIAGNOSIS OF THE SHAPE AND POSITION OF THE HEART.

CLINICAL MEASUREMENT OF THE HEART.

AFTER the anatomical notions just laid down have been carefully digested, the problem of the examination of the heart becomes much more easy. Nevertheless, no specialist in diseases of the heart has hitherto described a method of accurately measuring the organ in the living subject. It could be determined in an approximate manner whether the heart was large or small, but it was impossible to measure it in such a way that changes of size from year to year could be accurately ascertained, although all methods of exploration—inspection, palpation, percussion, and auscultation—were called into requisition.[1]

Gendrin was the first to secure more accurate data obtained upon the patient himself. He stated[2] that the first point of departure must be the apex-beat of the heart and that this must be measured upon the skeleton, *i.e.*, the ribs and intercostal spaces (which are the most fixed organs of the thorax), and not from the nipple, as is ordinarily done. The nipple is one of the least fixed organs. In females it varies in position according to obesity, the development of the breasts, the condition of pregnancy and lactation, age, etc. Even in man it may vary greatly, as is evident from the following observations which I have made:

Height of the Bi-mammillary Line, examined in One Hundred Men.

Third rib	2 times.
Third intercostal space	9 "
Fourth rib	28 "
Fourth intercostal space	41 "
Fifth rib	18 "
Fifth intercostal space	2 "
	100 times.

---

[1] Laennec: Traité de l'auscultation médicale, t. ii., p. 501. 1826.

[2] Gendrin: Leçons sur les maladies du cœur et des grosses artères, faites à l'hôpital de la Pitié pendant les années 1840 et 1841, p. 26. Paris, 1841–1842.

Distance of the Left Nipple from the Median Line, measured in One Hundred Men.

8	centimetres	0	11½ centimetres		10
8¼	"	2	12 "		8
9	"	9	12½ "		6
9½	"	9	13 "		3
10	"	17	13½ "		2
10¼	"	15	14 "		0
11	"	19			
					100

Unfortunately, this tendency to precision was not followed by Barth and Roger,[1] who state: "The dulness obtained in the præcordial region by moderate percussion does not give the real measure of the dimensions of the heart, but is related solely to the extent over which the organ is in immediate contact with the walls of the chest. It requires stronger and deeper percussion to recognize the parts concealed by the lungs, and the obscure sound then extends beyond the previous limits over a variable extent, on account of the differences in the size of the heart, according to age and the individual."

With Friedreich we see a fresh effort at precision with regard to the relations and measurement of the heart. This writer reduced still further the region of absolute dulness recognized by his predecessors. He properly remarks that percussion of the anterior surface of the sternum gives a clear sound. Consequently that portion of the right ventricle which is situated beneath this bone does not give any dulness, although it is not covered by lung. This observation of Friedreich is very just, and percussion should not be relied upon too much in measuring the size of the organ, since it is the percussion of the relative dulness which gives the closest approximations to the real dimensions.

Friedreich[2] describes the following plan of outlining the heart: "If there are no other signs of any special anomaly of the heart, its boundaries may be completed approximately by drawing a line, which starts from the sternal point of insertion of the right sixth costal cartilage and is directed to the left and a little downward toward that point of the sixth rib, which is situated immediately beneath the apex of the heart." This method is very ingenious, but it does not constitute a clinical procedure and is not obtained from a measurement taken upon the patient himself. One of these two points, the impulse of the heart, is readily determined by the apex-beat of the patient. But the right point of departure is imper-

---

[1] Barth and Roger: Traité pratique d'auscultation, suivi d'un precis de percussion, 10e édition, p. 709. 1880.

[2] Friedreich: Traité des maladies du cœur, translated by Lorber and Doyon, p. 108 et seq. 1878.

sonal and does not hold good if the heart is not healthy, if it has been lowered, pushed to the right or upward, etc.¹

Niemeyer,² relying solely on percussion, stumbles against the same obstacles as his predecessors, but he makes the following just and important reflection : " Under normal conditions the apex of the heart beats in the fifth intercostal space, in the fourth interspace when these spaces are very large and the abdomen is much enlarged, in the sixth space when the intercostal spaces are very narrow."

Burresi³ pretends to be able to detect a difference between cardiac and hepatic dulness, but every day experience does not permit us to accept this subtlety of observation.

Raynaud,⁴ following the early idea of Piorry, thinks that we must begin by establishing, by means of percussion, the upper border of the liver ; he then finds the apex-beat of the heart and, connecting these two starting-points, establishes the line of separation of the heart from the liver. After this, Raynaud, like his predecessors, ascertains by percussion the two zones of absolute and relative dulness.

By means of these various procedures we can ascertain whether the heart is large, medium sized, or small, but it is impossible to recognize any increase of size from year to year, and the amount of increment. I hope to show that these difficulties may be overcome by a new method.

*New Method of Measurement of the Heart.*⁵—Since 1872 I have adopted a method of measurement of the heart which has enabled me to determine its dimensions accurately.

The first step consists in determining the apex of the heart by sight, palpation, and auscultation. Upon stripping the chest—and this should always be done when a careful examination is to be made—we see the elevation of the thoracic wall by the apex of the heart at the moment of systole. The hand is then placed over this region and we can determine very readily by palpation in almost all cases the point at which the apex impinges. Finally, upon completing this examination by auscultation, especially with a flexible stethoscope provided with a small thoracic end, we can determine with great precision the situation of the apex during systole. This spot is then marked with a dermographic pencil.

The apex of the heart being thus marked, we count the intercostal space in which it is found. This requires certain precautions, because there is often such a large depression between the clavicle and the first rib that it

---

[1] Idem, loc. cit., p. 115.

[2] Niemeyer: Eléments de pathologie interne et de thérapeutique, translated by L. Culmann and Chas. Seugel, annotated by Cornil, t. 1er, p. 321. 1865.

[3] Burresi: Leçon clinique in Sperimentale di Firenze. 1871.

[4] Raynaud : Art. Cœur, in the Nouveau Dictionnaire de Médecine et de Chirurgie, t. viii., p. 385.

[5] C. Paul: Sur un nouveau procédé clinique de mensuration du cœur. Association Française pour l'Avancement des Sciences, Congrès de Paris, sitting of August 28, 1878.

may be mistaken by the inexperienced physician as the first intercostal space. We should therefore follow the edge of the sternum, remembering that the first intercostal space begins below the clavicle and first rib.

As we have already stated, the apex is usually in the fifth space, exceptionally in the fourth or sixth, if the thorax is very large or very small. In case such a source of error is suspected, measure the sternum and ascertain whether it differs much from the average, viz., 21 ctm. from the sternal notch to the base of the xiphoid appendix.

The distance of the apex from the median line is then measured, the average in the adult being 8 to 10 ctm.

After having determined the situation of the apex of the heart, the

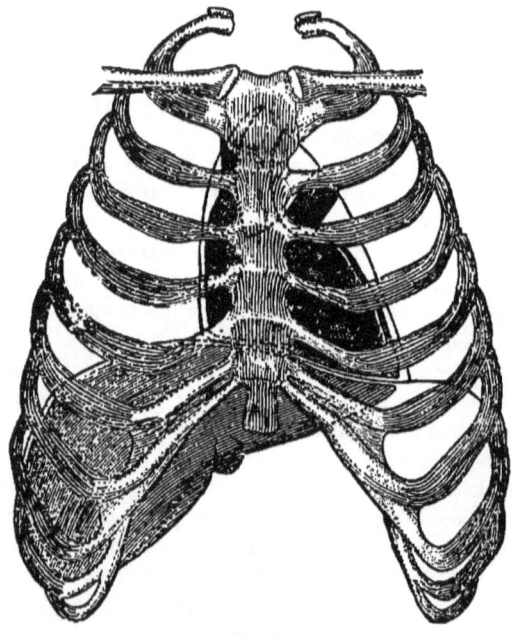

FIG. 3.

second point of departure is the upper border of the liver below the lung. Here percussion suffices to give exact information: the lung giving a clear sound and the liver a dull sound, it is easy to determine where one ends and the other begins. In order to take into consideration the slight convexity of the liver and the interposition of the lower border of the lung, I trace as the boundary the upper border of the finger which serves as a pleximeter in percussion, and thus obtain sufficient precision.

The level of the upper border of the liver should be fixed upon the skeleton. I have not chosen the corresponding rib, which is generally the fifth. because from the sternum to the mammillary line it describes a curve which is, by far, too descending. I have preferred to take the point of departure upon the edge of the sternum. I prolong the dermographic line of the upper border of the liver to the edge of the sternum and then ascertain the cartilage to the insertion of which it corresponds. In the adult this is generally the insertion of the right fifth cartilage. In connecting this line with the apex of the heart I thus obtain strictly the lower border of the organ.

In order to ascertain the length of the lower border, one end of which is already determined by the apex of the heart, it is sufficient to establish the vertical line representing the outer border of the right auricle.

This is furnished by percussion, being indicated by a change of the

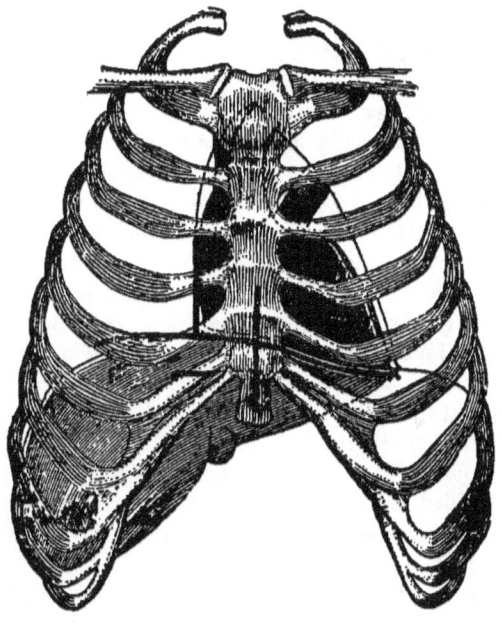

FIG. 4.

timbre of pulmonary resonance. For example, if we percuss from right to left at the level of the fourth left rib, we find at first the pulmonary sound, which is resonant, then that furnished by percussion of the sternum, which is also resonant, and the cardiac resonance, which is dull. We know that the anterior border of the right lung is more blunted than that of the left and encroaches little on the right auricle. I have also remarked that the most fixed part of the heart is the inferior vena cava, particularly its entrance into the right auricle. This auricle, then, is almost immovable, transversely it is only subject to greater or lesser dilatations, and its right border, which is normally 3 ctm. from the median line and 1½ ctm. from the right border of the sternum, varies very little.

Upon passing from the lung to the sternum, and on reaching about 1½ ctm. from the sternum, we do not find a dull sound, but a change of timbre with some diminution of resonance, which indicates the outer border of this auricle. This line is traced vertically and parallel to the sternum until it meets the hepatic line, and the right inferior angle of the cardiac triangle is thus determined. The distance from this angle to the apex of the heart gives the exact length of the lower border of the heart.

A third point of departure consists in indicating the obliquity of the lower border of the heart. In the normal condition, the apex of the heart is situated lower than the angle which corresponds to the right auricle, the difference in height being 1½ to 2 ctm. The level of the apex of the heart is therefore traced on the region of the liver below the right inferior angle of the heart, and the distance between these two points represents the depression of the apex of the heart.

Control upon the Cadaver of my Method of Measuring the Heart.—I have made several attempts to control upon the cadaver the dimensions which I have assigned to the heart during life by the aid of my method of measurement, and the results have been generally favorable. It must not be forgotten, however, that the situation of the heart after death is not absolutely the same as that during life. The ventricles are often retracted and the apex, consequently, elevated slightly; the right auricle is often found distended with black blood, especially in patients dying of heart disease.

On the other hand, the method of control is susceptible of certain errors. The method consists in marking upon the cadaver the points determined during life and there introducing steel needles, 10 to 15 ctm. in length. If the points of departure correspond to a cartilage or bone, a hole must be drilled in these parts through which the needles may be passed. Care must also be taken that the needles be introduced perpendicularly to the plane of the body.

The following will serve as illustrations of the results obtained by M. Letulle, one of my internes, in controlling upon the cadaver the measurements which I obtained:

EXPERIMENT I. *Senile Tuberculosis with Cavities.*—H——, aged sixty-five. Heart normal; apex in the fifth intercostal space, 8 ctm. from the sternum; upper border of the liver, right fifth costal cartilage (at its insertion); right border of the heart, about 1 ctm. from the right border of the sternum; transverse dulness, 6½ ctm. The length of the lower border of the heart, measured from the hepato-cardiac angle to the apex, is 11 ctm.

Death eight days after this examination.

Four needles are introduced as far as the posterior wall of the thorax.

The first needle enters obliquely in the fifth space, 8 ctm. to the right, upward and backward, in order to pierce the apex of the heart, and passes through it along the axis of the left ventricle. We find that it is buried in the left ventricle a little behind the apex. The inclination given to it has raised the apex before entering it. The apex of the heart corresponds exactly to the fifth intercostal space.

A second needle is passed through the fifth right costal cartilage, 5 mm. from its sternal insertion. It corresponds to the hepato-cardiac angle. Upon opening the thorax it is found to graze the upper border of the liver. More deeply it enters the hepatic parenchyma, so that it is found 1 ctm. below the inferior vena cava.

We had tried, in the living, to determine the transverse dulness of the heart by percussing upon a horizontal line passing across the third intercostal spaces; two needles were introduced corresponding to the extremities of the line thus determined.

One enters 1 ctm. from the right border of the sternum. It perforates the heart about 1 ctm. above the right auriculo-ventricular groove, thus leaving a notable surface of the right auricle outside of it. It must here be remarked that the auricle is distended with clots, which necessarily have modified its dimensions.

The other needle (the fourth), perforating 6½ ctm. from the first in the third left intercostal space, is imbedded in the left ventricle. It is at a distance of 14 mm. from the interventricular groove.

EXPERIMENT II. *Pulmonary Tuberculosis; Cavities at the Apex.*—F——, aged twenty years. Measurement of the heart: Apex in the fourth intercostal space at a distance of 8 ctm.; right border, right border of sternum; liver, right fifth intercostal cartilage (at its insertion). Rough anæmic murmur in left second intercostal space. Death a few days after this measurement.

Three needles are introduced: First needle: At 8 ctm. in the fourth space, in which it is introduced perpendicularly. It passes immediately above the apex, which it touches.

Second needle: Upon the right edge of the sternum in the third space. It grazes the right border of the right auricle and enters it a little in its deeper part.

Third needle: It was buried, by mistake, in the right fifth costal cartilage, 2 ctm. from its sternal insertion. It has entered the liver 2 ctm. below its convex border.

EXPERIMENT III. *Meningeal Hemorrhage.*—F——, aged thirty-five years. Measurement of the heart three hours before death: Apex in the fifth space at a distance of 8½ ctm.; the right border at the right border of the sternum; liver at the fifth costal cartilage. At the autopsy the heart is gorged with blood.

First needle: Has entered the very apex of the heart.

Second needle: Entering the third intercostal space at a distance of 8¼ ctm., is imbedded in the right auricle about 1½ ctm. from the right border of the auricle, which is filled with blood.

Third needle: Through the fifth costal cartilage at its insertion, literally grazes the inferior vena cava.

EXPERIMENT IV. *Aortic Insufficiency and Stenosis; Chronic Endaortitis; Angina Pectoris.*—G——, aged thirty-eight years. Apex, sixth space, at a distance of 8 ctm.; liver, sixth costal cartilage; right border, 1 ctm. from right edge of sternum. The patient died suddenly in an attack of angina pectoris. Three needles are introduced into the thorax:

1. At the apex: the needle has entered the apex of the left ventricle.

2. At the sternal insertion of the sixth costal cartilage: the needle is buried in the liver, a few millimetres from its convex border, in the vicinity of the inferior vena cava.

3. At 1 ctm. from the right edge of the sternum, in the third space: the needle traverses the right auricle.

EXPERIMENT V. *Atheroma of the Aortic Valves; Chronic Endaortitis; Considerable Hypertrophy of the Heart.*—For the clinical details of this observation we refer to the chapter on "Hodgson's Disease," and shall merely give the measurement of the heart:

Apex, sixth space at a distance of 11 ctm.; right border, $\frac{1}{4}$ ctm. from the right edge of the sternum; liver, seventh costal cartilage at its sternal insertion.

At the autopsy three needles were inserted into the thorax:

1. At the apex: the needle passes obliquely through the left ventricle, which forms a very round apex.

2. At the seventh costal cartilage: this enters the liver, grazing its most convex portion: the needle is a few millimetres distant from the heart.

3. At the right border of the sternum: taking into consideration the lowering of the hypertrophied heart, we introduced the needle, not into the third space, but into the fourth at a distance of $\frac{1}{4}$ ctm. from the right edge of the sternum. This needle was found imbedded in the auricle, which was distended by very abundant clots.

EXPERIMENT VI. *Atheroma of the Vessels; Mitral and Aortic Lesions; Hypertrophy of the Heart; Dry Pericarditis.*—V——, aged fifty-four years (the details of the case will be described hereafter). Apex, sixth space, at a distance of 12 ctm. from the median line; liver, fifth costal cartilage; vertical border (?) cannot be obtained on account of the considerable emphysema of both lungs.

At the autopsy, the first needle, introduced by mistake at a distance of 10 instead of 12 ctm., penetrates the left ventricle in the neighborhood of the inter-ventricular groove, about 2 ctm. from the apex.

The second needle, introduced at haphazard in the right third intercostal space, at the right border of the sternum, entered the right auricle near the right ventricular groove.

Finally, the third needle is passed through the right fifth costal cartilage at its sternal insertion, after determining upon the cadaver the evident dulness in this region. Upon opening the chest we notice with surprise that the needle passed through an encysted mediastino-diaphragmatic pouch of pleuritic fluid. This had given rise to dulness over the right fifth costal cartilage, and was a source of error which could be avoided with difficulty. The border of the liver was notably depressed, corresponding to the insertion of the seventh costal cartilage.

It is evident from these experiments, as well as from others which I have made, that the points of departure taken during life are really exact.

It is, then, sufficient, in order to correct the usual errors, to mark the level of the liver upon the upper border of the finger which serves as the pleximeter, and to trace the vertical border upon the external border of the finger. The thickness of the percussed finger compensates for the convexity of the organs.

CHAPTER III.

DIAGNOSIS OF DISPLACEMENTS OF THE HEART.

Diagnosis of Inversion of the Heart.—The diagnosis of this anomaly has always been easy, since it suffices to feel the apex of the heart beating in the right side of the chest instead of the left. But with the new method of measurement described in the last chapter, the diagnosis assumes new precision. I have had occasion to observe two cases:

OBSERVATION I. — A woman named D——, aged thirty-five years, presented herself at the Saint-Antoine Hospital, in April, 1878. She presented an arrest of development of the left side. The apex of the heart beat in the *right* fifth intercostal space 10 ctm. from the median line of the sternum. It must be mentioned, however, that the thorax is asymmetrical and that the capacity of the left side of the thorax is singularly diminished. In fact, the left second rib stops 3 ctm. from the sternum; this is also true of the third rib, so that the sternum is oblique downward and to the left, and if we take the distance from the apex of the heart to a vertical line passing through the sternal notch, the apex is distant only 8 ctm. from this median line. The liver is not reversed. The right angle of the heart and the entrance of the inferior vena cava are found at the level of the insertion of the left fourth cartilage, and the insertions of the left cartilages being very vicious, the vertical border, corresponding to the external border of the vertical or right auricle, appears to follow the left border of the sternum. The inferior border of the heart is thus calculated to have a length of about 12 ctm. The apex is lowered 3½ ctm.

The heart is, therefore, of normal size, with slight left hypertrophy.

The liver is not displaced; it is recognized by percussion, and does not pass beyond the false ribs. The atrophy of the left half of the body is manifest, the left breast is much less developed than the right. The two arms are equal in length; the left forearm and hand are much smaller than the right, by actual measurement.

OBSERVATION II.—Maurice M——, fifteen years of age, is imperfectly developed; his height is 146 ctm., the average at his age being 155 ctm. However, he appears to be developing more rapidly. The heart beats on the right side of the chest, the apex-beat being in the right fifth intercostal space, 7 ctm. from the median line. The edge of the liver corresponds on the left side to the insertion of the sixth cartilage. The vertical border of the heart is on the left side, 1½ ctm. from the edge of the sternum. The apex is lowered 1½ ctm. on the edge of the liver. Auscultation normal (reversed).

The liver, of normal size, is found on the left side. The spleen is on the right side, and appears small. The patient is not entirely left-handed;

he writes and draws with the right hand. He uses a knife with the right hand, but in throwing a ball, or making a powerful effort, uses the left hand. The right testicle hangs lower than the left.

I saw the patient eighteen months later. The heart had increased in size ; it presents the same relations to the ribs, but the apex is 8¼ ctm. from the median line.

As a rule, inversion of the viscera produces no bad effect.

There are other vices of position of the heart, which are scarcely compatible with life. I refer to those displacements in which the heart is situated in the abdominal cavity, the pharyngeal or cranial cavity, or even in front of the sternum outside of the chest. In one case which I observed with J. Franck, the heart was situated under the skin in the epigastric region.

DISPLACEMENTS OF THE HEART.—These are of two kinds ; the organ may either be displaced as a whole in one direction, or along its principal axes. In the latter event there is always, at the same time, a relative displacement in some other direction.

A. Displacement of the Heart Downward.—The heart may be lowered either by lesions which produce an increase in its weight, or by pressure from above downward. In the first case, the diaphragm finally yields under the influence of the increase of weight. I have shown above that the lower border of the heart is slightly oblique, with a difference in height of 2 ctm. between the hepatic angle, which corresponds to the insertion of the right fifth cartilage, and the apex of the heart, which is found in the left fifth intercostal space, 8 to 10 ctm. from the median line.

When the left heart begins to hypertrophy, it becomes heavier than the right side, descends, and then the obliquity of the inferior border increases ; the apex, instead of being 2 ctm. lower than the hepatic angle, descends to 3, 4, and even 5 ctm. below, and is then found in the sixth intercostal space. At the same time the ventricle becomes elongated, and, as the heart is fixed on the right side by the vena cava, this elongation can occur only to the left, and the apex is removed more and more from the median line (1 to 10 ctm.).

This descent of the apex of the heart, with increase in the obliquity of the lower border, is very easily recognized by the method described above.

On the other hand, if the right side of the heart is hypertrophied or dilated, the right angle of the heart is lowered, and the inferior border becomes horizontal. This occurs in cases of stenosis of the pulmonary artery, and in emphysema.

Finally, if both halves of the heart undergo hypertrophy, both angles are lowered, and the position of the heart is changed without any modification of its obliquity.

In pulmonary emphysema, the matter is more complex. According to Skoda, the heart increases in size in emphysema, and descends more on the left side than on the right, so that the apex approaches the median line. Bamberger and Schrœller entertain a different opinion, with which I agree entirely. They state that the impulse of the heart, which is felt under the sternum, and xiphoid cartilage, is due to the beat of the most distended part, viz., the right ventricle. This is readily demonstrated if the apex is sought, not alone by percussion, but also by auscultation.

A remarkable feature, in my opinion, is the fact that the right angle of the heart descends at the same time as the apex, and that the border of the heart becomes horizontal. The pressure of the heart upon the diaphragm is so great that the cardiac systole is indicated by an impulse in the epigastrium, even greater than in cases of hypertrophy of the right cavities.

In this case the horizontal position of the lower border of the heart is not alone due to its increased weight but also to the enlargement of the lungs, especially on the right side.

Furthermore, in emphysema, the heart undergoes a certain movement of torsion around its vertical axis. The left lung, in placing itself in front of the apex of the heart, separates the latter from the walls of the thorax. This causes a rotatory movement of the heart around the inferior vena cava, carrying the dilated right auricle and ventricle forward and increasing their area of contact with the anterior wall of the chest.

Finally, mention must be made of those cases in which the depression of the heart from increased weight is due to an abundant pericardial effusion and of those much rarer ones in which the increased weight is due to a neoplasm.

The depression of the heart, furthermore, may be caused by the traction of the central tendon of the diaphragm in cases in which the liver is hypertrophied (for example in hypertrophic cirrhosis), when the weight of the organ pulls down the diaphragm and consequently the pericardium and heart.

B. Displacement to the Left.—In enlargement of the heart the organ is always directed toward the left, on account of the fixation of the inferior vena cava, but this is not really displacement. On the other hand, if the patient lies on the left side the apex will be lowered 1 to 2 ctm. on this side; but the term displacement is applied only to those cases in which, the patient being in one of the positions recommended for examination of the heart, the heart as a whole is drawn or pushed toward the left.

If the right pleural cavity contains an effusion of fluid or gas (serous or purulent pleurisy, pyopneumothorax, aneurism), the heart may be pushed some centimetres to the left, and the apex-beat may even be felt in the axillary line in the sixth or seventh intercostal space. This is exceptional, but an increase of 2 to 4 ctm. in the distance of the apex from the median line is very frequent.

In other cases the retraction of the lung after pleurisy draws the pericardium and even the heart to the left, but this is more infrequent and usually less marked.

C. Displacement by being Pushed toward the Right.—This does not attain the same proportions as the corresponding displacement toward the left, because pressure starting from the left pleura (effusion of fluid or gas) will act chiefly upon the apex and cause the heart to revolve around its fixed point, the inferior vena cava. This is proven, even upon the living subject, by the fact that in proportion as the apex of the heart approaches the median line the cardiac dulness diminishes, and this diminution reaches its maximum when the apex has been pushed behind the sternum.

In these cases the cardiac pulsation appears most strongly to the right of the sternum, but it is never removed sufficiently from the median line to give the appearance of inversion of the heart.

Finally, in certain rare cases of diaphragmatic hernia the heart has been found pushed en masse into the right thoracic cavity.

Displacement by Traction toward the Right.—In the majority of the preceding cases the heart has been displaced by pressure. Displacement downward by traction has only been described from increase in the volume of the liver.

As a rule, displacement to the right by traction occurs only to a slight extent, but in rare cases it may attain considerable proportions. I recall one case in which this displacement was so marked that the patient, at first sight, appeared to be suffering from inversion. In this event, there is no torsion of the heart to diminish the cardiac triangle. The following case will serve as an illustration:

OBSERVATION III.—An engineer, forty-seven years of age, was seized suddenly, two years ago, with palpitation of the heart, which was declared to be functional by one of the best physicians in Paris. Soon after, the patient suffered from severe right pleurisy, which confined him to bed for three months, and was followed by considerable deformity of the thorax.

The right side of the chest is flattened, its antero-posterior diameter being 2 ctm. less than that of the left side; its circumference is only 1 ctm. less than that of the left. The right ribs are very oblique and the shoulder lowered. The right pectoral and intercostal muscles are very much atrophied.

In determining the situation and size of the heart, we are struck by the fact that it occupies mainly the right half of the thorax.

The left lung is resonant anteriorly and posteriorly as if the heart were inverted. Upon looking closely, the apex of the heart is found beating in the left fourth intercostal space, 6 ctm. from the median line. The upper border of the liver corresponds to the insertion of the right fourth costal cartilage, thus indicating a depression of the apex of 2 ctms. It follows that, on account of the retraction of the adhesions, the heart has been drawn toward the right, the apex being approximated 3 ctms. to the median line. The right lung is reduced to a third of its volume, the posterior border remaining permeable to air. The left lung, on the contrary, has increased in size; it is probably affected slightly by emphysema. The liver presents its normal position and size.

This is an example, therefore, of displacement of the heart toward the right by traction. It will be remarked in this case that the surface of the cardiac triangle has not diminished in extent, while in the corresponding displacement by pressure, in proportion as the effusion pushes upon the left side of the heart, the apex approaches the median line. At the same time the right auricle turns around the insertion of the inferior vena cava, and, finally, the surface of the heart in relation with the anterior thoracic walls does not increase on the right side in proportion to its diminution on the left side, so that the appreciable surface of the cardiac triangle diminishes accordingly.

D. Displacements Upward.—The heart may be pushed upward in certain cases of meteorism, ascites, abdominal tumors, and diaphragmatic hernia, and in the latter the apex may even beat in the third intercostal space. The latter cases are as exceptional as the former are frequent. I wish to add two cases which are less known, viz., cases of scoliosis in which the diaphragm, being stretched by the pressure, loses its curvature and gives a horizontal direction to the inferior surface of the heart. The other case, rare it is true, occurs in paralysis of the diaphragm when this paralysis is isolated; the diaphragm is then pushed upward by the pressure of the abdominal viscera. This is more marked during digestion, when the fermentation of alimentary matters in the stomach gives rise to the evolution of gases which produce stomachal meteorism, and consequently dyspnœa, by diminution of the respiratory field.

Flatulent dyspepsia and dilatation of the stomach often push the diaphragm and heart upward and occasion twitchings of the diaphragm, which the patients regard as pains in the heart. At the same time the apex of the heart beats upon the stomach as upon a drum and gives rise to palpitations. The patients often believe themselves to be affected with cardiac disease.

E. Displacement of the Heart Backward.—This is rare, and is due to diseases of the anterior mediastinum, especially aneurisms.

F. Displacement of the Heart along its Axes; Torsion of the Heart.—We have already seen that torsion of the heart along the vertical axis is almost always produced from the left side of the chest. Emphysema of the left lung pushes the apex backward and tends to diminish the lumen of the aorta and pulmonary artery.

Effusions into the left side (pleurisy, pneumothorax, aneurism) carry the apex forward and push the right auricle backward, the heart revolving upon its axis formed by the inferior vena cava. In this case, when the organ has been pushed to an extreme degree and the apex of the heart carried forward, the torsion of the aorta and pulmonary artery increases considerably. Blachez has shown that the compression of the left lung may cause the formation of a thrombus in the pulmonary artery. Labric has noted the presence of clots in the left heart, and Raynaud has indicated the possibility of degeneration of the myocardium.

Finally, in certain cases, the lower border of the heart may undergo peculiar rocking movements upon the transverse axis. Pleuritic exudations, some forms of pneumonia of the entire lung, and aneurisms may depress the hepatic angle more than 2 ctm., and then the inferior border of the heart no longer bends to the left, but to the right, the apex of the heart being more elevated than the hepatic border.

These facts, which have been observed upon autopsy, cannot be verified upon the living patient.

CHAPTER IV.

CLINICAL EXPLORATION OF THE MOVEMENTS OF THE HEART AND THE CARDIAC SOUNDS.

THIS exploration is effected by inspection, palpation, auscultation, and also by means of registering apparatus which furnish graphic traces of the movements of the heart.

INSPECTION.—In the normal condition the sides of the chest are sensibly though not absolutely symmetrical. We notice the expansion and retraction of the thorax during respiration, and a slight elevation in the fifth left intercostal space.

The absence of the cardiac impulse indicates that the apex is concealed either by the obesity of the patient, by pulmonary emphysema, or by a pleural or pericardial effusion. On the other hand, when the patient is lean and the cardiac contractions violent, a movement occurs, not alone over the apex but in the third, fourth, fifth, and sixth intercostal spaces.

If these palpitations are due to hypertrophy of some standing, elevation of the cartilages and ribs also occurs.

In other cases this movement is not an elevation, but a sort of aspiration toward the interior; it is the traction produced by the adhesion of the heart to the pericardium, *i.e.*, cardiac symphysis.

Furthermore, aneurisms produce rhythmical expansion of the ribs or merely of the soft parts, when the parts of the skeleton have undergone absorption. Finally, we may notice more or less varicose enlargement of the veins, indicating an obstruction in their course.

PALPATION.—We should early learn to recognize by touch the resistance given to the finger by the apex-beat in normal cases, in order to be able to determine whether it is increased or diminished. We can determine by palpation whether the impulse is regular or irregular, equal or unequal, sudden or slow; whether it is accompanied by thrill, and is single, double, or even treble.

PERCUSSION.—Most works give no other method of measuring the size and shape of the heart beyond percussion, especially deep percussion. In my opinion, this practice should be abandoned, for if the patient has an acute or advanced cardiac affection, this form of percussion is painful, and may even produce syncope.

In the method of examination which I have described, I have endeavored to avoid percussion of the heart, confining myself to the liver, lungs, and to a slight extent, the right border of the heart. Even in the pulmonary region percussion should be practised very gently if an aneurism is suspected.

AUSCULTATION.—This should first be performed with the unaided ear, in order to obtain a general idea of the normal or pathological sounds. The stethoscope is required to determine the exact topography of the sound, the period of the cardiac cycle during which it occurs, and finally its timbre. The patient may be examined in an erect, sitting, or recumbent position, but he should not be turned to the right or left.

I have made numerous experiments to determine the most suitable dimensions, etc., of the stethoscope, and have found that it should possess the following qualities:

It should be made of a bad conductor of heat, such as wood or hardened caoutchouc, so as not to impart a sensation of cold to the part auscultated. I have adopted an instrument with a calibre of 6 mm.

I believe that the best shape of the cap applied to the surface is that of the end of a trumpet. I have made a stethoscope composed of four tubes, sliding upon one another, so as to form an instrument varying from 10 to 35 ctm. in length; experience has shown that the most suitable length of wooden stethoscopes is 25 ctm. It has seemed to me preferable that the auricular ends of the instrument should be flat. This is the construction of the stiff stethoscope, but experience has taught me that the flexible stethoscope is preferable.

After numerous trials I have come to the conclusion that the following are the most suitable dimensions of the flexible stethoscope:

External diameter of the cap applied to the surface. 20 millimetres.
Height of the cap .... 40 "
Internal diameter of the tube.................... 6 "
Length... 450 "
Thickness........ .... 1½ "
External diameter............................... 9 "

The cap applied to the surface of the thorax should have the shape of the end of a trumpet, because this lends itself best to the reception of the waves of sound. Its diameter at the base, is 20 mm., in order that it may be introduced between the ribs and enter the intercostal spaces. The external diameter of the tube is 9 mm., *i.e.*, the average diameter of the external auditory canal. Those who have wider or narrower auditory canals should choose the tubes accordingly. The tube may be formed of vulcanized india-rubber; red English rubber, if soft, is preferable.

Even more advantageous than the monoauricular stethoscope is the biauricular stethoscope.

I have made recently, in my opinion, a considerable improvement in the flexible stethoscope. I had long endeavored to provide the instrument with a resonator, when chance led to the accomplishment of this object. Dr. Roussel, a Geneva physician, showed before the Therapeutical Society (March 21, 1881) an apparatus for the transfusion of blood, which presented a circular cup, intended to fix the apparatus to the arm of the individual who furnished the blood. The idea struck me of applying this circular cup to the receiving end of the flexible stethoscope, and I then had constructed the instrument as represented below, with the exception of the bulb producing the vacuum, which will be depicted later.

When the bulb is compressed, the cup adheres to the chest and the stethoscope is fixed automatically.

This new instrument is endowed with exceptional acoustic properties. The exact adaptation of the thoracic end to the tissues, the relative immo-

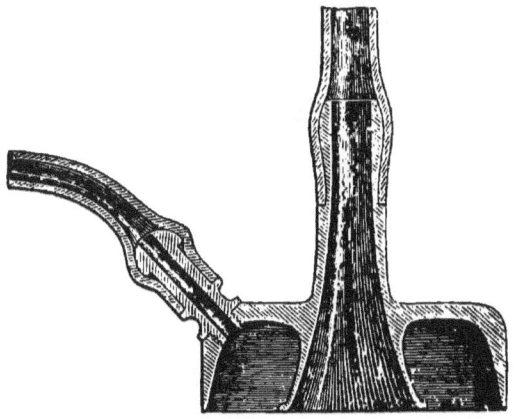

FIG. 5.

bility which results, and the cavity which surrounds it make the vibrations much more distinct. The cavity of the cup constitutes a resonant box and the fingers of the observer no longer interfere with the vibrations.

The circular cup has still another effect. As its cavity does not transmit the vibrations arising below; it does not permit the passage of the adjacent sonorous waves which interfere with those to which we listen. It isolates, in a striking manner, the sound to which we are listening.

NORMAL HEART-SOUNDS.—In auscultation of the normal heart two sounds are heard in succession: the first is dull and deep, the second is clearer, sharper, shorter, more superficial. The first sound is heard best at the apex of the heart; the second sound is heard best in the second intercostal space near the sternum, especially on the left side.

This is due to the fact that the pulmonary artery is more superficial than the aorta and approaches the left border of the sternum more closely than the right.

The first sound appears to be the result of the sounds produced by: contraction of the cardiac muscle (Laennec); impulse of the heart against the chest-walls (Magendie, Skoda, London Committee, Chauveau, Faivre, Marey); vibration of the blood (Gendrin); friction of the blood against

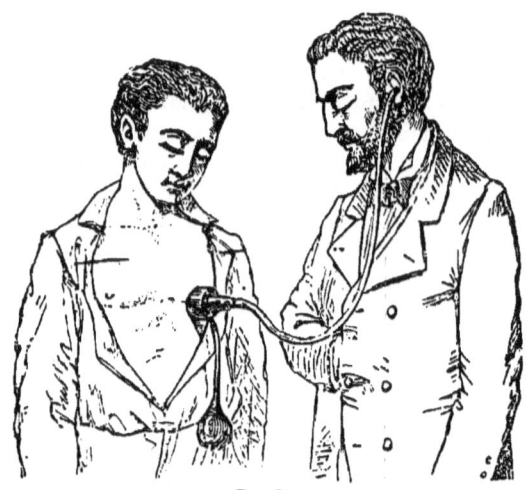

FIG. 6.

the walls and valves of the heart (Dublin Committee, Pigeaux); tension of the auricular valves (Hope, Renaud, Skoda, Philadelphia Committee, Chauveau, Faivre, Marey); lowering of the aortic and pulmonary valves (Bouillaud).

The second sound appears to be due to the sudden closure of the aortic and pulmonary valves. This theory is accepted by all.

CHAPTER V.

EXAMINATION OF THE MOVEMENTS OF THE HEART BY AUTOMATIC APPARATUS.

THE most perfect registering or graphic apparatus for the movements of the heart are those of Marey, and I refer to his works for the physiological details.[1] In this chapter I will recall merely those physiological data which will serve as the basis for clinical study.

Let us take a glance at the trace furnished by the contraction of the auricle and ventricle. The trace is that of the right auricle, but as we know by experience that the traces of both ventricles are sensibly identical, a single figure will enable us to recognize the succession and duration of each of the movements of the heart.

The trace represents several consecutive revolutions, the normal rhythm being represented as follows:

Fig. 12 represents the traces of the right auricle (No. 1), the right ventricle (No. 2), and the cardiac pulsation (No. 3), taken simultaneously during four complete revolutions of the heart, with a scale which permits the measurement, in fractions of a second, of the duration of the briefest movements of the organ. Those waves of the trace which are explained are marked with full lines instead of dotted lines.

We have resorted more than once to this fundamental trace for the interpretation of the different acts performed during a cardiac revolution. Let us confine ourselves for the present to the determination of the duration of each of them. We know that as the pulse beats about sixty times a minute, a cardiac revolution occupies about a second. The use of the registering apparatus permits the resolution of the different parts of this second with mathematical precision.

The trace in question, divided into tenths of a second, includes, as we see, eleven-tenths of a second; it has been taken upon a horse having 54 to 55 pulsations per minute. In this trace the following duration may be assigned to each act:

Contraction of the auricles	2 tenths of a second.
Ventricular systole	4 " " "
Closure of the semilunar valves	1 " " "
Diastole	4 " " "

[1] Marey: Physiologie médicale de la circulation du sang. 1863. La circulation du sang à l'état physiologique et dans les maladies. Paris: G. Masson. 1881.

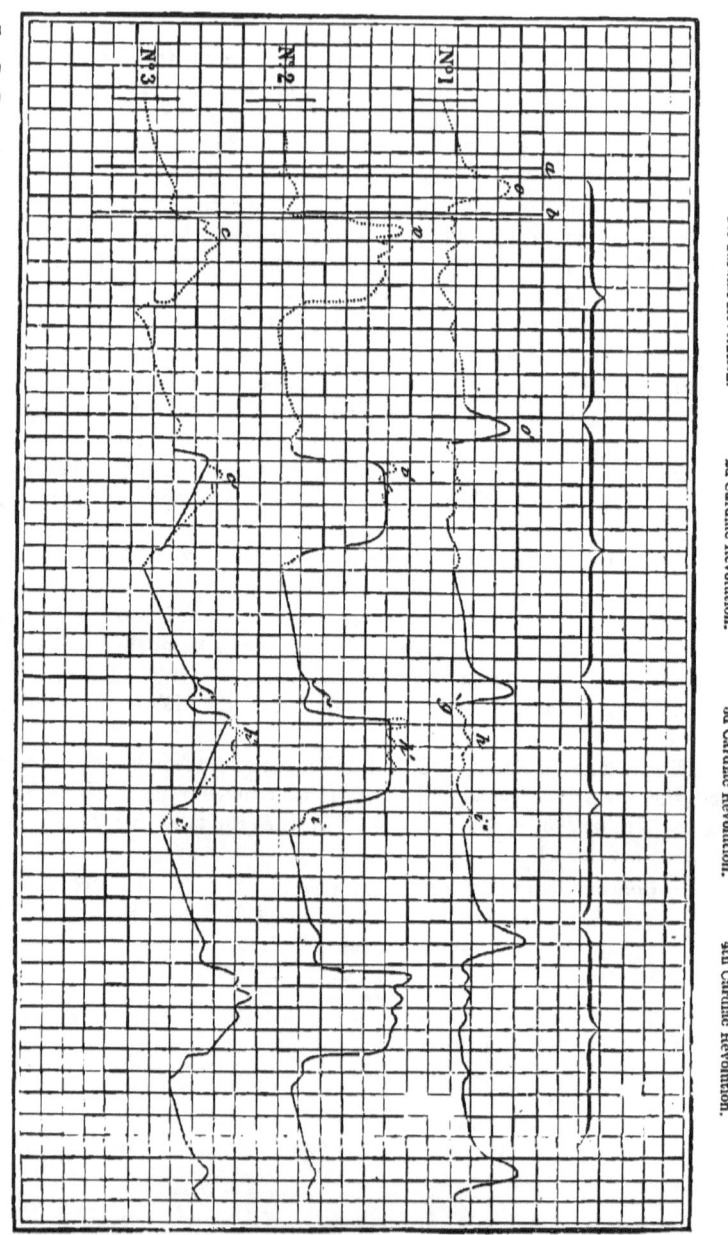

Fig. 7.—Determination of the Succession of the Cardiac Movements. Characteristics and Relations of the Auricular and Ventricular Movements and of the Cardiac Pulsation.

The period of auscultation of the two sounds may be divided as follows:

First sound.................... 3 tenths of a second.
Lesser period of silence............ 2 " " "
Second sound..................... 1 " " "
Greater period of silence............ 5 " " "

The entire cardiac revolution lasting eleven-tenths of a second, the first sound begins with the beginning of ventricular systole, and continues during a part of the systole, and the second sound takes place at the moment of the closure of the semilunar valves and of the impulse given by the entrance of blood into the ventricle, which ceases to contract. It follows, accordingly, that the schematic trace and auscultation do not begin the cardiac revolution at the same moment. The contraction of the auricles not producing any audible sound, the first sound does not begin until the ventricular systole, which includes the first sound and lesser silence, then comes the second sound, and finally the greater silence, which in-

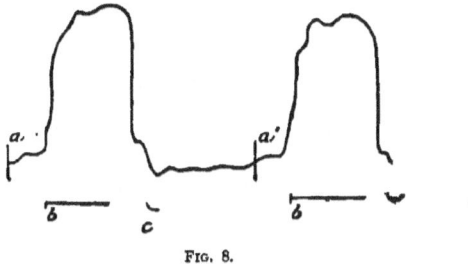

Fig. 8.

cludes not alone diastole but also the period of contraction of the auricle, and terminates at the moment when the following ventricular contraction begins.

This measure of the duration of a cardiac revolution is not absolute, it is merely relative; but Marey's apparatus gives it an approximation and precision not furnished by other methods.

We can conceive the difficulty of obtaining this measure in man. Suppose a man's pulse beats sixty times a minute, the entire revolution of the heart will last a second. Now, Marey's trace shows that the period which lasts from the beginning of the first sound to the end of the second being five-tenths of a second, the greater silence lasts six-tenths of a second. But if this period is five-tenths of a second for a pulse which beats fifty-four times a minute, it is not more than four-and-a-half-tenths of a second for a heart which beats sixty times a minute.

Volkmann[1] has attempted to measure these two intervals with a sort

---

[1] Volkmann: Zeitschr. f. rat. Med., iii., p. 321. 1845. Hæmodynamik, p. 177. Leipzig.

of metronome, formed of two pendulums beating half seconds, and which are modified in order to obtain a synchronous rhythm. He found that the ventricular systole and greater silence were equal in duration; it is evident that this approximates the truth, since the cardiograph gives the proportion of 5 to 6.

Hayden [1] was nearer the truth in stating that observations with the watch indicated that ventricular systole was a little shorter than the greater period of silence.

The investigations of Donders [2] have given analogous figures; the duration of the systole is 40.6 to 45.6, the duration of an entire revolution being 100; i.e., the relation of the systole to the greater silence is as 4.5 to 5.6.

Landois [3] has given analogous figures, obtained by Marey's polygraph.

In conclusion we may state that the pulse of the fœtus at full term varies from 112 to 160, the average being 135 to 140. In the new-born the pulse varies from 88 to 163, the average being 120 to 130. In the adult man, the average is 71 to 72; in old age it is often lowered to 56 pulsations.

Under physiological conditions, however, this number varies constantly with the position, condition of repose or action, the emotions, etc.

---

[1] Hayden: Dublin Quarterly Journal of Medical Sciences, xl., p. 456. 1865.
[2] Donders: Nederl. Archiv voor Genees en Naturk., ii., p. 184. 1868.
[3] Landois: Graph. Unters. ueber d. Herzschlag, p. 55. 1876.

CHAPTER VI.

THE MUSCLES AND NERVES OF THE HEART—THE INFLUENCE OF THE
NERVOUS SYSTEM ON THE CARDIAC RHYTHM.

I HAVE no intention of describing the muscles of the heart, but will refer for this purpose to treatises on anatomy. I will merely recall what is necessary to the comprehension of the pathology and treatment.

The heart is formed by a muscle situated between two serous membranes, the endocardium and pericardium. It constitutes a marked development of the muscular layer of the arteries. But instead of being composed, like the vessels, of smooth fibres, the heart is formed of muscular fibres which are striated like voluntary muscles, but, nevertheless, differ anatomically from the latter. We do not find the parallel and isolated arrangement of the voluntary fibres, but, on the contrary, very numerous anastomoses, which form a sort of reticulum. The connective tissue uniting the fibres is scanty; it may be overloaded with fat.

The heart muscle differs, then, from other striated muscles in various ways; not alone by the anatomical characteristics which I have enumerated, but also in the fact that it is constantly in action. It is evident from Marey's trace that the period of repose is almost nil compared with the period of action. The cardiac muscle is endowed also with a very peculiar nervous system, the anatomical and physiological conditions of which I shall now sum up so far as this is possible in the present state of our knowledge.

NERVES OF THE HEART.

The nerves of the heart are derived from two sources, the cerebrospinal system and the great sympathetic; they unite beneath the arch of the aorta to form one of the most important plexuses in the economy, the cardiac plexus. A large number of branches start from this plexus, following the vessels and entering the tissue of the heart to terminate in a very peculiar manner. In my description I shall follow modern anatomists, particularly Miot[1] and Reynier.[2]

---

[1] Miot: Recherches physiologiques sur l'innervation du cœur. Bruxelles, 1876.
[2] Reynier: Des nerfs du cœur, anatomie et physiologie. Thèse d'agrégation. Paris, 1880.

AFFERENT BRANCHES.

The afferent branches are derived from the cerebro-spinal system and the great sympathetic. The branches derived from the cerebro-spinal system are supplied by the pneumogastric nerve.

1. *Branches Supplied by the Pneumogastric.*—The pneumogastric starts from the posterior part of the medulla oblongata, from the restiform body, *i.e.*, from that portion of the posterior column of the cord which passes toward the cerebellum and forms the inferior cerebellar peduncle.

The groove of origin of the pneumogastric is in the domain of the posterior column of the medulla, but not in the continuation of the posterior lateral column (Farabeuf).

The roots, to the number of six or eight fibres, approximate one another in order to pass through the jugular foramen, and then form a ganglion. This ganglion anastomoses with the glosso-pharyngeal, facial, and sympathetic nerves. Beneath the jugular foramen the nerve forms the gangliform plexus and anastomoses with the internal branch of the spinal accessory, the glosso-pharyngeal, hypoglossal, and the cervical ganglion of the sympathetic.

The pneumogastric nerve, after all these anastomoses, supplies the cardiac plexuses with branches which start from its cervical part, others which start from the thoracic part, and finally branches which are detached from the inferior laryngeal nerve.

The branches which start from the cervical portion of the pneumogastric to pass to the cardiac plexus are two or three in number.

The *first cardiac nerve*, or superior cardiac nerve of Hirschfeld, is detached from the pneumogastric a little below the superior laryngeal, receives in its course some branches from the carotid plexus (Valentin), and unites below with the superior cardiac nerve supplied by the sympathetic. According to Valentin, it receives branches of the second cardiac nerve supplied by the pneumogastric. It passes into the cardiac plexus, in part directly, in part after having anastomosed with the other nerves of the heart.

The *second cardiac nerve*, according to Valentin, is larger than the preceding one; it often anastomoses with the third cardiac nerve before passing into the cardiac plexus.

The *third cardiac nerve* takes origin below the preceding one, anastomoses with the second cardiac nerve, and passes into the plexus.

According to Hirschfeld, the second and third cardiac nerves are not constant.

The *thoracic branches supplied by the pneumogastric* are called inferior cardiac nerves by Hirschfeld, and superior and inferior deep cardiac nerves by Valentin. These nerves start from the pneumogastric immediately below the origin of the recurrent nerve; they anastomose with branches supplied by the recurrent and sympathetic, and then pass into the cardiac plexus.

The *branches supplied by the inferior laryngeal nerve* or the middle cardiac nerves of Hirschfeld, arise from the convexity of the curve of these nerves. Those on the left side pass almost immediately into the plexus; those on the right side anastomose with the cervical branches of the pneumogastric and sympathetic before passing into the plexus.

2. *Afferent Branches Supplied by the Sympathetic.*—These are derived especially from the superior, middle, and inferior cervical ganglia. Hirschfeld states that, like the pneumogastrics, they are more marked on the right side than on the left, but this arrangement is not constant.

The *superior sympathetic cardiac nerve* is derived from the superior cervical ganglion, anastomoses in its course with the laryngeal plexus and the middle sympathetic nerve, then, in the thorax, with the fibres of the recurrent and pneumogastric nerves, before entering the plexus.

The *middle sympathetic cardiac nerve* arises from the middle cervical ganglion and enters the plexus at its upper and lateral part.

The *inferior sympathetic nerve* arises from the inferior cervical ganglion, sometimes from the first thoracic ganglion. It anastomoses in its course with the middle sympathetic cardiac nerve, and the recurrent laryngeal, but, according to Cruveilhier, this anastomosis is merely apparent and the nerves are simply in apposition.

To these nerves must be added the depressor nerve of Cyon and the accelerator nerve of the same author, but it will be better to consider them as efferent branches.

CARDIAC PLEXUS.

The cardiac plexus, formed of numerous branches, some of which are of large size, and of nervous ganglia, is situated in front of the bifurcation of the trachea, above the right branch of the pulmonary artery, to the right of the strand which results from the obliteration of the ductus arteriosus, and below and to the left of the angle formed by the ascending and transverse portions of the aorta.

The cardiac plexus is very constant in its arrangement; it is divided into three groups (Hirschfeld, Cruveilhier, Sappey, Reynier).

The *anterior group* is observed readily, without dissection, through the transparent pericardium. It spreads over the anterior surface of the ascending part of the arch of the aorta, between the latter and the pulmonary artery. This group furnishes filaments to the aorta, pulmonary artery, right auricle, and anterior groove of the heart; it also supplies branches to the lymphatic glands and the remains of the thymus gland. With this group is connected a small ganglion described by Andersch under the name thyroid ganglion, and another more important one, that of Wrisberg, situated between the superior vena cava and the trachea. This ganglion of Wrisberg, situated in the median line, serves to unite the right and left afferent cardiac nerves.

The *middle group* of the cardiac plexus is situated behind the arch of the aorta, between the latter and the trachea, and above the right pulmonary artery.

The *posterior group* is separated from the preceding by the right branch of the pulmonary artery; it is situated in front of the bifurcation of the trachea. It communicates freely with the anterior and posterior pulmonary plexuses, and with the superior tracheal and œsophageal plexus.

It supplies branches to the aorta and pulmonary artery upon their posterior surface and then passes into the tissue of the heart, being distributed to the auricles and ventricles.

EFFERENT BRANCHES.

The principal cardiac plexus gives off branches which go to form the anterior and posterior cardiac plexuses as well as the coronary plexuses.

The left and anterior cardiac plexus follows the anterior coronary artery and divides into two branches, one of which follows the groove separating the auricles from the ventricles, while the other passes along the anterior interventricular groove.

The posterior cardiac plexus follows the posterior coronary artery. It also divides into two branches, one following the auriculo-ventricular groove, the other the inferior or diaphragmatic interventricular groove and anastomosing at the apex of the heart with the anterior plexus. These plexuses distribute branches to the auricles and ventricles and also, according to Vulpian, to the coronary arteries.

Some of these branches present in their course minute ganglia, described by Remak in sheep, and the existence of which in man is admitted by Kölliker. According to Ranvier[1] these ganglia do not resemble the spinal ganglia; they are simple collections of more or less numerous peculiar cells. These cells contain a large nucleus, not situated in the centre, as is usually the case, but at one end. The cells are surrounded by a capsule containing nuclei belonging to endothelial cells, situated mainly near the hilus of the ganglion. According to Ludwig, Biddle, and others, these cells are never multipolar, but only bipolar or unipolar, especially the latter, with their prolongation directed toward the periphery.

Another peculiarity has been observed by Beale, Julius Arnold, and Ranvier. The nucleus contained in the cell situated beneath the capsule is connected with a straight nerve-fibre, around which is twined another and more slender fibre. The coils formed by this fibre become closer as they

---

[1] Ranvier: Gazette Médicale, 1878, No. 7, 16 février.

approach the cell and are lost in the enveloping substance. According to some, these intracardiac ganglia are derived from the sympathetic (Kölliker); according to others, from the pneumogastric nerve (Beale). Ranvier inclines to the latter opinion.

Termination of the Cardiac Fibres.—The cardiac fibres possess no myeline for some distance before arriving at their termination, and the absence of the medullary sheath in the latter part of their course renders the study of these fibres very difficult. The chloride of gold method of staining shows that they terminate in anastomosing meshes.

According to Ranvier,[1] when the nerve reaches a muscular fibre its diameter is exceedingly small, and measures 1, 2, or at the most 3 mm. The nervous fibrillæ then enter the muscular fibre, dividing and subdividing in its thickness, sometimes before; they penetrate into the muscle-cells and pass to the cells of the myocardium, appearing to anastomose with one another and forming a plexus, the meshes of which are a little larger than the muscle-cells themselves. It must be noted that each of the cells composing a muscular fibre does not receive directly a nervous fibre.

PHYSIOLOGY OF THE NERVES OF THE HEART.

ACTION OF THE INTRACARDIAC NERVOUS SYSTEM; THE PART PLAYED BY THE GANGLIA.—The first experiment showing the action of the ganglia was performed by Volkmann (1844). If the heart is cut into a number of pieces, those which do not contain ganglia cease to beat, while the others continue their rhythmical movement.

It must be added, as an offset to this experiment, that if the immobile parts are stimulated by the faradic current, they will again contract (Eckhard and Heidenhain), but with this difference, that they do not display a rhythmical movement, but merely a galvanic excitation at the moment of breaking the current (Ranvier).

There is still another offset to Volkmann's experiment. Stannius showed (1852) that the heart will be arrested in diastole if, instead of cutting it into several pieces, a ligature is tied around the sinus of the vena cava in order to separate it from the heart.

The experiment of Stannius being of fundamental importance, I will describe it in full:

1. He laid bare a frog's heart and applied a ligature to the venous sinus at the point where it empties into the auricle; the heart stopped beating at once. If the ligature is applied upon the auricles between the sinus and the ventricle, *i.e.*, a little below the situation of the ligature mentioned above, the heart remains immovable while the parts above continue to beat.

---

[1] Ranvier: Anatomie générale, p. 153. Recherches sur la grenouille et la tortue terrestre.

2. If the heart is thus arrested by ligature of the sinus, and a second ligature is placed around the auriculo-ventricular groove, the auricles remain at rest, but the heart recovers its activity.[1]

Ranvier[2] interprets Stannius' experiment upon the hypothesis of the existence of inhibitory and exciting centres in the heart. This hypothesis is not accepted by Vulpian, who admits the existence merely of exciting intracardiac ganglia. He adds this interesting fact, that, whatever may be the electrical stimulus employed, the cardiac muscles can never be placed in a condition of tetanus.

THE RÔLE OF THE EXTRACARDIAC NERVOUS SYSTEM; ACTION OF THE PNEUMOGASTRICS ON THE HEART; INHIBITORY NERVES.—The capital fact in this particular is that the pneumogastric nerve is a nerve of arrest. At first this property of the pneumogastric was observed solely under the influence of the electrical current, but Dastre and Morat have shown that the nerve responds to stimulation of every kind, mechanical or chemical, by an identical effect, viz., arrest or slowing of the heart's action.[3]

In order that this effect be manifested in mammals, it is merely necessary that the opposing system be enfeebled by section of the accelerator nerves, by freezing the animal, the condition of anæmia produced by a rapid hemorrhage, the injection of carbonate of soda, etc.

The inhibitory action of the pneumogastrics being no longer disputed, let us see the results of physiological experiments.

1. *Stimulation of the Intact Pneumogastrics.*—The Weber brothers have shown that faradic stimulation with a strong current produces, first, retardation or momentary arrest of the heart; second, lowering of the intracardiac pressure.

2. *Effects of Section of the Pneumogastric.*—At the present time it is believed that these effects are of two kinds: at first the mechanical irritation of the nerve, torn in its continuity by the action of the cutting instrument; then the separation of the heart from the centre of origin of the cut nerve.

In the primary irritation an action is produced on the one hand upon the centrifugal branches which pass to the heart and, on the other hand, upon the centripetal branches. Of these two effects, the second is much more prominent and appears earlier.

The ligature of one pneumogastric, the other being intact, produces retardation of two or three beats. F. Franck admits that this action is produced by a reflex which passes through the other pneumogastric, rather than by a direct action upon the lower end of the nerve. In fact, if, after the nerve has been tied, a second ligature is applied lower down, no action upon the heart is produced. But if the second ligature is applied

---

[1] Stannius: Zwei Reihen physiologischer Versuche. Mueller's Arch., 1852.
[2] Ranvier: Loc. cit., Gazette médicale, 1878, No. 7, 16 février.
[3] Dastre and Morat; Société de Biologie, Août, 1877.

above the first, the same phenomenon of temporary retardation is produced.

The second ligature placed below the first will act in exceptional instances when the animal is in such a condition that the pulsations of the heart are already notably lower than the average, either from section of the accelerator nerves, from freezing, or from a hemorrhage (Dastre and Morat).

If one pneumogastric has been cut and a ligature is applied to the other, temporary retardation will be produced, as in the preceding case. The explanation of this is not easy, since we can no longer rely upon the return of the irritation through the opposite nerve. We are forced to assume that the reflex passes through the same nerve before the ligature has had time to destroy the continuity of the fibres; but this is a very subtle explanation. All that can be said is that the production of this retardation is prevented by the administration of chloroform, curare, and atropine.

Finally, if a ligature is applied above the preceding one, no further effect is produced, as could be readily predicted.

After this temporary retardation, produced by the temporary irritation of the nerve, the real phenomena of section make their appearance, i.e., acceleration and increased regularity of the cardiac pulsations. The acceleration had been noted by Meyer in 1825. The increased regularity was not observed until later, and then in the dog, who habitually has an irregular pulse. The explanation of this fact has been given by von Bezold; it is the suppression of the inhibitory influence produced by the pneumogastrics. In fact, if the inhibitory action of this nerve has been destroyed by large doses of atropine or curare, section of the nerve does not give rise to acceleration.

Dastre and Morat have shown, in addition, that in asphyxia, the syncope is not a phenomenon of cardiac atony, but the result of the exaggerated activity of the pneumogastrics.[1]

The secondary action of this acceleration of the movements of the heart, without inhibition soon leads to the death of the animals from inanition, the food accumulating in the œsophagus without the power of being propelled farther. V. Aurey attributes the fatty degeneration of the heart found upon autopsy to the inanition rather than to the overwork of the heart.

The effect of *irritation of the central end of the cut vagus*, when the other nerve is intact, varies according to the intensity of the irritation. In

---

[1] We have still another proof of this interpretation in the fact that in emphysema with dilatation of the heart and a tendency to asphyxia, systole is prolonged and the two periods of silence being equalized, the rhythm in auscultation approaches that of a pendulum. Pidoux first observed this peculiarity and gave it the name pendulum bruit.

1856, Budge, who had produced vigorous stimulation, stated that stimulation of the central end allows the cardiac pulsations to continue, but arrests the respiratory movements; then the arterial pressure is raised and a sudden and violent systole is produced followed by retardation. In this case the pneumogastric acts like an ordinary sensory nerve; its irritation produces anxiety and suffering.

But if the irritation is moderate and the sensibility of the animal destroyed by a narcotic or anæsthetic, we observe an arrest, or at least retardation with lowered pressure. There is, then, an inhibitory reflex action, which pursues the following course: it ascends the central portion to the bulbar nucleus of origin, then passes to the opposite nucleus and descends in the opposite vagus, which has been left intact. The action is the same as if the opposite nerve were stimulated directly. This inhibitory action, produced by stimulation of the central end, is not produced if the other pneumogastric nerve has been cut.

Stimulation of the peripheral end of the cut pneumogastric by electrical currents produces retardation of the pulse, and this effect is more marked upon a cut nerve than upon an intact one; it also increases for some time after the section has been made. But it must be observed that this retardation is not produced immediately; one or two seconds elapse in warm-blooded animals and twenty-five or thirty seconds in cold-blooded animals (Legros and Onimus). Tarchanoff ascribes this variation in the retardation to the period of the cardiac revolution in which the operation is performed. The effect is more rapid if the stimulation is produced at the end of diastole or the beginning of systole; it is delayed if the irritation is effected during another period. The minimum retardation is one cardiac revolution, the maximum two revolutions. The inhibitory action is thus delayed, but, on the other hand, it is prolonged beyond the irritation; then the contractions make their reappearance, slowly at first, then more rapidly.

It is a peculiar phenomenon that if the stimulation which produces the arrest is prolonged, the arrest does not persist and the heart again begins to beat. The arrest is therefore a transitory phenomenon; it does not last more than fifteen to twenty seconds in a warm-blooded animal. After the arrest of the heart, the pulsations have an extraordinary amplitude, although the energy of the ventricular contraction has not increased (Arloing and Tripier); this is due to the fact that the resistance has diminished because the arterial system is relatively dilated during the arrest of the heart.

A last word with regard to the pneumogastric. This nerve contains various kinds of fibres, and those which bear the name of inhibitory fibres are derived from the spinal accessory through the internal anastomotic branch. This is proven by the fact that, when the spinal accessory is pulled out by the roots, the inhibitory action of the vagus is abolished.

THE PART PLAYED BY THE EXTRACARDIAC NERVOUS SYSTEM; THE GREAT SYMPATHETIC; THE ACCELERATOR NERVES.—Prochaska was the first who attributed the cause of the heart's action to the sympathetic system. In fact, stimulation of the sympathetic renders the pulsations of the heart more frequent, at the same time that the systoles are shorter (Donders, Baxt), so that, despite the much greater number of cardiac contractions, the arterial tension is increased very little or not at all. In this case the diastole is shorter than in the normal condition, and the heart is not filled completely. The quantity of blood received by the ventricles being thus diminished, the amount discharged by them is also lessened (Marey).

Stimulation of the sympathetic may accelerate the pulse considerably (500 per minute in the rabbit, Wendt), but it does not produce tetanus, as it would if the motor nerve of another striated muscle were irritated.

It follows from these experiments that it is probable that the sympathetic does not act directly upon the cardiac muscle, but rather upon the intracardiac ganglia.

This hypothesis is found to be verified more and more as experiments are multiplied.

F. Franck has shown, in addition, certain characteristics of the accelerator nerves. If an irritation is applied, separately, to one of the accelerator nerves—for example, the cardiac nerve which is derived from the inferior cervical ganglion (superior cardiac nerve of Schmiedberg)—it will be found that the stimulation undergoes a delay which is disproportionate to the short course of the nerve; the retardation is more than a second. Then, after the irritation has ceased, the acceleration of the heart is prolonged. In order to produce an effect a stronger stimulus must be applied to the accelerator nerves than to the inhibitory nerves. Moreover, the action of the accelerators, like that of the inhibitory fibres, may be interfered with by a reflex produced by the suffering experienced by the animal.

While the right pneumogastric often appears to have a greater inhibitory action than the left pneumogastric, the sympathetic of one side does not appear to have any greater accelerating action than that of the other.

The effects of section of the sympathetic are by no means so distinct as those of section of the pneumogastric, and this is readily understood. The sympathetic, by its numerous anastomoses, forms a sort of network, so that when the principal strand is divided, the anastomoses above and below re-establish the connection.

THE CENTRE OF ORIGIN OF THE CENTRIFUGAL NERVES.

As the nuclei of origin of the pneumogastric and spinal accessory nerves are situated in the medulla oblongata, the centre of inhibitory action has been sought in that region; while, as the sympathetic receives spinal nerves along its entire course, the centre of accelerating action has been sought in the cord.

In 1841, Budge noticed that upon irritating the lower surface of the medulla, or on passing an interrupted current through it, arrest of the heart was produced. In the same manner as in irritation of the pneumogastrics, the arrest occurs during diastole, and does not persist, even though the irritation continues. If the pneumogastrics are cut during the experiment, the arrest of the heart's action ceases. The same effects are produced by a blow upon the medulla, as by electrical stimulation (Vulpian). On the other hand, destruction of the medulla suppresses the inhibitory action, and, the accelerators having lost their antagonists, the pulsations become more frequent. The inhibitory centre is not very well defined; it appears to be less circumscribed than the respiratory centre, and corresponds to Budge's cilio-spinal region.

CARDIAC REFLEXES—SENSORY NERVES.

NERVE OF LUDWIG AND CYON.

Upon dividing the pneumogastric and stimulating the central end, the arterial tension is lowered. Ludwig and Thiry showed that this is not due to an action upon the heart, but upon all the vessels of the body, more especially those of the abdomen. They found that in cases of irritation of the cord, diminution of the vascular field was produced by the contraction of the small vessels. Ludwig and Cyon showed that this action is due to a nerve which starts from the heart, ascends in the pneumogastric, and passes to the medulla oblongata.

This nerve is very sensitive; if it is titillated or pricked in the rabbit it gives rise to pain or cries. If it is cut, and the peripheral end irritated, no effect is produced, but if the central end is stimulated, a diminution of pressure results. This depression is due to a dilatation of all the small vessels (Ludwig and Cyon) and to an increase of the respiratory movements (F. Franck).

Vulpian explains the action of the nerve in the following manner: "When the heart, under the influence of an obstruction to the pulmonary circulation, or some other cause, becomes distended with blood, and experiences difficulty in emptying itself at each ventricular systole a specific impression will be produced upon the peripheral ends of the cardiac nerves, either in the endocardium or the myocardium itself. The inhibitory nerves are thus brought specially into play. They conduct to the medulla the impression which they have received, and, through the medium of the bulbo-spinal centre and the vaso-dilator nerves, a general vaso-dilating action is produced, particularly in the mesenteric vessels through the action of the splanchnic nerves. If the heart, on the contrary, receives less blood than in the normal condition, the opposite effect will take place throughout the vaso-motor apparatus. The vaso-constrictor nerves will be stimulated, the majority of the small vessels will be more or less constricted, and the blood, being subjected to a greater *vis à tergo* in the veins, will flow more abundantly into the heart."

The sensibility of the inner surface of the heart was proven by an experiment made by Claude Bernard (1868). Upon touching the endocardium with a thermometer, the movements of the heart were accelerated.

From a recent thesis of Laffont, it appears that each depressor nerve has a nucleus of origin in the floor of the fourth ventricle near the origin of the pneumogastric. The nerves then descend through the first two pairs of dorsal nerves and pass to the dilating fibres of the abdominal viscera.

In 1880, F. Franck discovered other centripetal fibres. These fibres, starting from the endocardium, have no reflex action on the vessels, but rather upon the respiratory organs. An irritating injection when thrown into the right heart not alone gives rise to arrest or retardation of the heart, but also to respiratory disturbances, characterized most frequently by arrest of respiration.

CHAPTER VII.

VARIATIONS IN THE PULSE—PALPITATION—LIPOTHYMIA—SYNCOPE.

WE have seen in the last chapter that the heart possesses two sets of nerves, viz., inhibitors and accelerators.

We copy from F. Franck the following table, which sums them up:

CENTRIFUGAL NERVES.

Inhibitory Nerves. { Bulbar centre or rather bulbo-spinal. { Fibres of the pneumogastric (furnished in part by the spinal accessory).

Accelerator Nerves. {
  Bulbo-spinal centre (Schiff). {
    A. Fibres contained in the pneumogastric and coming from the spinal accessory. {
      1. Following the course of the nerve (Schiff).
      2. Following the superior laryngeal and Galen's anastomosis (Schiff).
    }
    B. Fibres contained in the sympathetic derived from its superior anastomosis.
  }
  Cervico-dorsal centre. {
    A. Filaments of the cervical sympathetic.
    B. Fibres derived from the last four or five cervical nerves, and which form the vertebral nerve of v. Bezold (Bever, Cyon).
    C. Fibres derived from the first two dorsal nerves (Cyon, Cl. Bernard, Stricker).
    D. Fibres derived from the third, fourth, and fifth dorsal nerves (Albertoni, Bufoldini).
  }
}

CENTRIPETAL NERVES.

The nerve of Ludwig and Cyon, visible in the rabbit; starting from the heart, passing toward the pneumogastric, with which it unites to reach the medulla oblongata, then descending in the first two dorsal nerves to join the sympathetic and terminating in the abdominal viscera.

Centripetal fibres of F. Franck, starting from the endocardium, passing to the medulla oblongata, and reflected upon the motor apparatus of respiration.

Fibres of Paglioni, starting from the pericardium and passing to the intracardiac ganglia (?).

The inhibitory and accelerator apparatus are not developed at the same time. Dastre and Morat describe their development as follows:

1. During fœtal life and immediately after birth, the extracardiac inhibitory apparatus does not act. At the same time the accelerator apparatus performs its function, in a word, the former has the better of the antagonism.

2. Some time after birth, from the second to seventh day in the cat (Aurep), the pneumogastric begins to acquire the power of inhibiting the heart.

3. A little later (during the third week in the cat, Aurep) the pneumogastric inhibitor exercises its function. Artificial irritation arrests the heart; section of the vagi accelerates it. But at this early age the advantage is still on the side of the accelerator system, hence the pulsations are much more frequent than in adult life.

4. In the adult, the inhibitory system finally assumes the predominance.

These data are in accordance with the results of clinical experience concerning the frequency of the pulse.

At the end of fœtal life the heart beats 135 to 140 times per minute; at the period of birth, 120 to 130 times a minute; from ten to fifteen years, 76 to 91 times; from twenty to twenty-five years, 69 to 73 times; from twenty-five to sixty years, 69 to 73 times. After the age of sixty the pulse diminishes in frequency; Richeraud has seen it fall to 29 in a man, aged eighty-eight years.

Apart from these averages there are individual variations. Sex possesses a certain influence. According to Quetelet, the pulse is more frequent in females than in males by 1 to 4 pulsations; according to Guy this increment is 7 to 8 per minute. It is said that the pulse is more frequent in large persons than in small ones.

Position often influences the frequency of the pulse. Guy gives the following averages:

When the individual is lying down .... 66 pulsations.
"   "   seated 71   "
"   "   standing 81   "

Exercise changes the rapidity of the pulse. If the exercise is not prolonged, the pulse is accelerated; if continued until fatigue ensues, the pulse becomes slower (Richtenfels, Froehlich).

The pulse also varies according to the meals, and becomes more rapid after each repast. A nitrogenous diet accelerates the pulse more than starchy food.

The pulse may be full or small, strong or weak, rapid or slow, hard or soft.

PALPITATION.

The term palpitation refers to frequent, rapid, sudden impulses of the heart, of which the patient has more or less knowledge, and which are accompanied by anxiety, sometimes even by dyspnœa.

G. Sée [1] mentions the following characteristics of palpitation :

1. The pulsations are more frequent.
2. The cardiac pulsations often are more rapid, *i.e.*, the duration of each impulse is less than normal.
3. The cardiac contractions appear to be more forcible and usually are perceived much more readily ; but it has not been proven that they are really more vigorous.
4. The pulsations are accompanied ordinarily by a modification of the timbre of the heart-sound.
5. In the majority of cases peculiar sensations are noticed by the patients.
6. Finally, the palpitations may coincide with disturbances of the rhythm or even with intermittency in the heart's action.

Palpitation may be continuous or intermittent, but it varies constantly in intensity.

Let us return to some of these characteristics. Palpitation may be temporary and only appear for a few seconds, minutes, or hours ; it may continue during sleep. At other times it is continuous.

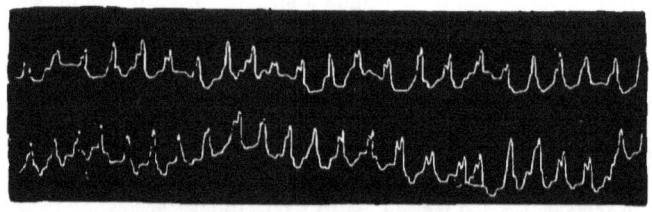

FIG. 9.—Trace of the Apex of the Heart in a Young Emotional Girl.

It may give rise merely to trifling sensations ; at other times, it is felt like the blows of a hammer, either in the cardiac region or the neck, and is often accompanied by anxiety or even by dyspnœa. The patients often are unable to lie upon the left side without experiencing dyspnœa.

Percussion affords negative results, auscultation merely shows that the sounds are produced more brusquely ; this suddenness often is more marked for the second sound, particularly the closure of the pulmonary valves. The force of the pulse may be retained, but it is often weak, sometimes even insensible.

Palpitation may be idiopathic, *i.e.*, the heart may be peculiarly emotional, apart from any morbid condition either of the circulatory or other organs. Thus we may mention the excitement produced by examination of a patient. This acceleration of the pulse is known generally under the term " medical pulse." It stops ordinarily at the end of five or ten seconds, but in some emotional patients it may last during the entire visit of the physician.

---

[1] G. Sée : Du diagnostic et du traitement des maladies du cœur et en particulier de leurs formes anomales, p. 123. 1879.

As a rule palpitation indicates a nervous or circulatory disorder. The nervous disorders are usually of a functional character, such as the congenital disposition known as nervosisme and which responds by palpitation to the slightest emotion (Fig. 9). Then follow hysteria, hypochondriasis,

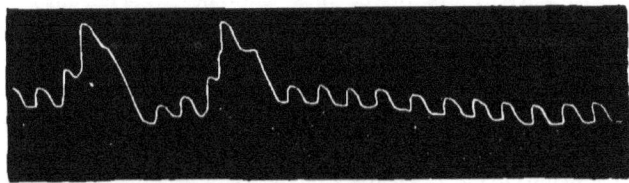

Fig. 10.—Trace of the Pulmonary Artery in an Anæmic Girl.

and the other psychical affections; then diseases of the brain and spinal cord or their membranes. It may be produced also by some local nervous affections, such as tumors along the course of the nerves of the neck or chest, neuritis or other nervous lesions.

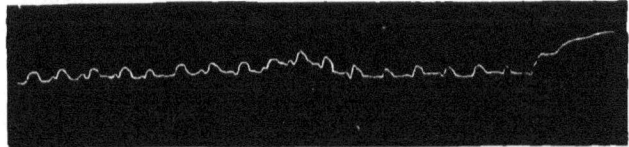

Fig. 11.—Trace of the Pulmonary Artery in a Young Girl suffering from Incipient Tuberculosis.

Palpitation may also be produced by changes in the blood, and at the head of these must be placed anæmia. In Fig. 10 I give the trace taken at the level of the pulmonary artery with the cardiograph in a chloro-anæmic girl. Palpitation may be produced also by intoxication from alcohol, coffee, tea, and tobacco.

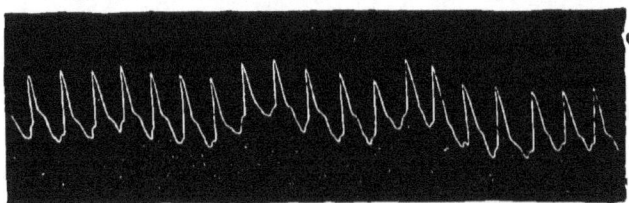

Fig. 12.—Trace of the Apex of the Heart, from the same Patient.

At the beginning of pulmonary phthisis, although the lesions may be scarcely indicated, reflex irritation of the accelerator nerves occurs and consequently palpitation. These patients, consulting us with regard to the cardiac disturbance, are confronted often by the diagnosis of beginning phthisis. An illustration of this form of palpitation is shown in Figs. 11 and 12, taken from a girl with beginning tuberculosis at the apex of the left lung.

Finally, palpitation may be symptomatic of dyspepsia, usually of the flatulent variety. In this form of the affection the patients complain of an intercostal pain at the level of the apex of the heart. They also complain of contractions in this region, of a sort of painful spasm. Examination of the heart reveals nothing but palpitation, but clinical examination enables us to recognize dyspepsia, usually of the flatulent variety, with dilatation of the greater cul-de-sac of the stomach. It is then found that the palpitation is produced particularly by the stomach, which is presented to the heart like a drum, upon which the apex beats. Careful examination shows that the pain is produced by stretching of the diaphragm at the point of its insertion into the ribs. When the primary disease recovers the palpitation disappears.

LIPOTHYMIA AND SYNCOPE.

As the accelerator system of the heart may be brought into play by various causes, so also may the inhibitory system. I have already mentioned the depressing emotions, such as fear, fright, physical pain, in this connection. The effects of heat and of pregnancy also belong to this category. There is another disease of this kind which I have described in the "Traité de Therapeutique," under the name "theatre sickness," and which should find a place here.

It occurs almost always in strangers in the city, who dine rapidly and reach the theatre in a heated condition. The patients, usually females, begin by suffering from dyspnœa and oppression; then, as the audience fills the room, the effects of the heat are felt more and more. Syncope occurs usually toward the beginning of the second act. This fact is so constant that at the Comédie Française, of which I am one of the physicians, I have, for a long time, not seen the second act.

Young women, at the beginning of pregnancy, are more exposed than others; thus, when young married couples come to Paris to spend the honeymoon, the "theatre sickness" may be the first indication of pregnancy. Attacks of hysteria with relaxation of the limbs may be distinguished from "theatre sickness" by the fact that the pulse remains unaffected in the former.

I have already quoted the experiments of Dastre and Morat upon the action of the blood in asphyxia. If asphyxia is prolonged in an animal, the pulsations become more infrequent and weaker, and syncope supervenes. In these cases this results from a reflex action upon the pneumogastrics; since, if these nerves are cut, the heart resumes its pulsations.

Among the causes of arrest of the heart's action must be mentioned the use of anæsthetics, particularly chloroform, which, apart from the pain produced by the operation, and even before the latter has been begun, may suddenly give rise to syncope. This syncope is produced, according to Franck, by stimulation of the inhibitory apparatus of the heart.

In addition, similar effects are produced by a large number of other drugs, the action of which will be considered at a later period.

But apart from these causes, which usually act accidentally, as it were, the most common cause of syncope is lesion of the pericardium or heart, usually of an organic character ; and, above all others, fatty degeneration of the fibres of the myocardium.

CHAPTER VIII.

ANÆMIA.

It often happens in boys and girls that anæmia produces palpitation of such violence that the physician may dread the development of hypertrophy. In addition, the murmurs produced by anæmia may possess the roughness of organic murmurs and, if they are not localized with great precision, the patients may be supposed to have cardiac disease.

I will here give the chief characteristics of anæmic murmurs, referring for further details to the paper which I have published on this subject.[1]

OBSERVATION IV.—Blanche H——, sixteen and a half years of age, a flower girl, entered the hospital April 15, 1877. She began to menstruate at fourteen years, the courses then appeared very irregularly for six months, after which they were suppressed and replaced by leucorrhœa. She also complains of gastralgia.

Upon auscultation an extremely loud blowing murmur is heard in both jugulars. A loud, hard, rasping systolic murmur is heard at the level of the second left intercostal space. This murmur extends over a space as large as a two-franc piece; it is followed by a clear, valvular click. In addition, the intercostal space is elevated at each systole and the finger distinctly perceives a thrill. The blowing murmur diminishes considerably when the patient is seated and remains in a sitting position.

She improves rapidly under the administration of arsenite of iron, and leaves the hospital at the end of a month, very much improved; the murmurs have diminished but are still present.

Every pathological murmur is recognized by three characteristics, its time, situation, and timbre.

The anæmic blowing murmur is situated in the second left intercostal space near the sternum, at the point which corresponds to the trunk of the pulmonary artery. It extends over a distance of 5 ctm. When the murmur is reduced to its smallest dimensions, it is always found in the second intercostal space.

But it often happens that the murmur is not confined to this space. It

---

[1] C. Paul: Sur le bruit de souffle anémo-spasmodique de l'artère pulmonaire désigné généralement sous le nom de bruit anémique de la base du cœur. Société Médicale des Hôpitaux, January 11, 1878, and in Union Médicale.

may extend to the third or first space or to both, but the maximum of its intensity is found always in the second space (Fig. 15).

In these cases the blowing murmurs are always intense, and in one or

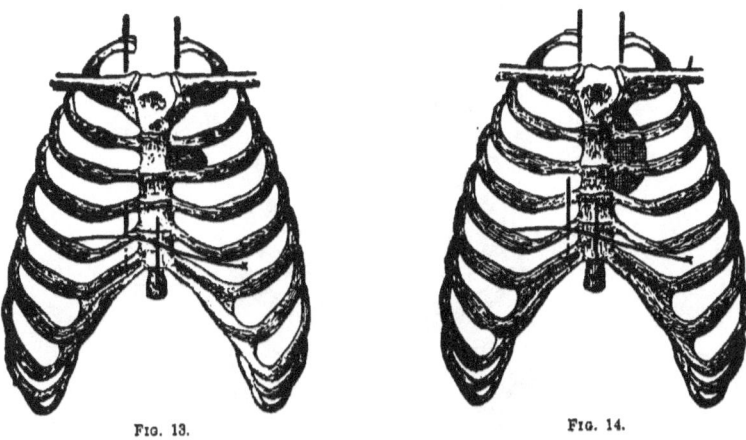

FIG. 13.   FIG. 14.

two cases I have found the murmur extending even into the second right intercostal space.

In a third type of localization of the anæmic blowing murmur, the maximum of intensity is found in the third left intercostal space; this occurs when the sternum is very short and the heart apparently situated

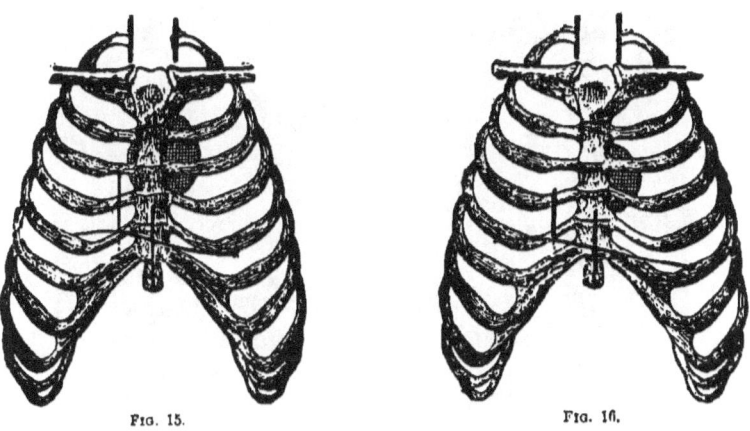

FIG. 15.   FIG. 16.

lower. From the third space the murmur may be continued into the second, and if the sternum is very short, it sometimes descends even into the fourth space (Fig. 16).

In a fourth class of cases the murmur is propagated to the apex or may there form a new locality. These cases are rare, but I have observed them twice (Fig. 17).

The anæmic blowing murmur, then, is situated always to the left of the sternum and, unless the latter is abnormally constructed, in the second

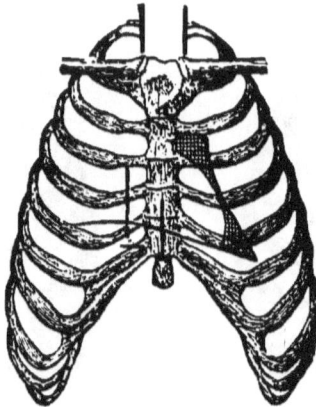

FIG. 17.

intercostal space at a point corresponding to the pulmonary artery. It is often propagated upward and downward, sometimes even forms a new localization at the apex.

The murmur in the pulmonary artery always coincides with a bilateral

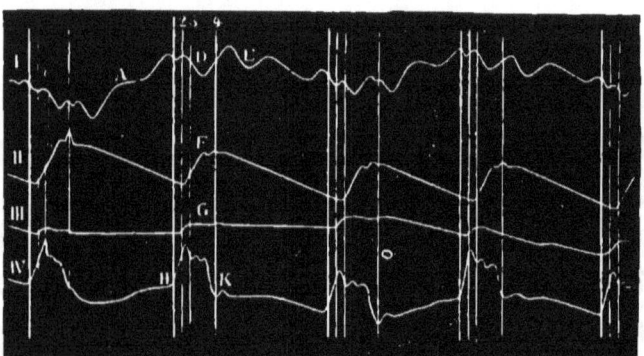

FIG. 18.—Reprinted from Potain.—I., Pulsations of the jugular vein; II., pulsations of the radial; III., pulsations of the carotid; IV., pulsations of the apex of the heart. A. progressive repletion of the vein; D, depression produced by the diastole of the auricle; E, depression produced by diastole of the ventricle; H, beginning of ventricular systole; K, end of the ventricular systole.

cervical murmur, or at least with a right cervical murmur. This murmur is situated very probably in the cervical veins, as is shown by the traces obtained by Potain[1] (Figs. 18 and 19).

---

[1] Potain : Du mouvement et des bruits qui se passent dans les veines jugulaires, Mémoires de la Société Médicale des Hôpitaux, p. 3. 1867.

4

We may say, therefore, that anæmic blowing murmurs have three principal sites:
1. A cervical site, particularly upon the right side.
2. In the pulmonary artery.
3. At the mitral valve.

The second site is never found without the first, nor the third without the other two.

The second characteristic of a cardiac murmur is the period of its occurrence. It is readily seen that the anæmic murmur begins at the beginning of systole and continues during a portion of this period; as a rule it persists during almost the entire systole and includes the lesser period of silence, so that the murmur terminates generally in a sudden

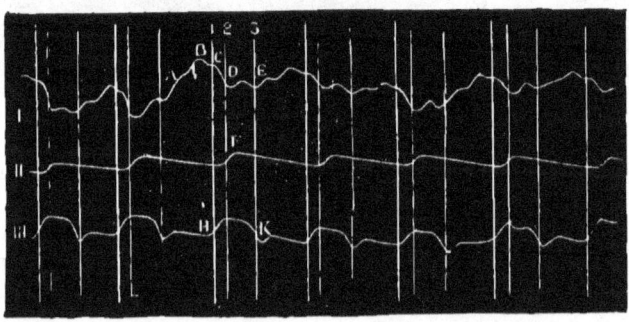

Fig. 19.—Reprinted from Potain.—I., Pulsations of the jugular vein; II., radial pulse; III., apex of the heart. A, progressive repletion of the vein; B, auricular contraction; C, ventricular contraction; D, auricular dilatation; E, ventricular dilatation.

manner, i.e., by the clicking of the semilunar valves. This click is usually superficial and distinct, and indicates the suddenness of the return flow and consequently the spasm of the vessels.

In some cases, however, the murmur is a little delayed and the normal first sound is heard before it. The rhythm then consists of three sounds, a click, a blowing murmur, and a click, the latter being particularly sharp.

OBSERVATION V. *Anæmia with Delayed Murmur.*—The patient, suffering from very marked chloro-anæmia, presents pallor of the integument and mucous membranes. Auscultation shows a bruit de diable in the vessels of the neck on the right side, slightly on the left. At the level of the pulmonary artery is heard a blowing murmur, occupying the second, first, and a part of the third left intercostal spaces. The murmur is systolic, lasts throughout the entire systole, and is followed from time to time by a sharp valvular click. The rhythm of the pulmonary artery is that of a bruit de galop composed as follows: A click, followed without an interval by a prolonged murmur ending with the semilunar click.

The blowing murmur diminishes in the sitting position, and is increased in the horizontal position; it is also modified by the cessation of respiration.

The contractions of the heart become less frequent, they fall from 92 to 72. The blowing murmur is prolonged and becomes more distinct.

When respiration is arrested with effort, the murmur diminishes without disappearing completely; then, when the effort ceases and the patient makes full inspirations, the murmur resumes its intensity and duration in proportion to the amplitude of the waves of blood.

During normal respiration the pulse beats 84; during the effort, 80; after the effort, 84.

The third characteristic of the anæmo-spasmodic murmur is its timbre. This is usually soft, like a puff of steam, but in some cases it assumes a

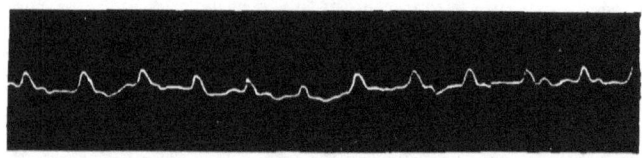

FIG. 20.—Trace of the Pulmonary Artery in a Young, Non-chlorotic Girl, suffering from slight Bronchitis without Fever.

roughness which approaches that of organic murmurs, and it is not rare to find it altogether rough and accompanied by a thrill.

We will not be astonished, therefore, to find that as the finger perceives the impulse of the artery, the cardiograph furnishes the corresponding trace (*vide* Fig. 22, compared with Figs. 20 and 21).

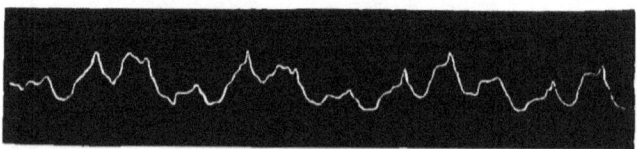

FIG. 21.—Trace of the Apex of the Heart in the Same Individual.

If we compare the trace taken upon the pulmonary artery in anæmic individuals, the spasmodic character of the systole will be found indicated by the much greater elevation of the needle (Fig. 23).

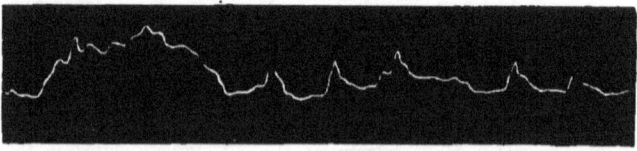

FIG. 22.—Trace of the Apex of the Heart in an Anæmic Individual.

The anæmic murmur presents two other important characteristics: it is modified by position and by the character of the respiration. The modification by position consists in this, that the murmur is less intense in the vertical than in the horizontal position. The explanation of this phenomenon appears to me to reside in the fact that the arterial pressure in the pulmonary artery is feeble. According to Marey it is only one-third of that in the aorta. Experiments made upon the horse have shown that the pressure in the right ventricle is 24 to 30 mm. of mercury, and in the left ventricle 95 to 140 mm. We can understand how gravity will possess a notable influence upon circulation of such feeble pressure, while it acts but little upon the left ventricle, the energy of which is three times as great.

Furthermore, if the patient ceases to breathe without making any effort, the first cardiac revolutions will give more distinctness to the blowing murmur ; then it remains stationary. If respiration is stopped while the patient makes an effort of expulsion, the murmur gradually diminishes, and often ceases altogether ; then, if the patient again begins to breathe, it gradually becomes more intense than it was at the beginning of the experiment.

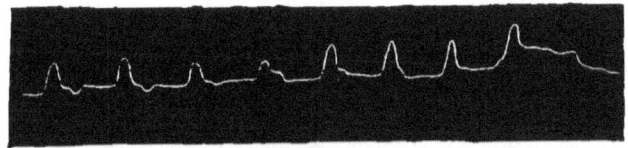

Fig. 23.—Trace of the Pulmonary Artery in an Anæmic Individual.

All these conditions, viz., its situation at the level of the pulmonary artery, its occurrence during systole, and the superficial character of the murmur, lead me to believe that the murmur is venous in its origin and situated in the pulmonary artery. The results of the change of position, and the modification of respiration have convinced me that the murmur is situated in the pulmonary artery. I know that this opinion has not been accepted by all, but I feel convinced that, when carefully examined, it will be adopted generally.

Differential Diagnosis between Anæmo-spasmodic Murmurs and Extracardiac Murmurs.—During the past ten years Potain has directed our attention to the interesting murmurs associated with the movements of the heart, but entirely independent of diseases of the organ, and due to certain accidental relations of the heart with adjacent organs.

These extracardiac murmurs present the following characteristics in common with anæmic murmurs:

1. They are heard at a point where the lung is situated in front of the heart and aorta and, in consequence, of the pulmonary artery.
2. They have a soft quality and sometimes are distinctly superficial.
3. The suspension of the respiratory movements modifies their intensity.
4. The intensity diminishes considerably when the patient passes from the horizontal to the sitting or standing position.

On account of these common characteristics it becomes necessary to make a differential diagnosis. I feel convinced, for my part, that some of the murmurs which Potain has regarded as extracardiac were merely anæmic murmurs in the pulmonary artery. This is proven by the following quotations copied from the works of Potain's pupils:

Choyan, in his thesis,[1] says: "Thus, in ausculting patients whose heart beat with extreme energy in consequence of excitement and exercise, I have often heard a blowing murmur which was no longer present after a quarter or half an hour of repose. Chlorotic females have often presented this peculiarity, and I am tempted to believe that certain blowing murmurs called anæmic are due to the same causes as those which are the subject of this thesis."

In an article by Cufer[2] is found the following passage: "According to Potain it is very probable that many of the murmurs which are regarded as anæmic and as occurring in the pulmonary artery, are merely extracardiac murmurs."

We will now see in what respects these murmurs differ from one another:

1. The anæmic murmur in the pulmonary artery is always accompanied by a jugular murmur.
2. It is sometimes propagated to the apex or it may possess a third site.
3. It is always systolic, while the extracardiac murmurs are sometimes systolic, sometimes diastolic.
4. The anæmic murmur is never converted into an interrupted murmur, as sometimes happens with the extracardiac variety.
5. The former is present at each cardiac cycle; the latter may be absent in certain cardiac cycles and only appear in those which coincide with the end of inspiration and the beginning of expiration.
6. The anæmic murmur is much less transitory than the extracardiac murmur; it is much less apt to disappear under the influence of slight cardiac modifications.
7. The former is never accompanied by moist or sibilant râles presenting the same rhythm.

In considering these characteristics, it is evident that the anæmospasmodic murmur is much more fixed than the extracardiac murmurs, which are essentially variable and modified continually by the respiratory rhythm. In order to modify to any considerable extent an anæmic murmur, vigorous efforts must be made.

The anæmic murmur is the resultant of two factors, the anæmia, and the spasm or sudden and violent cardiac contraction. Hence the name

---

[1] Progrès Méd., July 7, 1877. Des modifications des souffles extracardiaques, p. 522.
[2] Des bruits pleuraux et pulmonaires dus aux mouvements du cœur. Thèse, 1877.

anæmo-spasmodic murmur which I have applied to it. These two factors are present in all heart murmurs. Thus, when pathological murmurs are not well marked, the patients are made to walk in order to produce more marked contractions and to develop the abnormal sounds.

ANÆMIA IN ARTICULAR RHEUMATISM.

It has appeared to me so much the more necessary to discuss the question of anæmia in detail because it is mistaken very frequently for heart disease. I refer now to acute articular rheumatism, in which I have often seen anæmic murmurs mistaken for temporary endocarditis or even for organic affections of the heart.

In the first category I place those cases in which the rheumatism is not associated with any cardiac murmur, but merely with jugular murmurs. In such cases there can be no doubt.

Rheumatism with Jugular Anæmic Bruits.

OBSERVATION VI. *Anæmia in Acute Articular Rheumatism; Cervical Bruit.*—A woman, twenty-four years of age, entered the hospital January 25, 1876; she is nursing her child aged four months. January 22d she felt pain in the left knee, unaccompanied by swelling or redness. On the following day this had disappeared but was replaced by a pain in the left thigh. Pressure over Scarpa's triangle causes pain; moderate fever.

Auscultation of the neck shows a musical bruit of very great intensity in both jugulars; nothing abnormal in the heart.

Articular Rheumatism with Anæmic Jugular and Pulmonary Murmurs.

, In a second series of cases the rheumatism is accompanied by anæmic murmurs in the jugulars and the pulmonary artery.

OBSERVATION VII. *Acute Articular Rheumatism; Cervical and Pulmonary Bruit.*—A servant, nineteen years of age, entered the hospital on April 13, 1876. About two weeks ago she began to suffer from pain in urination and a discharge. Twelve days later—*i.e.*, three days ago—she was seized on waking with a pain in the calf of the left leg, then in the left ankle and knee.

Upon auscultation, a blowing murmur was heard in both jugular veins, and in the second left intercostal space near the sternum, a blowing systolic murmur identical with those found in anæmia.

As the affection advanced it became localized in the left wrist. The patient left the hospital on May 16th, her general health being very much improved, but the wrist is still a little painful and swollen.

The jugular murmurs and the blowing murmur in the pulmonary artery were still audible.

OBSERVATION VIII. *Slight Articular Rheumatism with Anæmia; Cervical and Pulmonary Bruits.*—T——, twenty-seven years of age, a carpenter, entered the hospital on June 28, 1879.

This patient is suffering from a general articular rheumatism. It is the first attack; the pains are moderate. A faint cervical murmur is heard, and another very distinct one in the pulmonary artery. The fever is moderate and oscillates around 38°. He left cured on July 29th.

Acute Articular Rheumatism with Jugular, Pulmonary, and Mitral Anæmic Murmurs.

In a third series I will class those cases in which the anæmia has three sites, cervical, pulmonary, and mitral, and in which the anæmic character of the mitral murmur is indicated by the fact that it disappears as soon as the anæmia improves, and that the pulmonary murmur diminishes.

Observation IX. *Acute Articular Rheumatism; Cervical and Pulmonary Bruit and Temporary Mitral Bruit.*—M. A——, eighteen years of age, entered the hospital July 23, 1880. Five days ago the patient was taken with the pains of acute articular rheumatism; the affected joints were reddened and swollen; profuse sweats. July 24th, a slight systolic murmur was heard at the apex upon auscultation. In the second left intercostal space near the sternum was heard a systolic blowing murmur, prolonged to the closure of the pulmonary valves; this murmur was propagated about 3 ctm. from this point. The "bruit de diable" was heard in the vessels of the neck. Four grammes of salicylate of soda were administered.

July 25th.—Diarrhœa; pains unchanged; temperature lowered; salicylate of soda, 6 grms.

July 26th.—Pains in the neck and upper limbs have disappeared; they persist in the knees and ankles. The murmur at the apex has disappeared; that in the pulmonary artery unchanged.

July 27th.—No pain; temperature normal. Pulmonary murmur slightly diminished.

August 2d.—Murmur weaker and extends over a smaller surface.

August 8th.—Patient leaves the hospital.

Observation X. *Acute Rheumatism; Anæmic Blowing Murmur in the Jugulars and Pulmonary Artery; Anæmic Mitral Blowing Murmur.*—Augusta B——, fifteen and a half years of age, entered the hospital September 9, 1877. The patient had fever, and complained of pains in the legs and arms. Salicylate of soda relieved the pains, but on the following days other joints became swollen and painful. October 1st the patient began to walk about.

During the attack auscultation showed an anæmic murmur in the vessels of the neck. Slight blowing, vibrating murmur in the pulmonary artery. At the apex of the heart was heard a systolic murmur, which, although very distinct at the beginning of the attack, rapidly diminished, and had disappeared by October 1st; the murmur in the pulmonary artery was still vibratile, and there was a slight blowing murmur in the jugulars.

Observation XI. *Acute Articular Rheumatism, accompanied by Anæmia; Cervical, Pulmonary, and Mitral Murmurs.*—Caroline I——, eighteen years of age, domestic, entered the hospital January 18, 1879. Was attacked with articular rheumatism January 15; the knees, feet, elbows, and hands were involved. The joints are painful, hot, slightly reddened, and swollen. The anæmia accompanying the fever is shown by the following signs:

The patient is pale, and a very rough blowing murmur is heard in the vessels of the neck on both sides. A very rough systolic blowing murmur, followed by a sharp valvular click, is heard in the left second intercostal space. A circumscribed, very soft systolic murmur is also heard at the apex. The heart is small, the apex-beat being felt in the fifth intercostal space, 7 ctm. from the median line.

There is no doubt that the murmurs in the neck and pulmonary artery are anæmic in character. With regard to the murmur at the apex, the course of the disease shows that it is not due to endocarditis, but to anæmia. In fact, it disappeared on January 27th, and nothing was left but a slight murmur in the neck.

ACUTE ARTICULAR RHEUMATISM WITH ANÆMIA AND PROBABLE ENDOCARDITIS.

In a fourth series I place those cases in which the persistence of the mitral murmur renders endocarditis probable but not absolutely certain.

OBSERVATION XII. *Acute Articular Rheumatism, Anæmia, and slight Endocarditis.*—J. D——, twenty-eight years of age, entered the hospital March 21, 1878. The patient suffered from acute articular rheumatism for the first time at the age of seventeen years, and since that time has been subject to palpitation of the heart. Four days ago she was seized with severe pains in the wrists and limbs, with abundant sweats. Examination of the heart gave the following results: The apex beats in the fifth intercostal space, 12 ctm. from the median line ; the upper border of the liver corresponds to the insertion of the sixth cartilage (short sternum) ; the vertical border is near the sternum. The apex is lowered 15 mm.

A slightly rough intermittent blowing murmur is heard in the vessels of the neck. A rough systolic murmur of two pitches and followed by a distinct valvular click is heard in the second left intercostal space. It is propagated in the second intercostal space to a distance of 5 ctm. from the sternum, and in the third space to a distance of 3 ctm. The area of this murmur forms a triangle, the base of which is situated upon the second rib and the apex directed toward the apex of the heart. The murmur at the apex is systolic, rough, covering the first murmur and the lesser period of silence, and more intense than the pulmonary murmur. The question was whether the third murmur was due to anæmia or endocarditis. Four days after entrance into the hospital a friction murmur made its appearance upon the left side, and the dyspnœa soon increased. A blister was applied and the dyspnœa diminished somewhat, but intense congestion developed in the right upper lobe, with an almost tubular murmur at the level of the infraspinous fossa ; pulse, 120.

For two months the phenomena varied in intensity without disappearing ; finally, on May 10th, she left the ward convalescent. July 10th she presented herself in good condition ; the cough had ceased ; the heart beat regularly, and there was merely a slight systolic prolongation at the apex.

In a fifth series I place those cases in which anæmia and endocarditis coincide.

OBSERVATION XIII. *Anæmia and Heart Disease occurring at the same time during Rheumatism.*—P——, twenty years of age, entered the hospital November 23, 1877. Upon admission, most joints of upper and lower limbs were implicated ; there was moderate fever.

Auscultation showed blowing murmurs in the vessels of the neck on both sides; there is also a systolic blowing murmur to the left of the sternum in the pulmonary artery. At the apex is heard a soft systolic murmur, which is not propagated toward the apex but toward the mitral orifice. All the murmurs gradually diminished, and the patient left the hospital in the early part of December. The patient returned December 6th; the bruits in the neck and pulmonary artery have diminished very much; the blowing murmur at the apex, which had also diminished, nevertheless remained most marked.

OBSERVATION XIV. *Rheumatism with Anæmia and Mitral Lesion.*—Ernestine D——, twenty-seven years of age, entered the hospital March 27, 1879. The patient stated that she took cold on the previous Tuesday, and on the following day had pain in the side and lumbago. To-day she presents the following symptoms: lumbar pain and slight dyspnœa; nothing noticeable in the lungs on auscultation and percussion. Apex-beat of heart in the sixth intercostal space; a loud systolic blowing murmur is heard at the apex, the maximum of intensity being in this situation. The heart does not appear to be markedly hypertrophied.

The patient had an attack of articular rheumatism a year ago, and has often had palpitations since that time; a blowing murmur is heard in the vessels of the neck.

OBSERVATION XV. *Rheumatism, Mitral Endocarditis and Anæmia.*—Louise G——, eighteen years of age, domestic, entered the hospital January 20, 1876. This patient had an attack of rheumatism, lasting a month, in September, 1875. The second attack began January 8, 1876. It involved first the left knee, then the right leg, then the upper limbs. Upon admission, the right knee was painful and swollen; the phalangeal articulations were enlarged transversely. Auscultation showed a blowing murmur in the jugulars, and a systolic blowing murmur at the apex; nothing heard over the pulmonary artery. In this case it must be held that the jugular murmur is anæmic in character, while the mitral murmur is due to slight endocarditis.

OBSERVATION XVI. *Rheumatism accompanied by Anæmia and then complicated by Endocarditis.*—The patient, P. M——, eighteen years of age, was seized on December 23, 1880, with chills, fever, and finally double tonsillitis. A week later polyarticular rheumatism developed. Auscultation showed a double blowing murmur in the vessels of the neck on the right side. In the left second intercostal space, near the sternum, was heard a systolic blowing murmur, occupying the entire systole, and followed immediately by a sudden, metallic click of the semilunar valves. These two sounds indicated clearly the anæmic condition of the patient.

Five or six days later, auscultation revealed a third bruit at the apex and propagated toward the axillary line. This bruit occupies the entire duration of systole and is prolonged into the greater period of silence. At first it creates the impression of a double murmur, since the cardiac revolution terminates in a soft blowing murmur, and not in a distinct click like that of the semilunar valves. This is the prolonged systolic bruit to which I have applied the term paradox bruit. This murmur is an evidence of endocarditis and not of anæmia. This is also proven by the fact that since the development of the endocarditis the murmur in the pulmonary artery has markedly diminished.

Finally, in a sixth series I place those cases in which the anæmia is accompanied by articular rheumatism, complicated with endocarditis and pericarditis.

OBSERVATION XVII. *Acute Articular Rheumatism complicated by Endopericarditis and Anæmia.*—L. B——, twenty-three years of age, entered the hospital January 10, 1878. Upon admission she had acute articular rheumatism with fever. Auscultation revealed blowing murmurs in three places: first, in the jugular veins, especially on the right side; secondly, a blowing systolic murmur in the pulmonary artery, in the second intercostal space to the left of the sternum; thirdly, a soft systolic murmur at the apex.

The first two murmurs were regarded as clearly anæmic; with regard to the third I was unable to determine positively whether it was due to anæmia or endocarditis. But the course of the disease soon decided the question. Five days later a pleurisy began, then pericarditis, and on January 18th auscultation was characteristic.

The murmur in the pulmonary artery had disappeared. The blowing murmur at the apex had disappeared and was replaced by a double friction sound occupying the inferior third of the cardiac triangle, the maximum being behind the sternum, at the level of the insertion of the fifth costal cartilages. Two days after, the exudation had diminished, and the friction murmur occupied the entire pericardial region.

If the murmur at the apex had been anæmic in origin, it would have disappeared before the murmur in the pulmonary artery.

It follows from the study of these various illustrative cases that all patients who present blowing heart murmurs during the course of an acute articular rheumatism are not affected necessarily by an organic cardiac affection, but that some of the murmurs may be attributable to anæmia.

If the blowing bruits are found in the jugulars and pulmonary artery, and not at the apex, they are due to anæmia. If a murmur is found at the apex, but is merely temporary and disappears before that in the pulmonary artery, or even at the same time, it may also be attributed to anæmia.

But if the murmur at the apex is present without a bruit in the pulmonary artery, or if it persists after the latter has disappeared, mitral disease is really present. The pericardial murmur will be rarely mistaken for an anæmic bruit, except in the event (which will be discussed later) that the pulmonary artery and infundibulum give rise, during dry pericarditis, to a systolic friction murmur which accompanies the dilatation of the vessel at the moment of the passage of blood through it.

DISEASES OF THE HEART AND ITS MEMBRANES.

CHAPTER IX.

THE PERICARDIUM AND PERICARDITIS.

Anotomical Data.—Following the arrangement which I have adopted in this work, I will begin with a few details concerning the anatomy of the pericardium.

The pericardium forms a hollow cone, situated immovably in the thoracic cavity; it is perhaps the most fixed part of the human body. Its base corresponds to the tendinous centre of the diaphragm, its apex to the base of the first part of the sternum in the median line of the body, as is readily seen by forcing the finger up into the pericardial cul-de-sac; the pericardium is therefore a triangular pyramid.[1]

Its anterior surface is in contact with the sternum merely over a small space upon the median line, because the anterior cul-de-sac of the pleura covers it on each side. It follows, accordingly, that paracentesis of the pericardium is made always through the left pleura, and care must be taken not to traverse the lung. The inferior border of this triangular surface begins on the right at the level of the insertion of the fifth costal cartilage, then passes a little obliquely to the left into the fifth intercostal space.

The two lateral surfaces will not occupy our attention. It is sufficient to know that the left border slopes from the apex of the pericardium toward the apex of the heart, and, on its way, meets the two points where the aorta and pulmonary artery escape from the serous pericardium.

The right border descends to about $1\frac{1}{2}$ ctm. below the sternum; at the level of the first rib it meets the superior vena cava, which passes out of the pericardium at this place. The right auricle extends a little beyond this line, and the inferior vena cava passes vertically into the auricle, but more posteriorly.

A few words with regard to the structure of the pericardium. The parietal layer is thick, the visceral layer thin and separated from the myocardium by subserous connective tissue which is often loaded with fat.

---

[1] Vide page 2 with regard to its situation and means of fixation.

Each layer is composed of two strata, an internal one formed of pavement epithelium in a single layer; the external one is composed of bundles of connective tissue interlacing with meshes of elastic fibres. Neither the vessels nor the nerves of the pericardium present any special features; the latter, according to Luschka, are derived from the right pneumogastric, recurrent, and phrenic nerves.

The serous cavity always contains a certain quantity of fluid. According to Gorup-Besanez, this fluid is alkaline, and has the following composition:

Water	96.2 to 95.5
Solid parts	37.17 to 44.87
Albumen	21.62 to 24.68
Fibrin	0.81
Extractive matters	8.21 to 12.67
Salts	7.34 to 6.69

Anatomo-Pathological Data.—The following are the different lesions which may be found in the pericardium:

The most common and most difficult to establish is *hydro-pericardium*. In fact, a serous fluid is found constantly in the cavity upon autopsies. This fluid, examined under the microscope, is found to contain isolated epithelium cells forming shreds. This dropsy of the pericardium is attributed often to an asystoly, occurring at the close of life.

The *milk spots* of the pericardium are composed of condensed connective tissue, and may or may not be secondary to false membranes, as their surface is often covered by a layer of pavement epithelium.

Finally, the most common change consists in the presence of the *inflammatory exudation* which characterizes the false membranes of pericarditis. When these false membranes persist, they become organized and form papillary projections. New-formed vessels develop, and the new membrane is thus formed. Certain varieties of pericarditis are very peculiar. Tubercular pericarditis is characterized by the presence of tubercular granulations upon the upper portions of the pericardium and by exudation at the level of the large vessels. These tubercles may become cheesy, detached, and form gray or ochre-colored masses capable of undergoing calcareous degeneration.

Purulent pericarditis is very rare except in infancy.

THE DIAGNOSIS OF PERICARDITIS.—Pain is the first symptom to which we will devote our attention.

In primary pericarditis, beginning with violence, the pain is situated in the præcordial region below the nipple, or toward the lower end of the sternum. It radiates throughout the entire præcordial region, toward the axilla, into the left arm, the epigastric region, and left hypochondrium. It is felt very rarely on the right side of the thorax.

The pain is spontaneous, lancinating, tearing, sometimes atrocious. It is increased by the various methods of examination, by coughing, by the movements of respiration. At times no spontaneous pain is present; it is produced only on pressure.

What is the cause of this pain? Laennec attributes it to pleurisy of the adjacent parts, but I think that a distinction must be made. All the painful irradiations in pericarditis are situated in other parts than the pericardium, but real pains do occur in this structure. I admit the occurrence of central and peripheral pains.

Peter has called attention to pain in the phrenics, which is characterized by two signs, viz.: pain upon pressure over the cervical and costo-xiphoid points. The former is found between the two insertions of the sterno-mastoid muscles, immediately above the clavicle, sometimes a little higher. The costo-xiphoid point is situated at the level of the termination of the right and left phrenic nerves, on each side of the xiphoid appendix. The same writer has also isolated certain pains irradiated along the intercostal nerves, below the nipple.

With regard to the central pericardiac pain, on account of its extreme intensity Peter believes in the occurrence of an acute neuritis of the cardiac plexuses, accompanied by a tendency to syncope, to circulatory and respiratory disorders, which appear to be due to an extension of the process to the pneumogastric and sympathetic. This view, however, is purely hypothetical.

Bochefontaine and Bourceret[1] have experimented recently upon the sensibility of the pericardium under normal and pathological conditions. They have come to the conclusion that the healthy pericardium is sensitive, and that the inflamed pericardium presents a lively sensibility, at least upon its external surface and in its thickness.

But the future must decide whether we must admit the existence of a neuritis of the cardiac plexuses, or of a parenchymatous pain, properly speaking, in acute painful pericarditis. Another group of important phenomena consists of the disturbances of rhythm of the heart. The pulsations are more frequent, and this is, perhaps, the disease in which the most marked acceleration of the pulse is observed. In the beginning the pulse is somewhat stronger; as a rule, it soon becomes irregular and intermittent; a reduplication of the second sound is sometimes observed.

When the pericarditis is serous in character, when the effusion is considerable and the patient is emaciated and young, the enlargement of the pericardium is visible sometimes to the naked eye. Graves has even noticed, in infants, the pushing back of the left lung, the apex of which may project into the supraclavicular fossa.

Apart from these local functional disturbances, there are general symptoms which impart peculiar features to the disease. The fever cannot be

---

[1] Acad. des Sciences, Octobre 17, 1877.

judged, in this instance, by the frequency of the pulse, because the conditions are peculiar. The disturbances of circulation are not reflex sympathetic disorders, but the circulatory centre itself is affected.

With regard to the temperature, Bouillaud had noted an important fact. Sometimes, while the skin is very hot, the limbs are notably cool. At the same time there is a very great tendency to syncope. Charcot noticed a depression of the central temperature in certain cases of acute pericarditis in old age. Lorain observed depression of the temperature in the rheumatic pericarditis occurring in the course of acute rheumatism. Finally, Brouardel twice noted a remarkable fall of temperature during typhoid fever, at the moment of the development of a pericarditis.

These facts must be taken into consideration. In secondary pericarditis the depression of an originally febrile temperature may sometimes put us on the track of a diagnosis which is often difficult.

In addition there are a certain number of sympathetic disturbances in pericarditis which must be considered.

The dyspnœa is often extreme; it is not always diaphragmatic. In certain cases, in fact, it is a nervous dyspnœa, and is accompanied by violent respiratory efforts, by profound inspirations. An important sign, in my opinion, is the dilatation of the nares, which is often as marked as in pleurisy. Orthopnœa has not been noted by authors, but it appears to me to be an important sign. The anxiety, pallor of the face, and hiccough are also useful signs.

Finally, apart from these nervous phenomena, obstinate insomnia is usually present, and often a mild delirium.

But in certain cases, perhaps I ought to say in the larger proportion of cases—particularly in secondary pericarditis—the functional phenomena are absent or are masked by other disturbances, so that the disease is latent and recognized only upon physical exploration.

The term latent pericarditis has a very precise signification; it refers to a concealed, insidious disease, not to one which is unrecognizable. The term latent disease signifies that the disease is not revealed by functional disturbances of the affected organ, but is recognizable by the physical phenomena. We must rely chiefly, therefore, upon the physical signs of the disease.

The following are the physical signs of pericarditis:

1. *Palpation* of the præcordial region sometimes furnishes certain data, viz., the sensation of rough friction, which Broussais compared to the crackling of parchment. This is felt at the onset of the disease, when the fibrinous exudation is in process of formation. It disappears when the effusion occurs, but reappears at the close, though more feebly. As a rule, in examining for this sign the patient should be placed in a semi-recumbent position.

2. *Percussion* is useful, in a considerable number of cases, as the præ-

cordial dulness is increased; but it does not possess the same importance in adults as in children. Differences in percussion are also observed in changing the position of the patient, but these signs are delicate and unimportant.

I now arrive at the consideration of those signs which have a considerable importance as they enable us to predicate the existence of pericarditis.

3. *Auscultation* furnishes the most important indications. The pericardial friction sound, the only pathognomonic sign of the disease, is both systolic and diastolic. It often gives the sensation of coming and going. In addition, it is superficial, it is not propagated, and increases upon pressure. But the most important feature of the friction sound is its situation.

A pathological bruit is determined, upon auscultation, by three characteristics—its position, time, and timbre.

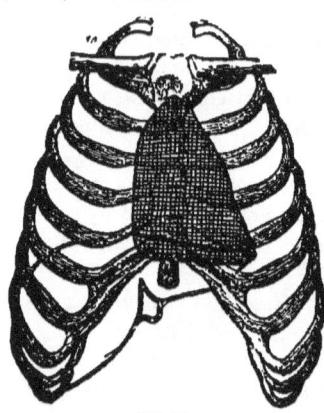

Fig. 24.

Let me say at the outset that when pains are present they often precede the production of pathological bruits; this is especially true of peripheral or extrinsic pains. If we suspect pericarditis at an early period and perform auscultation from the beginning, the murmur will be found to develop at the lower border of the heart. It soon rises and invades the entire pericardial surface and maps out the pericardial triangle with remarkable precision, so that with the exception of one point, it can even be recognized with the eyes closed.

The sides of the pericardium may be mapped out with the precision which I have indicated, but this does not hold true of the inferior border. If this border is mapped out carefully, in the manner which I have described above, it will be found that the friction sound overlaps it by 1½ ctm. This is due to the fact that the liver is a solid body and a good conductor of sound.

OBSERVATION XVIII. *Acute General Pericarditis; Friction Murmur giving the exact Shape and Dimensions of the Pericardium.*—The patient entered the hospital in 1876 for a chronic bronchitis with purulent secretion.

Careful examination showed, in the cardiac region, a leathery bruit, giving the sensation of a double friction. In marking out the region in which the bruit is heard, a triangular figure is obtained. The apex of the triangle is situated upon the median line, behind the first portion of the sternum. The right border, almost vertical, corresponds to the right border of the sternum; it is situated about 1 ctm. from the latter. The left border descends obliquely, following the normal border of the pericardium. The inferior border or the base of the triangle is approximately horizontal, but it is situated 2 or 3 ctm. below the border of the pericardium. In seeking an interpretation of this phenomenon, it must be remembered in the first place, that in emphysematous individuals the heart is lowered and presses downward upon the diaphragm, which pushes against the liver. The latter organ is solid, transmits the pericardial murmurs and enables them to be heard a little below the inferior or diaphragmatic border of the pericardium.

The murmur is a friction murmur, and readily heard; it is double and corresponds to systole and diastole. It is soft in the upper half of the triangle, louder and rougher in the lower half. Its maximum of intensity does not correspond to the apex of the heart but to the xiphoid appendix. The murmur ceases abruptly at the borders of the triangle.

The patient died on the following day, and a diffuse pericarditis was found upon autopsy. The visceral and parietal layers of the pericardium was covered with vegetations; there was very little fluid in the pericardium. The lesions were diffused uniformly. The heart was slightly hypertrophied and presented no valvular lesion.

OBSERVATION XIX. *Acute Pericarditis in the Course of an Organic Affection of the Heart.*—A patient who had long been treated in the hospital for an organic affection of the heart, with mitral lesion, muscular degeneration, and anasarca, was attacked two days ago with pericarditis.

The affection is characterized by a leathery murmur, recalling the friction sounds of inflamed tendinous sheaths. The murmur is double and gives the sensation of coming and going; its intensity is not sufficiently great to mask the pathological sounds at the apex, which are sharply defined. The murmur extends over the entire cardiac triangle. In defining the surface over which it is heard we obtain a triangular figure corresponding exactly to the anterior surface of the pericardium. The apex of this triangle is situated in the median line of the sternum, a little above the insertion of the cartilages of the second ribs. The right border descends obliquely to the insertion of the right fifth costal cartilage into the sternum. The left border descends to the apex of the heart. The lower border or base of the triangle corresponds to the lower border of the cardiac triangle, but is situated 1 ctm. below it. The murmur occurs in systole and diastole; its timbre is distinctly that of a friction sound. The maximum of intensity corresponds to the aortic orifice, behind the sternum, at the level of the insertion of the third costal cartilages.

OBSERVATION XX.—The following case occurred in the service of my colleague Dr. Mesnet:

Jean B——, sixty-four years of age, entered the hospital June 19, 1878. The patient was suffering from rheumatic pains, particularly in the fingers of the right hand. Upon auscultation of the cardiac region, a very rough and superficial double friction sound is heard, increasing on pressure. This murmur gives a distinct sensation of coming and going. It covers the entire surface of the cardiac triangle, and has its maximum of intensity in the fourth intercostal space, to the left of the sternum, very near this bone. The heart sounds are very distinct at the apex.

The vessels are atheromatous and show the visible pulse of Corrigan; no murmur of aortic insufficiency. Six days later, the pericardial friction ceased.

OBSERVATION XXI. *Acute Articular Rheumatism with Pericarditis and Pleurisy.*—X—— entered the hospital January 2, 1879. In this case the pericarditis developed in a patient who had already suffered from three attacks of acute articular rheumatism, which had not left any cardiac affection. The present attack began five days before admission to the hospital. Upon admission the rheumatism was found to be general, without any cardiac complication.

Careful daily examination of the heart revealed abnormal sounds on the twelfth day of the disease. A double friction murmur was first heard over the diaphragmatic border of the heart and extending upward to the level of the third rib. This superficial murmur sounds like crackling snow and clearly indicates pericardial friction. On the following day the murmur extended over the entire pericardial triangle, mounting to its summit behind the first portion of the sternum.

In the beginning, the maximum of intensity was heard at the level of the xiphoid appendix, on the following day it was heard at the level of the infundibulum of the pulmonary artery. January 26th, the murmur had disappeared. A left pleurisy developed during the attack of pericarditis, and disappeared at the same time as the latter.

February 1st, the patient was convalescent.

It is not sufficient to state the extent of the region occupied by the pericardial friction murmur, but its maximum and minimum of intensity must also be determined.

The most important fact, and one which has not been noticed hitherto, is that the minimum of intensity is at the apex of the heart, which has the briefest contact with the thoracic walls.

The maximum of intensity is found on the left border of the sternum at the level of the third and fourth intercostal spaces. When the friction murmur is circumscribed it is heard in this situation alone.

In addition to this constant location of the maximum of intensity of the pericardial friction bruit, another isolated spot is often found at the inner part of the second intercostal space, at the level of the exit of the pulmonary artery. This upper location is very remarkable, as it gives to the murmur a very peculiar rhythm, which I will consider in discussing the period of the cardiac revolution at which it is produced (Fig. 25).

It is remarkable to notice that the pulmonary artery may give rise to a systolic friction murmur, while, when a partial pericarditis (like that which surrounds the superior vena cava and aorta in tuberculosis) is present, the adhesions are such that the friction murmur is hardly ever heard. In such a case, in which the pericarditis is really latent and the local phenomena are also mute, the diagnosis cannot be made, and clinical experience alone will enable us to suspect it at times. In tuberculosis I have often searched unsuccessfully for a friction murmur, localized at the apex of the pericardium and which would permit the diagnosis of this adhesive pericarditis of the apex which is so frequent in this disease. In such cases I have found only extracardiac bruits, *i.e.*, pleural or pulmonary bruits with a cardiac rhythm.

As a rule the pericardial friction murmur is systolic and diastolic, it gives a characteristic coming and going sensation. When the murmur begins to grow shorter, it first disappears during diastole, then at the beginning of systole, and finally is heard only during the second part of the systole, covering the lesser period of silence. In such a case the short murmur may give to the cardiac rhythm the appearance of a rhythm of three murmurs, as if the second sound were reduplicated.

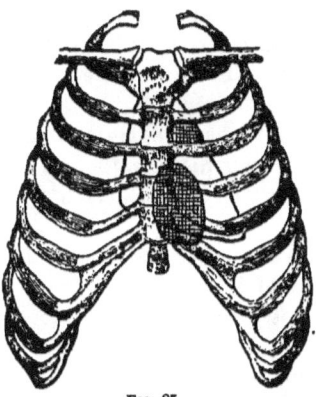

Fig. 25.

The isolated bruit in the upper location is never double, it corresponds to the systole and is due to the increase in the volume of the pulmonary artery at the moment of the arrival of the wave of blood. This is proven by the fact that the murmur does not always begin with systole and sometimes is not produced until the arterial dilatation has reached its maximum. Hence follows a retardation of this murmur, which enables us to hear the first sound of the heart before the murmur, and thus imparts the sensation of a murmur immediately preceded and followed by a valvular click.

Unlike the anæmic murmur, this friction sound is not diminished by the sitting position, but, like the former, it is modified by arrest of respiration and by exertion. But the condition of dyspnœa and the dread of syncope will not permit us to insist upon such experiments.

The third characteristic of the pericardial friction murmur is its timbre. It is rough, like the creaking of new leather (Collin), rubbing or rustling (Bouillaud), scraping or rasping. It is superficial and varies according to position and the pressure exerted upon the præcordial region.

The recumbent posture intensifies the sound if the exudation is not very abundant (Stokes, Corrigan), but it increases in the sitting position if there is very little or no effusion. It is also increased upon pressure or forced expiration.

Finally, there is no morbid vascular bruit which is modified more rapidly by cups, leeches, and especially vesicants.

Let us now sum up the important characteristics of these murmurs:

The pericardial friction murmur does not begin at the very onset of pericarditis (Mayne, Stokes) and is often preceded by extrinsic pains. It appears at first at the lower border of the heart, mounts rapidly along the sternum, and may soon reach the apex of the pericardium if the pericarditis remains dry. If effusion occurs the murmur disappears from below upward, and reappears from above downward when the effusion disappears.

Its maximum of intensity corresponds to the inferior part and to the left border of the sternum. There is sometimes a second locality at the level of the pulmonary artery.

The murmur is often double, systolic and diastolic, then it becomes shorter, ceasing at first during diastole, then at the beginning of systole, and only continues during the second half of the systole, covering the lesser period of silence.

When the murmur is isolated at the level of the pulmonary artery it is reinforced at the moment of systole. It may exist only during systole and may be delayed to such an extent as to allow the first sound of the heart to be heard previous to it.

With regard to timbre, it is more frequently rough than soft, and is modified readily by position, pressure, and treatment.

CHAPTER X.

COMPLICATIONS OF PERICARDITIS—VARIOUS FORMS OF THE DISEASE—PERICARDIAL ADHESIONS—COURSE—TERMINATION.

AT a later period I will detail the symptoms of each of the affections with which pericarditis may be complicated; but even now it will be easy to recognize, for example, when another affection is superadded to the pericarditis or when the latter occurs after a previous disease of the heart.

The following observations will show how this diagnosis can be effected.

OBSERVATION XXII. *Acute Articular Rheumatism, with Endopericarditis.* —L. H———, fifteen years of age, entered the hospital February 28, 1878. The disease began, three days before admission, with profuse perspiration and vague pains in the lower limbs. On the sixth day an anæmic blowing murmur was heard in the pulmonary artery and the vessels of the neck. At the same time a murmur appeared at the apex, but it was questionable whether this was anæmic or due to endocarditis of the mitral valve. On the seventh day, the respiration became anxious and on the ninth day pericardial friction was heard and the temperature was lowered 1° C. The friction sound covers the entire pericardial triangle; it is dry and gives the sensation of coming and going. It has two situations of maximum intensity, one at the level of the xiphoid appendix, the other over the pulmonary artery. It is very evident that the intensification at the latter site is caused by the addition of the friction sound to the systolic blowing murmur in the pulmonary artery.

On the following day the friction sounds had disappeared. On the eleventh day the friction sound reappeared, though feebly; it accompanied the systolic blowing murmur and did not present the double timbre of coming and going.

On the twelfth day the friction sound was limited to the level of the infundibulum of the pulmonary artery; it constitutes a rhythm of three sounds—the click of the auriculo-ventricular valves, a systolic bruit, and finally the click of the semilunar valves.

On the fifteenth day everything had disappeared, and the general condition was good.

OBSERVATION XXIII. *Aortic Stenosis, Cardiac Hypertrophy, Final Pericarditis.*—D. L——— entered the hospital May 31, 1877. Examination showed that the apex of the heart was beating in the sixth intercostal space, 17 ctm. from the median line. The upper border of the liver was $3\frac{1}{2}$ ctm. lower than normal, *i.e.*, at the insertion of the right fifth costal cartilage. The vertical border of the right auricle is 2½ ctm. from the sternum. The lower border of the heart is almost horizontal.

The entire cardiac surface is covered by a bruit of new leather, sibilant and double; the first sound is harsher, the second more prolonged and sibilant. They have the same timbre over the entire pericardial surface, but the intensity diminishes over the apex.

The murmur covers the entire pericardial surface and forms a triangular figure corresponding exactly to the pericardial triangle. It must be remarked that the base of the triangle is situated about 2 or 3 ctm. below the inferior border of the heart. The murmurs scarcely extend beyond the right and left borders.

The leathery bruit has its minimum of intensity at the apex, its maximum at the xiphoid appendix. There is another locality in which the blowing murmur has considerable intensity, viz., to the right of the sternum, in the second and third intercostal spaces and ascending toward the brachio-cephalic trunk.

The pulse is full, strong, regular, and possessing manifest locomotion, particularly when the artery is not tense and the arm is raised.

The patient states that he has only been sick since the previous January, i.e., for five months.

He states that he was suddenly seized one morning with a chill and that he entered the Necker Hospital a few days later, where the existence of pericarditis was recognized.

Upon his second admission to hospital, a general dry pericarditis is again recognized. The heart is hypertrophied; the cardiac orifices are healthy, with the exception of the aorta, which is dilated like the vessels of the limbs. The rough murmur, which has its maximum of intensity to the right of the sternum, in the second and third intercostal spaces, diminishes at the level of the brachio-cephalic trunk.

At the autopsy, a general dry pericarditis is found with slightly developed vegetations, an excentric cardiac hypertrophy and stenosis of the aortic orifice from adhesion of the valves.

OBSERVATION XXIV. *Acute Articular Rheumatism, complicated with Anæmia, then with Pericarditis and Left Pleurisy, then with Right Pleurisy.*—L. B——, aged twenty-three years, entered the hospital January 10, 1878. Since the age of seventeen the patient has suffered every winter from pains in the joints. Four days prior to admission she was affected with pains in the knees and high fever.

An anæmic murmur is found in the vessels of the neck and the pulmonary artery. This murmur is delayed; the closure of the auriculo-ventricular valves is first heard, then the murmur is heard during systole and is followed by the click of the semilunar valves. At the apex there is a faint, soft systolic, blowing murmur. The murmurs at the apex present a galloping rhythm.

On the eleventh day of the disease right pleurisy develops and on the fourteenth day pericardial friction murmurs appear, the maximum of intensity being found at the xiphoid appendix. On the fifteenth day the anæmia diminishes, the bruit in the pulmonary artery disappears, leaving only the double murmurs in the jugulars. There is a systolic murmur at the apex, but it is extracardiac and only appears with each inspiration. The pericardial bruits have increased, and a double friction sound is heard which occupies the entire lower border of the heart, the apex being at the insertion of the third right costal cartilage. On the eighteenth day the pericardial bruits diminished in intensity; at the same time a systolic murmur, like a jet of steam, appeared at the apex and is probably due to an endocarditis. The right pleurisy improved when the left pleura became affected. On the thirty-first day the pericardial friction has become very loud and extensive. On the fortieth day it has disappeared completely, and the blowing murmur of the endocarditis alone remains at the apex. This also improves gradually and the patient enters into convalescence.

I will not dwell upon the differential diagnosis from pleurisy, as the cessation of respiration will suffice to abolish the pleuritic murmurs.

But if the abnormal sounds do not cease upon the cessation of respiration, can we say that they are cardiac sounds? By no means. The impulse of the heart itself may give rise to abnormal bruits in the præcordial region; it matters little whether these sounds are pulmonary blowing murmurs or systolic pleural friction sounds produced by the apex. It is certain that this abnormal bruit does not map out the pericardium. If it is partial and exists only at the apex, we are very probably justified in stating that it is not a pericardial friction sound.

In the preceding pages we have discussed pseudo-membranous or ordinary pericarditis. There is another very interesting variety whose diagnosis is often difficult, viz., serous pericarditis; this may be transformed into the sero-purulent variety. The latter form is met with especially in infancy.

Pericarditis with effusion is characterized by more or less marked prominence of the intercostal spaces, and may even be accompanied by projection of these parts. The præcordial dulness is very extensive, and the pericardial sac, distended with fluid, pushes back the lung. The heart sounds are feeble and the impulse of the apex may disappear. Finally, the orthopnœa is pushed to its maximum.

The diagnosis of this variety is often made only by way of exclusion. But there is one symptom which often facilitates the diagnosis, viz., the tendency to syncope, which may terminate in sudden death.

What is the cause of this accident? Stokes attributed it to the condition of the cardiac muscle itself, to the existence of a concomitant myocarditis. This view, however, is purely hypothetical. F. Frank solved the question by the result of his experiments upon compression of the heart. A tube inserted into the pericardium conveyed into the cavity air subjected to an increasing pressure, and the death of the animal was readily produced in this manner. But the sudden death in pericarditis must not be attributed exclusively to the intra-pericardial pressure. There is no doubt that the nerves of the heart also play a part in the production of this accident. We have taken it for granted hitherto that the pericarditis was primary or protopathic, but in the large majority of cases it is secondary or symptomatic and readily escapes notice under such conditions. The functional disturbances are then modified or are even absent.

Age is one of the most important etiological factors. In the new-born pericarditis most frequently passes unnoticed. In old age the remarkable independence of the different organs is also shown in pathological conditions, and we should not be surprised to find old people walking about while suffering from purulent pleurisy or pericarditis.

At other times the pericarditis is latent because it is concealed by other and more marked phenomena referable to adjacent organs, such as a serious affection of the thoracic viscera, or an osseous lesion in the præcordial region.

There are certain diseases in which pericarditis occurs so frequently that it should always be looked for. In the foremost rank stands acute articular rheumatism, then scarlatina, rubeola, variola, purulent infection, puerperal infection, typhoid fever. I will also add erysipelas, which, according to Jaccoud, is complicated now and then by pericarditis. It has also been known to occur during the course of cancer, tuberculosis, and Bright's disease. It may also be due to extension of inflammation from inflammatory lesions of the heart, endocardium, and aorta.

An abscess of the liver may also open into the pericardial cavity. Stokes and Graves have reported cases of rupture of a hydro-pneumothorax into the pericardium.

This accident gave rise to a peculiar murmur, bruit de moulin (a mill), produced by the cardiac contractions. This murmur should not be mistaken for another very peculiar extracardiac bruit, which has been described by Stokes and which I also observed in a phthisical patient. This individual was suffering from pyo-pneumothorax. At a certain time the level of the fluid reached a point where the impulse of the heart gave rise to a systolic succussion sound.

Pericarditis may recover entirely or it may pass into the chronic condition, characterized by the persistence of pericardial adhesions.

Stokes called this form pericarditis obliterans, Bouillaud applied to it the term ankylosis of the heart. At the present day it is known as pericardial adhesions.

PERICARDIAL ADHESIONS.

The pathological anatomy of pericardial adhesions is interesting. These adhesions may be total or partial. Andral reports a case in which they formed during acute pericarditis. They are observed under three conditions. Bands are found which are stretched especially around the large vessels, more rarely in the lower region of the heart. These bands may break and result in long filaments which are free at one end and usually attached to the visceral pericardium.

At other times lamellar and partial adhesions are scattered over the surface of the heart to a variable extent. Or finally the adhesions may be general. The adhesions which cover the entire surface of the heart present different aspects according to circumstances. Sometimes a hard, thick tissue is found, which may even be 1 ctm. in thickness. At other times the layers are thin and loose and can be torn off readily.

Vessels ramify in these new membranes and adhesions, which may sometimes undergo calcareous, tubercular, and even cancerous degeneration, as has been observed by Cruveilhier.

These chronic adhesions occur at all ages, even as early as eleven months, as was observed in Betrier's service at the Hôtel Dieu.

Morgagni, Haller, and later Bouillaud believed that hypertrophy of the heart with dilatation of its cavities resulted from pericardial adhesions. Stokes opposed this view and stated that he had seen a considerable number of cases without hypertrophy or dilatation. I can also recall a remarkable case of cardiac adhesions with a small heart.

Nevertheless, it is very positive that a certain number of cases are accompanied by hypertrophy without valvular lesion.

On the other hand, dilatation of the heart, either with or without hypertrophy, often results from the lesion in question. At other times atrophy may be the final result, as is admitted by a certain number of authors. This atrophy may be simple or the result of calcification of the pericardium and heart. But Liouville has published a remarkable case of calcification with hypertrophy.

Passive dilatation of the orifices of the heart may also be a final result.

Diagnosis.

The symptoms of pericardial adhesions are very often misconstrued, always obscure, sometimes unnoticeable. Pulsation of the epigastrium below the xiphoid appendix has been attributed to it, and Barth, Roger, and Skoda regard this as a pathognomonic sign. It is probable that this sign has often been mistaken for hepatic palpitations due to lesions of the tricuspid valves. Skoda, Freidreich, and Potain have described cases of reduplication of the heart sounds in pericardial adhesions, but no importance is now attached to the symptom in the diagnosis of this disease.

Finally, Williams has described agitation with systolic retraction of the intercostal spaces at each cardiac revolution. But, according to this author, an adjacent pleurisy with adhesions must also be present in order to produce this phenomenon. As a rule, however, pleurisy is not present in pericardial adhesions.

It is evident, accordingly, that the symptoms of the disease are very vague and the diagnosis doubtful.

In my opinion, however, there is a very rational manner of regarding the symptoms. If cardiac hypertrophy is present, with a violent impulse of the heart as a whole but a feeble impulse of the apex, we are justified in suspecting pericardial adhesions. Under all circumstances, however, it must be admitted that the diagnosis remains hypothetical unless the patient is markedly emaciated.

Among one hundred and fifteen cases of sudden death, Aran observed nine of complete adhesion of the pericardium.

CHAPTER XI.

HYDROPERICARDIUM.

This lesion has lost much of its importance since pathological anatomy differentiated serous effusions from inflammatory products.

It is rare that the pericardium does not contain from twenty to thirty grammes of fluid upon autopsy. This quantity is much greater in certain cases, for example, in emphysema, phthisis, organic heart disease, thoracic deformities, etc., in a word, in all those conditions which result in a stasis of blood in the coronary veins. In a general way, it may be stated that the amount of pericardial fluid is greater the longer the death agony has lasted.

The fluid, as a rule, is a clear, brownish or greenish serum, rarely rendered cloudy by epithelium cells. A small quantity of the coloring matter of the blood sometimes imparts to the fluid a reddish or brownish discoloration. Its reaction is always alkaline, and has the following constitution:

	Wachsmuth.	E. Wagner.	Gorup-Besanez.
Water	95.27 to 97.34	96.51	95.51
Albumen	1.43 to 3.01	2.01	2.46
Fibrin	……	……	0.08
Extractive matters	……	……	1.27
Inorganic salts	……	……	0.95

No traces of inflammation are to be found upon the pericardium. If the dropsy has lasted for a long time the subserous cellular tissue becomes œdematous.

Hydropericardium may be due to local affections in the neighboring parts, or to general affections. The local affections which give rise to it are such as produce stasis in the veins or lymphatics of the heart and its membranes, or the adjacent parts. In other cases the general cause may consist of diffuse dropsy, parenchymatous nephritis, tuberculosis, cancer, and other cachexiæ.

Hydropericardium is always a latent affection, and when not abundant, escapes detection. It is only recognizable when the pericardium is found distended by fluid. Hydropericardium may then be suspected, if no friction murmur has been detected.

As a rule, the affection occurs as a final phenomenon of grave local or general diseases. The most difficult part of the diagnosis in such cases consists in determining whether the phenomena observed are due to the asystoly from muscular degeneration or exhaustion, or to the hydropericardium.

The following are the symptoms observed in such cases : The pulse is small, there is marked venous stasis, cyanosis, diminution of the secretion of urine, compression of the lungs, and depression of the diaphragm.

HÆMOPERICARDIUM.

Hæmopericardium is the result of a rupture of the heart, an aneurism of the first part of the arch of the aorta, or even an aneurism of the coronary arteries. The rapidity with which death occurs, corresponds to the amount of blood escaping at one time ; if the pericardium only becomes filled after the lapse of half an hour or an hour, death is delayed in consequence.

PNEUMOPERICARDIUM.

The presence of gas in the pericardium is extremely rare. Some authors have observed it when the pericardium contained sanious fluid which had undergone decomposition. At other times, the gas is derived from abscesses of adjacent organs which have perforated the pericardium.

In the cases of Chambers, Beckers, and Tuetel, an ulceration of the œsophagus had perforated the pericardium. In Dowel's case the ulceration started in a cavity, in Eisenlohr's case in a pyo-pneumothorax. Sæxinger found an ulcer of the stomach ; Graves an abscess of the liver. The affection may be due also to penetrating wounds of the chest. Förster states that in certain cases the gas may be the result of post-mortem decomposition.

The general symptoms of pneumopericardium are the same as those of hydropericardium, superadded to fever and delirium.

The local symptoms present some peculiarities. The distention of the pericardium gives rise to projections, which are visible if the patient is not too obese. Palpation may reveal a sort of gurgling, synchronous with the heart sounds. Percussion gives a characteristic sound ; it is a clear, tympanitic bruit, with a metallic timbre. This sound, occupying the upper portions of the pericardial cavity, is displaced according to the different positions of the patient. The area of absolute cardiac dulness is considerably diminished. Feine states that he heard the cracked pot note, although there was no opening into the pericardium.

Auscultation reveals some curious phenomena. The heart sounds are so loud that they are heard, not only by the patient himself, but also by those surrounding him. The heart sounds have an exceptionally high pitch, and sometimes have a clear, metallic quality.

Exudation into the pericardium occurs at the same time, and the friction sounds also present a high pitch and a metallic quality. Metallic tinkle has sometimes been heard.

Death always ensues rapidly, except in the traumatic cases, in which recovery may occur.

Tumors and Foreign Bodies in the Pericardium.

The pericardium is sometimes the site of carcinomatous growths, which are due to the propagation of cancer from the mediastinum or œsophagus.

Finally, foreign bodies known as cardioliths have been found in the pericardium. They are usually of a whitish color, smooth, and as large as a pea or bean; they are fibroid in their structure, being formed of concentric layers and calcified either throughout their entire extent or only in the centre. Hydatids have also been observed rarely within the pericardial cavity.[1]

---

[1] Stein-Lein: Diss , Erlangen. Hyrtl: Sitzungsber, der k.k., Akademie zu Wien, 51, Bd. 23, March, 1865. Klob: Zeitschr. der k.k., Gesellsch. d. Aerzte zu Wien, 49, 1860. Rokitansky: Lehrbuch d. path. Anatomie.

CHAPTER XII.

THE ENDOCARDIUM AND ENDOCARDITIS.

Anatomical Data.—The endocardium is composed:

1. Of an endothelial layer consisting of pavement epithelium.
2. A sub-epithelial layer, formed of large flattened cells, directed parallel to the inner surface of the endocardium; they present a lenticular nucleus and are separated from one another by hyaline basement substance.
3. A fibro-elastic deep layer, which varies in thickness according to its situation. It is ten times thicker in the left auricle than in the cavity of the ventricle.

The valvular folds are formed by the reduplication of the endocardium. The ventricular surface of the aortic valves is thinner than the aortic surface.

The vessels and nerves of the endocardium are very numerous; they are situated in the deepest or fibrous layer. In the auriculo-ventricular valves, however, a certain number of capillaries pass between the two deeper layers and some even make their way to the surface, while none are found in the semilunar valves.

The nerves are derived from the pneumogastric and sympathetic; some fibres have been traced to the vicinity of the endocardium. The latter does not possess much sensibility.

This sensibility is peculiar in character; it is unconscious, reflex, and responds only to its normal stimulant, the blood. If the heart of an animal is removed and rhythmical contractions have ceased, they may be made to reappear by the passage of a current of blood. Faradic stimulation of the endocardium also produces reflex contractions.

ENDOCARDITIS.

Anatomo-pathological Data.—Redness constitutes the initial lesion of endocarditis. According to Bouillaud, it may be rosy or scarlet, general or partial, but is much more marked upon the valves. Laennec showed that in a certain number of cases these changes were due to post-mortem imbibition of blood. Bouillaud then insisted upon the signs above mentioned, but states that, in order to constitute evidences of inflammation, they must be associated with the following conditions: 1. The cadaver must not be in a state of decomposition ; 2, the endocardium must be thickened ; 3, a certain amount of pus must be present; 4, or, at least, other inflammatory lesions such as softening or friability of the endocardium, ulcerations, or erosions. Förster[1] adds that the redness of imbibition is superficial.

Cadaveric imbibition is confined to those parts which correspond to the red portions of the heart-clot, and the limit of redness is bounded by a line parallel to the horizon, whatever may have been the position of the body.

Finally, inflammatory redness differs from post-mortem imbibition, inasmuch as it extends deeper than the latter.

The second characteristic of inflammation is thickening of the endocardium, accompanied by friability and softening. If the disease has lasted a week the thickening is marked upon the valves, where it is characterized by a cloudy, unpolished, corrugated appearance, and by adhesion to the subjacent tissues. No exudation can be detected. Förster states that this appearance of endocarditis is due to proliferation of the cells of the connective tissue.

Small fibrinous cysts, due to the disintegration of fibrin (Lebert, Charcot, Virchow), are sometimes found. It is generally admitted that no pus can be detected, but Lancereaux states that he has observed it. In such cases the small abscesses are found in the deep connective tissue layer, and are, so to speak, sub-endocardial.

The third characteristic of endocarditis is the production of vegetations.

They are so small at times, that they result merely in a rough, shagreened appearance of the serous membrane. In this shape they extend often over a large part of the left auricle and ventricle. They are sometimes much larger and attain the size of a pea. In the horse I have seen them an inch in length. As a rule, they are conical, sometimes are arranged in nummular patches, but generally in raspberry-like projections.

They are situated commonly upon the auriculo-ventricular or semilunar valves, more rarely at the orifices. Those upon the semilunar valves are situated, in general, near the corpus Aurantii ; they are not situated upon the edge, but follow the vascular network. Cornil and Ranvier have shown that they are due to proliferation of the cells of the flattened layers with infiltration of a certain number of white blood-globules and a hyaline layer of fibrin. According to Klebs,[2] the vegetations generally contain micrococci. We will see that these organisms play an important part in ulcerative endocarditis.

In chronic endocarditis, on the other hand, the lesion assumes the character of an interstitial inflammation, with a tendency to cirrhosis or sclerosis of the valve.

---

[1] Förster: Handb. d. path. Anatomie. 1863.

[2] Rosenstein: Krankheiten des Herzens, p. 99. Leipzig, 1870.

As the lesions are in contact with a rapid current of blood, it follows that they will be swept away if not very adherent to the subjacent tissues. Thus the fibrin which has been removed from the blood by projections, and the vegetations themselves when their base is not firmly attached, are carried away by the blood. The detached particles constitute emboli, which pass to the different viscera.

According to Sperling, among 84 cases of embolism from endocarditis of the left heart, the various organs were affected in the following proportions:

Kidney	57	times.
Spleen	39	"
Brain	15	"
Skin	14	"
Liver and intestines	1	"
	126	"

When the emboli are large and visible to the naked eye they produce anæmia of the organs, and then necrobiosis; when very small or microscopic they produce infarctions and abscesses.

These infarctions are recognized by their pyramidal shape, representing the ramifications of the circulation. They are red, black, brown, or cheesy, according to age. Virchow states that, unlike clots, they cannot be dissolved in potash, a characteristic since confirmed by other observers (Charcot, Chalvet, Lancereaux, Duguet, Hayem).

In the fœtus, in whom the right heart acts more vigorously than the left, endocarditis affects the former more frequently. Rauchfuss collected 192 cases of endocarditis on the right side in the fœtus, and only 15 cases on the left side.

After birth, this relation is reversed, the left heart being affected much more frequently than the right.

According to the majority of writers, pericarditis is much more frequent than endocarditis. The frequency of the latter affection in acute articular rheumatism has also been investigated. Bamberger gives the proportion as 20 to 100; Lebert, 17 to 100; Wunderlich, 15.7 to 100; Rolle, 12.6 to 100.

Bouillaud alone has given a much larger proportion, viz., 80 to 100. I am convinced, from my investigations on anæmia, that Bouillaud has classed among the endocarditides all the blowing murmurs produced by anæmia, and I have shown how frequent these murmurs are.

Sperling furnishes the following figures with regard to the location of endocarditis in 300 cases: mitral valve, 255 times; aortic valve, 129 times; tricuspid valves, 29 times; pulmonary valve, 3 times.

We also furnish the following statistics:

Willigk, among 4,567 autopsies, noted 238 valvular lesions, or 5 per cent.

Förster, among 639 autopsies, noted 72 valvular lesions, or 11 per cent.
Chambers, among 2,161 autopsies, noted 367 valvular lesions, or 17 per cent.

With regard to sex:
Willigk, among 238 cases, observed 86 males, 152 females.
Cason, among 41 cases, observed 28 males, 13 females.
Bamberger, among 230 cases, observed 118 males, 112 females.

DIAGNOSIS OF ENDOCARDITIS.

The onset of endocarditis is always insidious, especially when it occurs in the course of other serious affections. As a rule, however, a dull pain is experienced, rather a malaise or feeling of anxiety than a severe pain. When the pain is intense, it can be explained almost always by a coexisting lesion, either pericarditis, pleurisy, or intercostal neuralgia.

Upon palpation the impulse of the heart is found to be more vigorous, or at least it is felt over a greater extent of surface than normally.

Is it necessary to conclude, therefore, that the heart is increased in volume? Some authors have thought so; but the conclusion will probably be reached that this is a purely theoretical view, if the means at our disposal for measuring the heart be resorted to with any degree of care. Since I discovered the plan for accurate measurement of the heart, I have observed nothing of this nature.

Auscultation furnishes very valuable data in simple acute endocarditis. Bouillaud had noted the presence of a blowing murmur, which he describes as sometimes rough, sometimes soft, sometimes short, sometimes prolonged. In order to describe this murmur with the requisite precision, I will examine it under its triple relations of place, time, and timbre.

The site of the murmur depends upon the site of the lesion, and on account of the much greater frequency of the lesions at the mitral valve, it is heard particularly at the apex (Figs. 26 and 27). If the region occupied by the bruit in this locality is very small, it is heard at the very extremity of the apex where the shock is felt, and a little to the outside rather than within this point.

If the region is more extended it covers the right ventricle and approaches more or less to the median line, but does not pass the middle of the distance between the apex and the median line.

The region in which it is audible then forms a triangle, the apex of which is situated between the heart and the axilla; the base of the triangle is vertical and parallel to the median line; the inferior horizontal border is parallel to the lower border of the heart, and about 1 ctm. below it; the upper border of the triangle follows the right border of the heart.

The maximum of intensity of the murmur is usually at the apex of the heart, but is sometimes at the upper angle of the schematic triangle. The murmur begins exactly at the beginning of systole, it is longer than the clicking of the mitral valves, and lasts during the entire period of systole.

It covers, therefore, the lesser period of silence and continues until the second sound. But it is not always arrested at this period, and may continue even during diastole. The murmur is then reinforced during the closure of the semilunar valves, and is prolonged into the greater period of silence, thus assuming the appearance of a systolic followed by a diastolic murmur. For this reason I have given to this prolonged systolic murmur the name "paradox bruit." It will be considered in detail under the head of mitral insufficiency. This prolongation of the systolic murmur into the period of greater silence is only heard over a very small space, usually toward the outer angle of the schematic triangle; as we approach the median line the bruit becomes shorter and soon purely systolic, followed by a very clear semilunar click, thus showing that these valves are healthy.

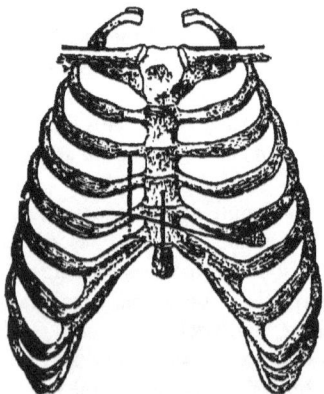
FIG. 26.—Schema of the Blowing Murmur of Mitral Endocarditis.

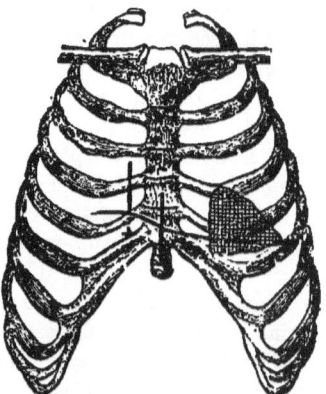
FIG. 27.

The murmur varies in timbre, being sometimes soft, at other times being more metallic.

I stated above that the murmur presents its maximum of intensity, as a rule, at the apex, but never at the inner angle of the schematic triangle, thus differing entirely from the pericardial friction sound, which always presents the minimum of intensity at the apex. When the two affections coexist the schematic triangle assumes a peculiar shape; it first covers the pericardial triangle, and then shows a prolongation toward the axilla.

How is the murmur affected by changes in the position of the patient? Sidney Ringer, Hardy, and Behier state that it increases when the patient is seated; Cuffer asserts that it is intensified in the recumbent posture. For my own part I have never noticed any great difference.

The pulse is frequent in endocarditis and may rise to 140 or 160. The rhythm is irregular, with intermissions and inequalities in successive pulsations.

The temperature is elevated, even when the endocarditis develops during the course of a febrile disease : at times a chill is observed.

The venous circulation is interfered with ; a bluish tinge of the face and hands and œdema of the lower limbs may be noticeable. On the part of the respiratory organs we notice anxiety, oppression, orthopnœa, which may lead to fainting and lipothymia.

When the endocarditis occupies the aortic orifice, the symptoms differ to a remarkable degree, not alone with regard to the local but also the general symptoms.

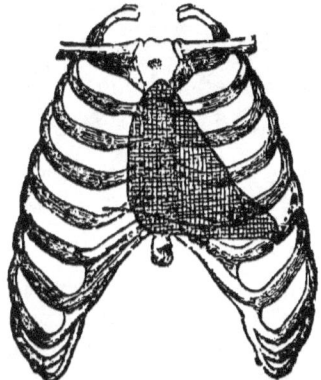

Fig. 28.— Schema of the Blowing Murmur of Pericarditis with Mitral Endocarditis.

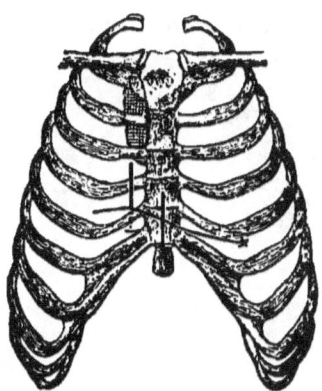

Fig. 29.

In this event the inflammation is localized at first upon the nodules of Arantius. If the vegetations acquire a certain size they result in narrowing of the orifice, and a murmur is found which forms an ascending column along the right border of the sternum. This column usually starts from the third right costal cartilage and rises to the clavicle (Fig. 29).

The murmur is systolic and covers the lesser period of silence ; it is followed immediately by the semilunar click, but it never passes into the greater period of silence. It remains purely systolic. With regard to timbre, it is usually rough and superficial ; it is not affected by a change in the position of the patient.

If the lesions are such that insufficiency is produced, the murmur is entirely different. It then forms a descending column which starts from the third right cartilage ; this column, which is almost as large as the sternum, occupies the right border and largest portion of this bone, and often descends below the inferior border of the heart (Fig. 30).

This bruit is diastolic and replaces the click of the aorta, but is not prolonged very far into diastole, since it hardly ever suppresses entirely the greater period of silence. It is soft and deep, and is increased by standing, which enables it to flow back more completely into the ventricle.

Not alone the local phenomena but the general ones also differ. It does not give rise to cyanosis but to pallor and anæmia, a sort of temporary chlorosis with anæmia.

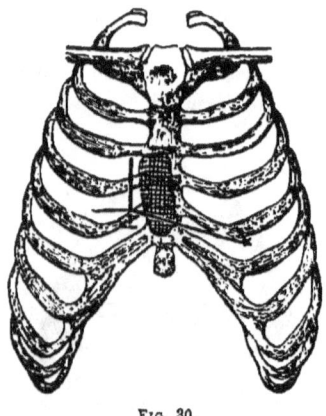

Fig. 30.

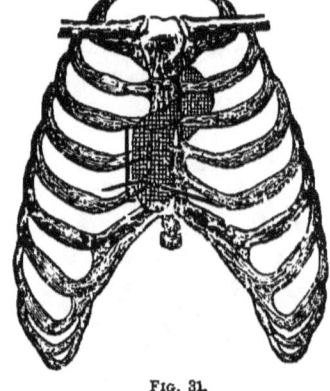

Fig. 31.

These two systolic and diastolic murmurs are readily distinguished from anæmic bruits, as they are situated on the other side of the sternum. However, they are often associated with one another, as is shown in Fig. 31, which shows the location of an anæmic murmur in the pulmonary artery and a diastolic murmur of aortic insufficiency.

DIFFERENTIAL DIAGNOSIS OF ENDOCARDITIS FROM ANÆMIA AND PERICARDITIS.

There is no necessity of dwelling long upon the differential diagnosis of anæmia and endocarditis. The evident localization of the murmur either at the apex of the heart or at the right of the sternum shows that the pulmonary artery is not affected.

Anæmia, on the contrary, gives a blowing murmur which is evidently in the pulmonary artery. I need merely refer to the chapter on anæmia, in which I have shown that it is always possible to distinguish anæmia from endocarditis.

If the murmur is situated at the apex alone, there can be no question of anæmia, since this is always associated with the murmurs in the pulmonary artery and right jugular. If there are three localities in which the bruit is heard, it may be due to endocarditis or anæmia; as anæmia is modified more rapidly than endocarditis, it may be concluded, if the mitral murmur disappears first, that we have to deal with anæmia, not with endocarditis. If the mitral bruit is accompanied by a jugular bruit, without one in the pulmonary artery, there is evidently a coincidence of anæmia and endocarditis.

Finally, if the three murmurs coexist, and that in the pulmonary artery disappears first, anæmia and endocarditis are both present.

The following are three new illustrations of these combinations:

OBSERVATION XXV. *Rheumatism, Endocarditis with a Mitral Murmur, and Anæmia with a Cervical Murmur.*—Ernestine D——, aged seventeen years, entered the hospital March 28, 1877.

The patient suffered from acute articular rheumatism a year ago, and since that time she has been subject constantly to palpitation of the heart.

Five days ago she was seized with lumbago. Upon admission she complains of lumbar pains and dyspnœa. The impulses of the heart are vigorous; the apex beats behind the sixth rib. Auscultation shows a blowing murmur, with reinforcement, in both jugulars; nothing in the pulmonary artery. A systolic blowing murmur is heard at the apex.

OBSERVATION XXVI. *Acute Articular Rheumatism with Anæmia, without Endocarditis.*—Marie G——, entered the hospital for a second attack of acute articular rheumatism. A bruit de diable is heard in the jugulars and a marked bruit in the pulmonary artery; nothing at the apex.

OBSERVATION XXVII. *Acute Articular Rheumatism with Anæmia, without Endocarditis.*—Marie R——, entered the hospital March 4, 1881, suffering from acute, monoarticular, gonorrhœal rheumatism. April 15th, a fresh attack occurred, involving a large number of joints. The patient complained of palpitation, and auscultation revealed two anæmic murmurs, one in the vessels of the neck on the right side, the other in the pulmonary artery. The latter murmur is situated in the left second intercostal space; it is systolic and is followed immediately by the second sound of the heart; its timbre is rough. The murmur increases upon the cessation of respiration.

I do not speak of the diagnosis of an endocarditis situated in the pulmonary artery and which may be mistaken for anæmia, since endocarditis in this locality is extremely rare. However, Dujardin-Beaumetz has observed one case. In such an event the bruit is located in the second left intercostal space, but is not accompanied by a jugular bruit, as is the rule in anæmia.

The diagnosis between tricuspid endocarditis and anæmia will be rendered easy by the difference in the location of the murmurs, the former giving rise to a systolic bruit situated at the level of the fifth costal cartilage, near the sternum and near the border of the heart.

Differential Diagnosis between Endocarditis and Pericarditis.—These two affections have certain common characteristics, but their bruits differ notably from one another:

1. *Topography.*—The topography of endocardial murmurs depends upon the orifices affected. They are situated almost always at the periphery of the cardiac triangle, and pass beyond it on various sides—toward the axilla in mitral disease, toward the right clavicle in aortic disease, and toward the left clavicle in affections of the pulmonary artery.

The bruits of pericarditis are limited to the pericardial triangle, particularly to the middle and lower portion along the left border of the sternum. This site may be occupied also by a tricuspid murmur, and the characteristics of the latter will be discussed at a later period.

2. *Time.*—The bruit of pericarditis is often systolic and diastolic, with a sensation of coming and going which is not met with in endocarditis. If the bruit is only systolic, it is recognized nevertheless by its timbre and modifications.

3. *Timbre.*—The bruit of pericarditis may be soft like that of endocarditis, but this is exceptional; it is much more frequently a friction sound.

4. *Depth.*—The bruit of pericarditis is much more superficial than that of endocarditis.

5. *Evolution.*—As the exudation increases the friction sound rises; extends above as it diminishes below, and reappears when the exudation is absorbed.

6. Pericardial murmurs are modified readily by pressure upon the sternum, change of position, etc.; this does not hold good of endocardial murmurs.

Differential Diagnosis between Endocarditis, Pleurisy, and Pneumonia.—At first sight this may seem puerile, but we know that the central organs of circulation may give rise to extracardiac murmurs which are singularly deceptive. In such cases, cessation of respiration will almost always suffice to recognize their true character. I refer the reader to the chapter on anæmia, in which the characteristics of these murmurs have been considered in detail.

Diagnosis of Endocarditis and Myocarditis.—Logically this diagnosis should not be made until the symptoms of myocarditis have been described.

For the present I will suppose that the patient presents in a marked manner the symptoms attributed to endocarditis, and that anæmia, pericarditis, etc., have been eliminated. Are we, therefore, certain that we have to deal with an attack of endocarditis?

Palpitation of the heart and a murmur at the apex are observed very frequently during the course of an acute articular rheumatism, and the diagnosis of endocarditis is made at once. But if this murmur disappears after a few days without leaving any effects upon the action of the heart, can we be certain that it has been produced by an exudation? Furthermore, if this murmur, as sometimes happens, disappears for a few days and then reappears, the theory of a permanent lesion becomes problematical.

I have supposed that a bruit which acts in this manner may be the result of a temporary paresis of the cardiac muscles, particularly of the columnæ carneæ, *i.e.*, a form of localized, temporary myocarditis.

COURSE OF ACUTE OR SUBACUTE PLASTIC ENDOCARDITIS.—Plastic endocarditis usually lasts one to three weeks and then ends in resolution. It often terminates in chronic sclerotic endocarditis. At other times it proves fatal by complications which are situated either in the myocardium or surrounding organs (myocarditis, pericarditis, pneumonia, pleurisy, embolism). In these cases death may not occur until the end of two or three months. We may even ask, with Fuller, whether vegetations which have once formed, may not disappear.

The diagnosis of endocarditis also includes its cause. It may be primary, due to cold, and not followed necessarily by articular affections, although in such cases it is usually the first manifestation of acute articular rheumatism. It is said that it is never traumatic in its origin, as pericarditis may be.

Endocarditis, then, is usually symptomatic of rheumatism, particularly of the acute articular variety. It may also be symptomatic of gonorrhœal rheumatism. Some writers have observed endocarditis during the course of gonorrhœa without articular complication.

The other diseases which predispose to endocarditis are: pregnancy, and particularly the puerperal condition, scarlatina, variola, rubeola, diphtheria, erysipelas, nephritis, and finally syphilis (Ricord, Lebert, Virchow).

CHAPTER XIII.

ULCERATIVE ENDOCARDITIS.

ULCERATIVE endocarditis is remarkable for the loss of substance it occasions. It appears to affect particularly those parts in which the connective tissue is abundant, the mitral and aortic valves, even the tricuspid and pulmonary valves. The lesions are more frequent in the left heart than in the right, though the difference is not so great as in plastic endocarditis.

The lesion is deeper than in the latter form, it is situated in the fibrous layer. The ulcerations are found especially upon the mitral valve, at the edge of one or both leaflets, sometimes in the midst of vegetations, sometimes on the opposite surface; at times small abscesses are found under the endocardium.

The aortic valves are affected usually upon their upper surface. The ulcerations are surrounded by vegetations, and perforation of the leaflets may occur.

At other times the ulceration spreads laterally, separates the layers of the valves and forms cavities with narrow openings, which are filled with blood and form small aneurisms.

At the onset, the protoplasm of the cellular elements becomes swollen and cloudy; then the endocardium in the affected region becomes opaque, grayish, and less smooth, and, if the process continues, it attacks the nuclei of the elastic elements. The endothelial covering begins to disappear in places, and the lymphatic corpuscles make their way into the interstices.

Fibrinous deposits then appear upon the affected spots, and the subjacent tissue becomes soft and friable. The small granulations and the deposits covering it are detached, the tissue softens, and ulceration is produced. In addition to these ulcerations foci of myocarditis are found, and these may empty their contents into the endocardium.

The blood presents lesions corresponding to the two forms, the septic and purulent.

In the septic form, Virchow found in the blood the debris of the endocardium; in the infarctions, Charcot and Vulpian found fine granulations which resisted acid and alkaline reagents, and the fibrinous detritus of pus-globules, in fine, putrid capillary emboli.

In the form which resembles purulent infection, the fluid is not really pus (Virchow and Charcot) but disintegrated fibrin, and the secondary abscesses, therefore, are not true metastatic abscesses.

The myocardium is profoundly changed. The muscle is soft and friable; in places, the fibres present fatty granulations. The nuclei of the interstitial tissue are increased. At the surface of the muscle, either beneath the endocardium or pericardium, are found small foci of granulo-fatty granulations which appear to be altered leucocytes. The muscular fibres in the affected parts have lost their transverse striæ. Finally, the myocardium contains true abscesses. The pericardium is covered sometimes by false membranes. In addition there are certain lesions produced by the detachment of particles from the central lesions (infarctions), though these are not constant. The infarctions may be present in the spleen, kidneys, brain, and liver.

In this case the spleen is softened, and one portion is found to have a different color from the rest of the surface. This part corresponds to an obliterated vessel. At the outset of the lesion, a deep redness, a simple hyperæmia, is found; but at a later period, an indurated mass is found of a yellowish chamois color. At a still later period, the tissues may be disintegrated and become diffluent, but true pus is never observed.

In the kidneys, the infarctions occupy the cortical substance and form a yellowish pyramidal mass, the base being situated at the periphery, and the apex toward the centre. The infarction is pale yellow, surrounded by a red or violet circle. In addition scattered ecchymoses are noticed. Duguet and Hayem have sometimes found very small infarctions, which can be enucleated and are formed of leucocytes and red blood-globules.

In the brain, the spots of softening occupy the optic thalami and corpora striata. A thickened false membrane is found in the fissure of Sylvius, corresponding to an obstruction of the middle cerebral artery. A certain number of small uniform foci are found and, upon the meninges, thickening of the membranes and ecchymotic spots.

Infarctions in the liver are rare. The organ is found enlarged, friable, soft and yellowish. The hepatic cells are altered, they contain fatty granulations which are sometimes the cause of the jaundice.

As a rule the lungs are but little affected; congestion of the lower parts and a few puriform patches may be observed.

Hemorrhages are found disseminated throughout the intestinal tract, the muscles, parotid gland, retina, choroid, and finally the skin. Deposits of tyrosine and leucine may be present in the bones.

Pathogeny.—Senhouse-Kirkes thought that the cardiac products are detached, obliterate the small vessels, and lead to necrosis of the corresponding field of circulation.

Virchow has supported this view by showing the analogous chemical composition of the fragments found in the vessels at the centre of the lesion and the changes of the endocardium.

Charcot and Vulpian accept this theory, but do not believe that these fragments possess a peculiar septic nature.

Jaccoud and Lancereaux regard this affection as representing sometimes putrid infection, sometimes purulent infection.

Finally, Dupuy, Daguet, Hayem, and Martineau think that the peripheral lesions are not due to embolism, but that they may be produced under the sole influence of the general condition.

DIAGNOSIS OF ULCERATIVE ENDOCARDITIS.—The symptoms may be divided into two classes: 1. The central or cardiac disturbances and the general symptoms produced by the a dynamic condition. 2. The symptoms produced by lesions of the different viscera.

The cardiac symptoms are often slight, and lost in the general grave disturbances presented by the patient. At other times they are marked, but the condition of the patient does not permit a methodical examination.

The pain is sometimes very acute, and the patient may be tormented by such excessive anguish that he is unable to rest. He suffers from painful palpitations which are increased on the slightest movement. The heart beats irregularly, sometimes with violence and rapidity, at other times there are periods of syncope in which the pulse scarcely can be felt. The examination of the heart is difficult, as percussion is painful and the patient cannot endure pressure upon the cardiac region. Blowing murmurs are heard which vary in situation and intensity, some being due to endocarditis, others to pericarditis.

The general symptoms always present a grave appearance: anxiety, orthopnœa, lipothymia, tendency to syncope, frequent pulse accompanied by elevation of temperature (up to 40°), finally the prostration, delirium, and even convulsions. In the adynamic form, prostration predominates.

The patient lies in a condition of low delirium, is constantly muttering incoherently, but often returns to the same ideas and subjects. The tongue is dry and black, the face is often jaundiced, the abdomen tympanitic; diarrhœa is usually present. The lungs are congested and disseminated crepitant râles are heard. The temperature is high; the pulse is often soft and dicrotic.

In the pyæmic form all these symptoms are present, but they are interrupted from time to time by violent and repeated chills, followed by fever and sweat; these symptoms sometimes return at regular intervals. Jaundice is almost always present.

The visceral phenomena are marked by congestion of the base of the lungs with dyspnœa and the râles of pneumonia, tubular breathing and muco-purulent expectoration. The spleen is painful and enlarged; the urine contains albumen.

The most frequent causes of ulcerative endocarditis are overwork, and then the puerperal condition. O. Dür reports that among 6,848 parturient women endocarditis was noted 63 times, and that 11 of these cases assumed the ulcerative form. Winge observed it as a complication of rupture of the urethra produced by a false passage ; Eisenlohr, in a case of vertebral caries. Virchow regards as a predisposing cause the existence of small arterial vessels, such as he has found in chlorosis. Klebs asserts that it is due to the introduction of micrococci, which enter through the medium of wounds, as, for example, in the puerperal condition ; that these microbes, circulating in the blood, form embolic obstructions in the capillary vessels of the valves, leading to necrobiosis of the parts, and consequently to ulceration and its sequences.

The diagnosis reduces itself then to two points. If the endocarditis is primary, the characteristics of the disease will be manifested and its nature established by the nature and gravity of the general phenomena.

In the other event, we have to deal with a grave puerperal, eruptive, typhoid, or other disease, in which serious symptoms appear, particularly with reference to the heart. A careful examination of the patient must then be made in order to determine whether the complications are caused by a cardiac lesion.

Ulcerative endocarditis lasts five or six weeks when it is primary or engrafted upon an old valvular lesion. The duration of that form which occurs during acute articular rheumatism is not more than two to four weeks. Finally, the variety which appears during the puerperal condition or grave fevers, does not last more than four to six days. The termination is always fatal.

CHAPTER XIV.

CHRONIC OR SCLEROTIC ENDOCARDITIS.

Chronic endocarditis is rarely general; it has even a greater tendency than acute endocarditis to localize itself upon the valves of the heart, the orifices, or their vicinity. The localization varies with age. The right heart is affected most frequently in the fœtus; in adult life and adolescence, the mitral valve; in old age, the aortic valve.

The lesions of chronic endocarditis consist mainly of hyperplasia of the connective tissue, which forms, as in the other viscera, an induration which causes atrophy of the neighboring elements; the parts then become retracted and finally undergo fibrous or calcareous transformation.

If the chronic endocarditis follows the acute form with vegetations, the latter become hard and form a sort of saw, either at the edge of the valve or a little higher, near the orifice. The lesion often is not limited to the valves, but may affect the chordæ tendinæ and the free extremity of the papillary muscles. The following table shows the frequency of the occurrence of the lesion in the different valves.

	Mitral.	Aortic.	Tricuspid.	Pulmonary.
Barclay	51	61	..	..
Ormerod	133	118	17	4
Bamberger	150	50	..	..
Forget	10	9	..	..
Kaulis	22	13	..	..
Sperling	255	129	19	5
	621	380	36	9

Left heart....1,001   Right heart....45

I will study the lesions of the valves and orifices in their order of frequency.

LESIONS OF THE MITRAL VALVE.

Anatomical Data.—The left ventricle forms the posterior half of the heart. It has the shape of a triangular pyramid, and consequently has three surfaces, three borders, a base and apex. The inferior surface, like that of the right ventricle, rests flat upon the diaphragm. The ventricular groove divides the lower surface of the heart into two equal parts, each ventricle presenting an equal surface to the diaphragm. The posterior surface of the left ventricle, forms the posterior or pulmonary surface of the heart.

The base is formed by the aortic and mitral orifices. The aortic orifice forms a plane which is directed outward and upward; the mitral looks upward, outward, and backward. But in those cases in which insuffi-

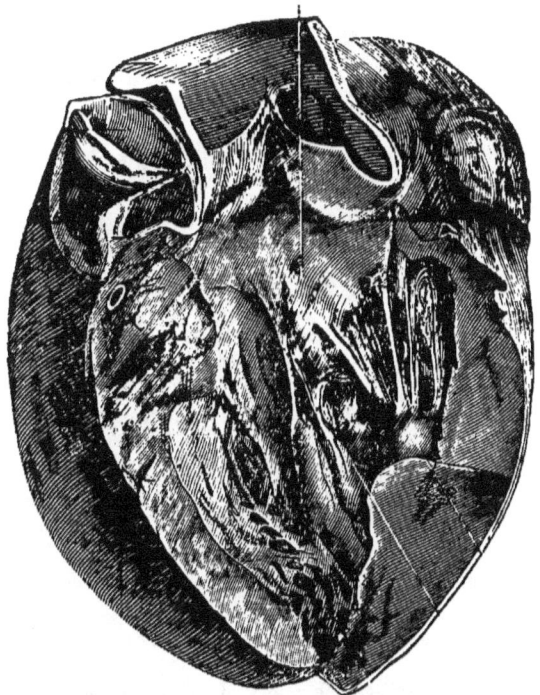

FIG. 32.—Inner Surface of the Left Ventricle (Cruveilhier).

ciency arises from excentric hypertrophy and dilatation of the heart, the posterior border of the mitral orifice is directed more and more to the left, and the orifice is directed, finally, from before backward.

The aortic orifice is closed by the semilunar valves, the venous orifice by the mitral valve.

The capacity of the left ventricle has been determined by Legallois, by means of the weight of mercury which it will hold. This was found to be 10.68 for the left ventricle, and 11.72 for the right ventricle.

The cavity of the left ventricle is ovoid in shape. The inner surface is smooth in the region near the aortic orifice in the arterial portion; it is bounded behind in its auricular part and toward the apex by a trellis, the meshes of which become closer as they approach the apex. In this region the elongated meshes form a sort of spongy tissue (Fig. 32). These meshes are formed by the three classes of columnæ carneæ (Fig. 33).

The apex of the left ventricle constitutes the apex of the heart.

The mitral valve is composed of two leaflets, an anterior larger one which is parallel to the anterior surface of the heart, and a posterior one which is much smaller (Figs. 34, 35, 36).

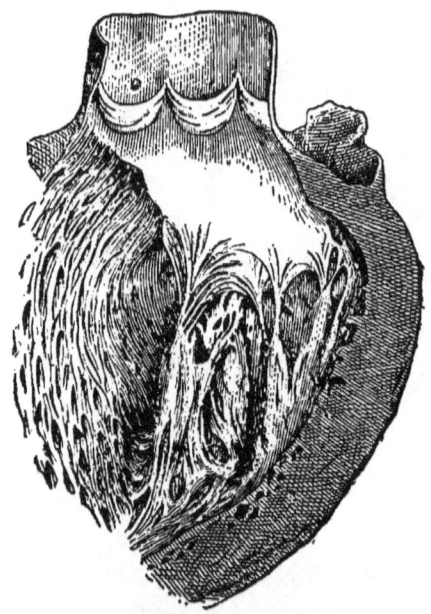

FIG. 33.—Inner Surface of Left Ventricle stripped of the Endocardium (M. Sée).

The anterior valve divides the cavity of the ventricle into two unequal parts, an anterior larger part, known as the aortic or arterial cavity of the ventricle, and a posterior smaller one, known as the venous cavity. Each papillary muscle is inserted into both leaflets, the anterior one into the inner portion, the posterior one into the outer part of the valves.

The mitral valve is connected by its adherent border with the circumference of the mitral orifice, and at its free border with the chordæ tendineæ.

1. The chordæ tendineæ of the first order, the strongest ones, are continued upon the outer surface of the valve, and inserted into the fibrous rings; some are free, others adherent.

2. Those of the second order are arrested upon the sides of the valve, without being continued to the fibrous rings of the orifice.

3. Those of the third order are arrested at the free border of the

valve; they are derived usually from the columnæ and rarely from the papillary muscles.

To recapitulate, the right mitral valve divides the cavity of the left ventricle into two unequal parts. The anterior or arterial cavity is smooth

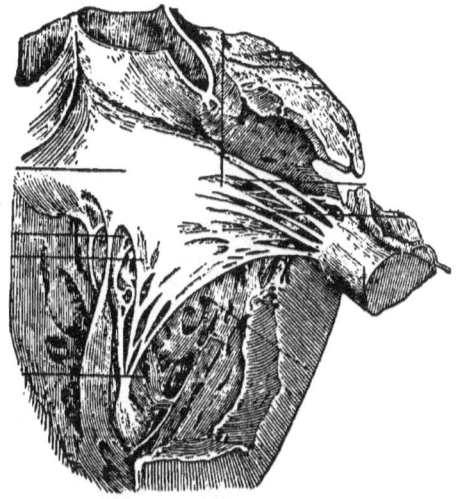

FIG. 34.

upon its valvular surface and upon the anterior surface formed by the septum. It presents no columnæ carneæ of the first or second order and none of the third order, except in the lower part, so as to offer no rugosity which may interfere with the exit of the blood during systole.

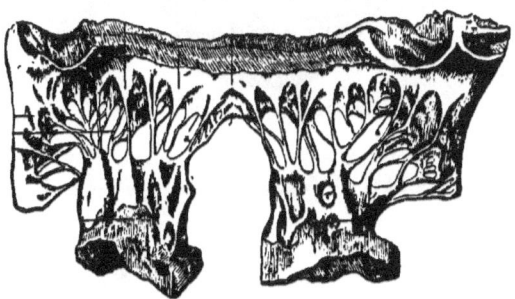

FIG. 35.—Mitral Valve, separated (M. Séc).

Behind the right mitral valve is found superiorly the mitral canal, which passes from the auriculo-valvular orifice to the valvular border, then the trellised work of the tendon which serves as a communication between the arterial and venous portions of the ventricle, and lower down the columnæ carneæ of the first order.

Physiological Data.—All are agreed with regard to the condition of the heart during diastole. The heart, in dilating, stretches the columnæ carneæ and the venous valves and orifices are open.

A number of theories have been advanced with regard to systole. According to one theory, which was advanced by Lower in 1679 and has received a number of modifications, the occlusion of the orifices by the valves

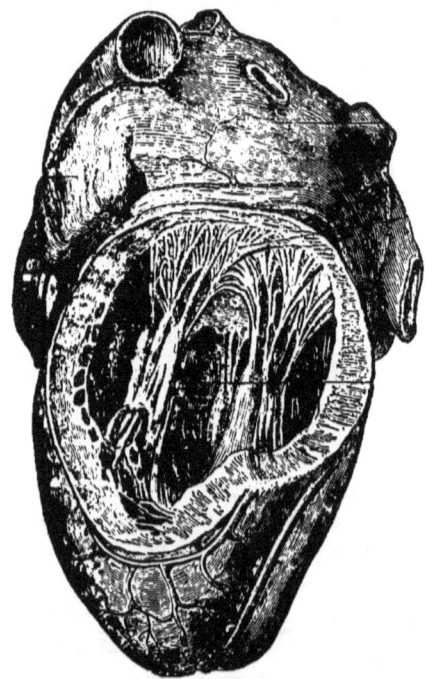

FIG. 36.—Smaller Mitral Valve, seen from behind (Cruveilhier).

is effected passively by the pressure of the blood. But all theories of this nature leave out of sight the cardinal fact that systole is an active muscular contraction.

The second theory is that of active occlusion. As early as 1825 Meckel stated that the various parts of the valves are approximated to one another and the auriculo-ventricular orifice closed with force by the energy of the ventricular contraction.

Burdach (1827) observed that if the stretching of the valves was a passive phenomenon, the presence of the papillary muscles was needless. According to him, when the papillary muscles contract they stretch the valves like sails, separating them from the walls. They are drawn in the direction of the axis of the ventricle; the occlusion is completed, he says, by the pressure of the blood.

It is difficult to understand this view, as Bouillaud calls the columnæ carneæ tensors or elevators of the valves. But his next remarks are very just. He states that, during systole, the left or auricular half of the left ventricle is almost entirely effaced, while the right or arterial half throws into the aorta the column of blood which it has received from the left auricle.

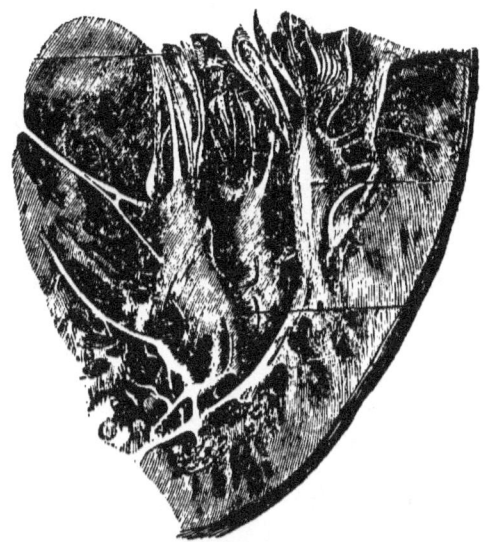

FIG. 37.—Respective Position of the three Papillary Muscles.

Marc Sée[1] has developed the theory which was first advocated by Parchappe. He states that the systole is an active function and that, while the entire ventricular walls contract, the papillary muscles also contract, and that the effect of the contraction is to produce tension of the chordæ tendineæ and lowering of the valves despite the shortening of the columnæ carneæ.

The papillary muscles of the left ventricle are arranged in such a manner as to fit into one another and to fill up the left portion of the cavity (Figs. 37 and 38). In contracting they draw to the left both leaflets of the mitral valve which are applied against one another and against the ventricular wall. The right leaflet plays the essential part in the closure of the auriculo-ventricular orifice. But neither the left leaflet nor the two acces-

---

[1] Marc Sée: Sur le mode de fonctionnement des valvules auriculo-ventriculaires du cœur, Archives de physiologie, 2e serié, t. 1er, p. 381.

sory valvular tongues are useless. They prevent the **reflux** of blood along the wall into the ventricle.

The theory of active occlusion may then be summed up as follows. At the moment of systole the contraction of the columnæ pulls upon the valves and maintains them in the direction of the axis of the heart. The pressure of the blood forces the two leaflets against one another (Burdach).

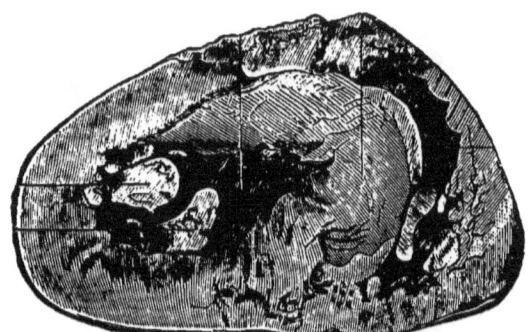

FIG 38.—Figure showing the Interlacing of the Columnæ Carneæ.

The ventricular contraction, by approximating the columnæ carneæ, fills up the auricular portion of the cavity of the left ventricle (Cruveilhier). The valves are folded and the tendons come in contact in such a manner as to close the space which separates the valvular borders of the columnæ carneæ (Parchappe). The papillary muscles are applied to one another and close the cavity in the region of the apex (M. Sée) (Fig. 38).

7

CHAPTER XV.

LESIONS OF THE MITRAL VALVE AND ORIFICE—NARROWING—INSUFFICIENCY.

Anatomo-Pathological Data.—We have seen that acute endocarditis affects the mitral valve much more often than the other regions of the endocardium. These lesions usually occupy the auricular surface of the valve and generally its free border. If the process is mild, proliferation of the connective tissue of the valve occurs. This sclerosis also attacks the tendons, which become shortened, and reaches the free extremities of the papillary muscles, which are indurated. At a later period calcareous incrustations in the valve develop, especially upon its free border. In some rare cases the lesions are situated at the orifice.

The theory of the action of the mitral valve which we have described, forces us to admit that every lesion of the valves produces both narrowing and insufficiency. But it is true that at a later period one of these lesions will predominate.

In such cases there is both stenosis and insufficiency. If the orifice is small it admits very little blood and the obstruction to its entrance is considerable. But if the orifice is immobile the insufficiency is in reality total, since the opening will admit as much blood as entered. In such cases the insufficiency is as complete as if the valve were entirely suppressed in a healthy subject.

When the mitral lesion is accompanied by dilatation of the heart, with or without hypertrophy, it will be found that the papillary muscles are separated from one another, that the trellis formed by the chordæ tendinæ has large meshes, and that the lower border of the valve forms the section of a tubular orifice.

In the first case the diastole is retarded and the quantity of blood conveyed at each systole is diminished; in the second event the diastole is easy and systole inefficient.

After the mitral lesion has been established, the secondary disturbances will vary a little according as stenosis or insufficiency predominate. These changes are summarized in the following table:

LESIONS OF THE MITRAL VALVE AND ORIFICE.

	Stenosis.	Insufficiency.
Left auricle	Enlarged cavity. Hypertrophic walls. Atheromatous walls.	Enlarged cavity. Hypertrophic walls. Atheromatous walls.
Left ventricle	Diminished cavity. Thinner walls.	Enlarged cavity. Hypertrophic walls.
Papillary muscles.	Degenerated before the walls.	Hypertrophic.
Aorta	Smaller.	Thickened walls.
Right ventricle	Enlarged cavity. Thickened walls.	Enlarged cavity. Thickened walls.
Tricuspid valve	Stretched. Insufficient.	Stretched. Insufficient.

DIAGNOSIS OF MITRAL DISEASE.

The general symptoms of mitral disease are the ordinary symptoms of cardiac affections, dyspnœa on effort, then palpitation and small, often irregular pulse.

Examination shows that the apex is lowered and removed farther from the median line; it descends to the sixth intercostal space and is 10, 12, 15, even 17 ctms. from the median line. It follows that the lower border of the heart becomes more and more oblique. The depression of the apex corresponds to the increased weight of the heart, *i.e.*, to the hypertrophy of its walls.

The right angle remains at first in its place at the level of the insertion of the right fifth cartilage, and the right border is 3 ctm. from the median line, but when dilatation of the right heart occurs toward the close of life, the vertical border is 4 ctm. from the median line and the angle descends so as to correspond to the right sixth cartilage.

If the hand is applied to the præcordial region, the impulse of the heart will be found to be more diffuse and less distinct, and a purring thrill is often felt.

Potain and Rendu have given the following description of this thrill: "The thrill of insufficiency never begins before the moment of ventricular systole, it prolongs the impulse of the apex and disappears before the closure of the semilunar valves. It is propagated toward the axilla and is not present at the origin of the large vessels. The thrill, accordingly, begins with systole and continues throughout its entire period, covering the systolic bruit and the lesser period of silence.

"The thrill of mitral stenosis begins in the middle of diastole, is reinforced at presystole and ceases completely at the moment of the præcordial impulse. This thrill is peculiar, inasmuch as it is felt at the end of the repose of the heart during the second half of diastole and ceases abruptly at the moment when the præcordial impulse occurs. It is, accordingly, a diastolic, or to speak more correctly, presystolic thrill, since it has its maximum of intensity immediately before the ventricular systole."

Two things must be distinguished here, the fact and the theory. The fact is real if we restrict ourselves to the statement that the thrill precedes the præcordial impulse, but I do not think that it occurs during the contraction of the auricle.

Sphygmographic tracings have demonstrated that in mitral insufficiency, and particularly in mitral stenosis, the impulse of the apex is delayed beyond the beginning of systole, and that consequently the moment which immediately precedes the impulse of the apex is not the presystole, but the beginning of systole.

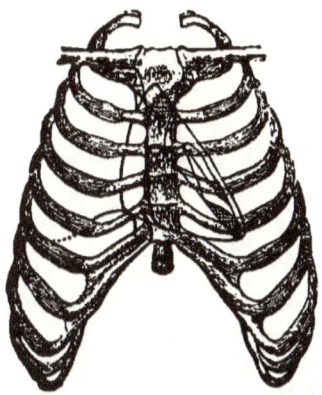

FIG. 39.

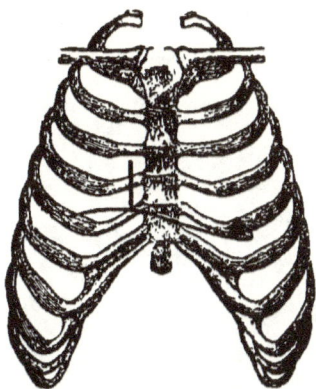

FIG. 40.

Auscultation. A. Topography of Mitral Murmurs.—The soft or rough blowing murmurs produced by lesions of the mitral valve have a very definite situation. When their area is very limited they are situated at the level of the apex of the heart; when their area increases they spread in three directions, outward, inward, and upward toward the insertion of the cartilage of the left third rib, *i.e.*, a point corresponding to the mitral orifice.

The mitral orifice corresponds exactly to the sternal insertion of the third left cartilage; the valve is directed from this point directly toward the apex and corresponds to the inner part of the third intercostal space (this is represented in Fig. 39). When the area of the murmur is very small, accordingly, it is limited to the apex; when it grows larger it forms a triangle parallel to that furnished by measurement of the heart, though not exactly. Its apex is approximated more or less to the mitral orifice, the right inferior angle passes beyond the apex toward the axilla, and the internal inferior angle is directed toward the sternum. Of the three sides, the inferior descends a little below the border of the heart, as the sound is conducted through the liver. The external border is oblique, parallel to the border of the heart; the inner border or base is parallel to the sternum (Fig. 41).

When the murmur rises toward the mitral orifice it is heard to a much less extent toward the axilla. It then occupies a very extensive area, but scarcely passes below the inferior border of the heart (Fig. 42).

The point of maximum intensity of the bruit also varies. When its area is very small its maximum is situated naturally at the apex. In the second case, in which the bruit passes beyond the apex to the outside, inside, and upward, its maximum is also at the apex, and in both these cases the murmur is soft. But when the bruit increases in intensity, when it becomes rough and there is marked stenosis and roughness of the valves, the point of maximum intensity is elevated more and more and is found in the fourth, third, and, in certain cases, even in the second left intercostal

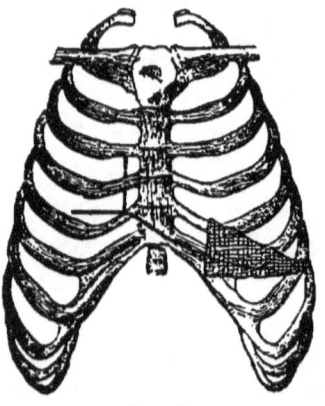

FIG. 41.

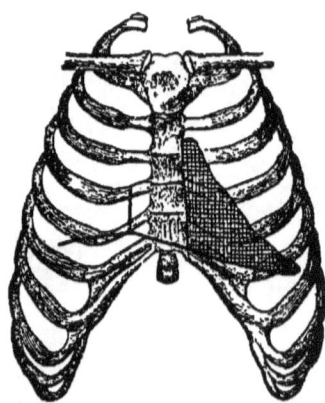

FIG. 42.

space. When the valves are calcified it may be stated, in general, that the maximum of intensity of the murmur is found at the upper angle of the triangle.

In certain exceptional cases the murmur may be limited exclusively to the second intercostal space. I have then found the leaflets normal, but the vegetations and rugosities were situated upon the adherent border of the valve, at the orifice itself.

The question arises, Why does the bruit extend to the apex and not to the vertebral column? According to the general law, it should be heard near the fifth spinous, but this does not obtain unless there is considerable hypertrophy of the heart. It must be concluded that the conditions for the transmission of sound through the organs of the posterior mediastinum are defective. The transmission of the sound to the apex in the opposite direction to the current of blood must be due to the fact that the hypertrophied walls of the left heart are a better conductor. It will be noticed that although the maximum of intensity of the bruit is situated at the apex in cases of insufficiency, it approaches the mitral valve in cases of stenosis and is transmitted directly to the adjacent parts, the third left intercostal space and even the second.

B. Period.—In mitral insufficiency the first sound of the heart is suppressed at the mitral valve and is replaced by a blowing murmur. This murmur commences with systole, covers the tricuspid sound, and also the lesser period of silence in part or entirely, and is followed by the second sound or the click of the semilunar valves. Potain and Rendu state that this bruit is very loud at the beginning, reaches its maximum at once, and then diminishes. It is not affected by the respiratory movements.

The bruit corresponds evidently with the systole and the reflux of blood from the ventricle into the auricle, and in so far theory is in accord with clinical observation. But another point must be considered. Theoretically, this murmur should terminate with the systole. Now, if another postsystolic murmur is heard at the beginning of diastole, is it due to stenosis?

Let us consider the facts. In examining a patient suffering from mitral disease, it is not uncommon to hear at first a soft murmur lasting throughout the entire systole, then a reinforcement of the bruit furnished by the click of the semilunar valves, and finally a soft bruit which terminates the revolution and thus covers a part of the diastole.

An inexperienced observer might believe, in such a case, that two murmurs, systolic and diastolic, are present, and that the latter murmur is caused by the blood flowing into the ventricle through a narrowed and roughened orifice.

This phenomenon may, indeed, deceive us if the auscultation of the murmur is confined to the region of the apex of the heart. But if auscultation is performed over the entire heart, the following fact will be observed: in proportion as the stethoscope is brought nearer to the median line the murmur is prolonged less and less, and if we have reached the middle of the lower border of the heart before arriving at the sternum, it will be noticed that the bruit becomes shorter and shorter and remains confined to the systole. Therefore we have to deal with a prolonged systolic murmur which assumes the appearance of a diastolic murmur. I have applied to it, accordingly, the term paradox bruit.

OBSERVATION XXVIII.—C. A—— entered the hospital in 1877, suffering from a mitral affection. Measurement of the heart gave the following results: The apex is situated in the sixth intercostal space, and 12 ctm. from the median line; the right or hepatic angle of the heart has not moved; the upper border of the liver corresponds to the insertion of the fifth costal cartilage; the vertical border is 1 ctm. from the sternum; there is evident hypertrophy of the left ventricle.

Auscultation at the apex reveals a prolonged systolic murmur covering the first sound, the lesser period of silence, the second sound, and prolonged into the greater period of silence. This murmur has its maximum of duration at the apex. It becomes shorter as the median line is approached along the lower border of the heart. About half way between the apex and the median line the duration of the murmur diminishes to such an extent that the click of the semilunar valves is heard distinctly. The systolic murmur diminishes still more after leaving this point and disappears at the level of the tricuspid. It is evident that this was a prolonged systolic murmur.

OBSERVATION XXIX.—V——, aged thirty-five years, entered the hospital in 1878. This woman was attacked six years previously with severe bronchitis. About the same time she had an abortion and premature labor. Since then she has merely suffered, from time to time, from slight palpitations.

For the past six weeks she has emaciated considerably, coughs a little, and the sputa contain streaks of blood. At the right apex is found dulness, rude respiratory murmur, but no râles. Measurement of the heart gave the following results:

Apex beat in the fifth intercostal space 11½ ctm. from the median line; liver dulness ascends to insertion of right fifth cartilage. The vertical border of heart is 3½ ctm. from median line.

A bruit is heard over an area forming a triangle, the apex of which is 4½ ctm. outside of the apex of the heart. The murmur is systolic, followed immediately by the semilunar click. Pulse small, with false intermissions.

Under the influence of digitalis the murmur becomes more distinct. It is prolonged into the greater period of silence, giving the appearance of a double systolic and diastolic murmur. On placing the stethoscope nearer the median line, the murmur is found to grow shorter and to be purely systolic.

OBSERVATION XXX. *Mitral Affection; Paradox Murmur.*—J. D——, aged twenty-seven years, entered the hospital in 1879, suffering from heart disease. Examination of the heart: Apex beat in the fifth intercostal space, 10 ctm. from the median line; the upper border of the liver corresponds to the insertion of the fifth cartilage; the vertical border of heart is 4 ctm. from median line.

At the apex is heard a soft, prolonged blowing murmur, the area of which forms a triangle which passes about 1 ctm. beyond the apex and internally corresponds almost to the edge of the sternum. The bruit begins with systole, lasts throughout the entire period of systole, and is prolonged into the greater period of silence, with a reinforcement at the semilunar click, giving the appearance of two successive bruits.

At the level of the tricuspid is heard a murmur which is merely the prolongation of the mitral murmur. The murmur is soft, like a jet of steam. In fine, it is a blowing murmur, whose maximum is found in the fourth space, 5 ctm. from the median line, and indicates a mitral lesion with insufficiency.

I have observed a number of similar cases, but it appears to me to be needless to report them.

To recapitulate, in ausculting a patient suffering from mitral insufficiency a double murmur may be heard, the first covering the entire systole, the second beginning at the click of the semilunar valves and covering a part of the greater period of silence. Upon careful examination, however, the bruit, which appears to be diastolic, is found in reality to be systolic in character. In fact, it is sufficient to find that the murmur shortens as we approach the sternum in order to assure ourselves that there are not two sounds, but a single prolonged systolic murmur. Furthermore, in proportion as the patient grows better the murmur is shortened and becomes purely systolic. For this reason I have applied to it the term paradox bruit.

I know of no diastolic bruit at the apex which belongs to a mitral lesion. Every murmur of this character heard at the apex is either a paradox bruit or is a transmitted bruit of aortic insufficiency.

Last year I observed, with my friend Dr. Blachez, a man suffering from mitral stenosis with considerable hypertrophy, in whom the blowing murmur, situated in the second intercostal space, began with a rough timbre during systole and finished with a soft timbre in diastole ; it was a paradox bruit.

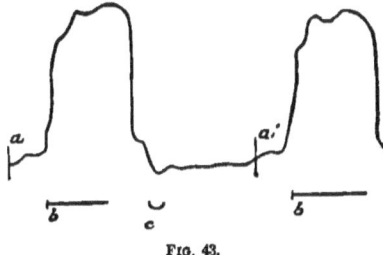

FIG. 43.

It would be surprising if the passage of blood from the auricle into the ventricle, propelled by the feeble contraction of the auricle or aspired by the diastole, should produce a murmur, while the blood, flowing back into this orifice with all the energy imparted by the ventricular contraction, should not produce a murmur.

THE SO-CALLED PRESYSTOLIC MURMUR.—In 1843 Fauvel rendered a valuable service to science in declaring that mitral stenosis is characterized not by a diastolic but a presystolic murmur, or rather a murmur preceding the ventricular systole and corresponding to the auricular systole. It is well to repeat here that the physiological cardiac revolution begins with the contraction of the auricles and ends with the close of diastole ; while when examined clinically by auscultation the cardiac revolution begins with the ventricular systole and ends with the auricular systole.

The first sound and the lesser period of repose correspond to the systole ; the second sound to the closure of the semilunar valves ; the beginning of the diastole and the greater period of silence include the diastole plus the auricular systole.

The ventricular systole perceived by the ear is preceded, therefore, by a mute systole, the auricular. This difference is well shown by Fig. 43, the physiological systole lasting from *a* to *a'*, the systole of auscultation from *b* to *b'*. The greater period of repose, from *c* to *b'*, includes the diastole plus the auricular systole.

Fauvel and all those who have followed him have said with regard to the presystolic murmur: This bruit precedes the impulse of the apex of the heart and the first sound. It is therefore presystolic in relation to the ventricular systole; it corresponds to the passage of the blood from the auricle into the ventricle through the narrowed orifice. All this is very simple.

But let us see in how far the observation is correct: 1, It is true that the pathological bruit in question precedes the impulse of the apex; 2, when separate it precedes the systolic bruit; 3, upon autopsy this murmur is found to correspond to a mitral lesion in which stenosis predominates. On the other hand, it must be remarked that—1, This murmur,

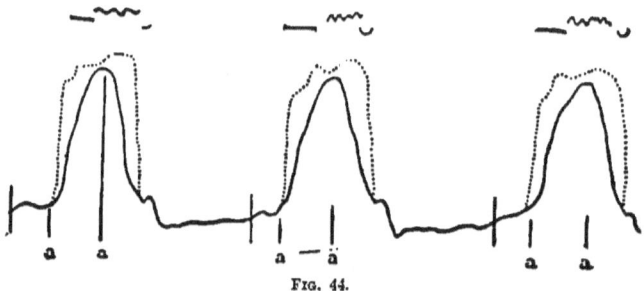

FIG. 44.

which precedes the systolic murmur, is never a more or less variable rough murmur, like the lesion which it is reputed to represent—it is always a dry click; 2, it does not always precede the systolic murmur, but is often confounded with it, being distinguished only by a roughness of timbre when the systolic murmur is soft; 3, the trace furnished by the cardiograph reveals an important fact, viz.: that the impulse of the apex no longer corresponds to the beginning of systole, but to its middle, so that what immediately precedes the impulse of the apex is not the auricular but the ventricular systole. Accordingly, the murmur which immediately precedes the impulse of the apex is not a presystolic murmur with relation to the ventricular systole, but is a systolic murmur.

In Fig. 44, which represents the difference between the trace of the mitral lesion and the normal trace of the ventricular systole (in the dotted line), the delay in the impulse of the apex is indicated by the letters *a—á*.

Now, what occurs during this period? The closure of the tricuspid valve and the sound which it produces; the mitral systolic murmur then follows with the impulse of the apex and lasts until the closure of the semilunar valves. The first sound which is heard, which precedes the impulse of the apex and the mitral bruit, therefore is not a presystolic bruit—it is a systolic bruit, the normal sound of the tricuspid valve; then comes the pathological mitral bruit, and finally the semilunar click. In this way we comprehend readily:

1. Why the so-called presystolic bruit is always a click and not a blowing murmur.

2. Why it is not always followed by a short period of silence, as the greater or less retardation of the impulse of the apex causes the pathological bruit to cover the tricuspid sound more or less.

3. Why the reflux produced by the ventricular contraction gives rise to a murmur, but the flow produced by the auricular contraction does not cause it. The classical theory admits that the auricular contraction produces a blowing murmur, but the ventricular contraction, although it is four times more vigorous, and causes the blood to pass through the same orifice, does not produce a murmur.

There is, therefore, no abnormal presystolic murmur; that which is heard at the beginning of the cardiac revolution is the closure of the tricuspid valves. On account of the obstacle to the flow of blood produced by the stenosis, the systole requires a certain time to overcome this hindrance, and consequently the impulse of the apex is retarded, and at the same time the ventricular systolic bruit.

DUPLICATION OF THE MURMURS.—Alteration in the rhythm of the bruits is another symptom of mitral stenosis. This change in rhythm may consist either of duplication or reduplication of the murmurs.

When patients enter the hospital we find, very often, that the action of the heart is so confused that it is impossible to determine its rhythm. In proportion as the patient is rested, the frequency of the contractions diminishes and the rhythm becomes better marked. A peculiar phenomenon is then observed quite frequently in mitral stenosis. Two cardiac revolutions follow one another immediately, indicating that the second revolution has followed a first incomplete one. Now, in cases of stenosis in which the entrance of blood from the auricle into the ventricle meets with a certain amount of difficulty, it may happen that so little blood has entered the ventricle during the first revolution that the systole, so to speak, has shot wide of the mark.

Another phenomenon is shown very clearly upon the cardiograph, *i.e.*, after two or three complete revolutions the ensuing one is weak, insufficient, followed by a period of silence which indicates a sort of cardiac rest; then follows another series of contractions of which the first is the most vigorous. The following ones grow weaker and weaker until another arrest of the heart occurs. In such cases we may readily observe that in the first pulsation of this series the systole is brusque and complete from the beginning, as in the normal condition, then at each following revolution the retardation of the impulse of the apex grows more and more marked.

LESIONS OF THE MITRAL VALVE AND ORIFICE. 107

In the same manner, if a trace of simple mitral insufficiency is taken in a young subject, the heart is found to contract precipitately at the beginning and to give an abnormal trace, then, as the emotional excitement subsides, the trace loses the characteristics of insufficiency.

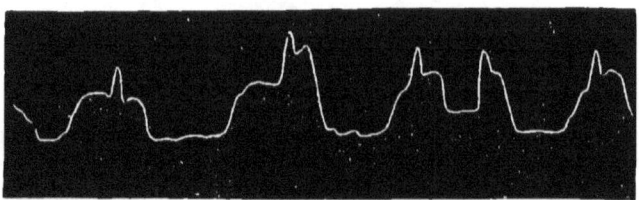

FIG 45.

Auscultation in these cases may reveal different murmurs. At first a sort of rolling[1] which lasts throughout the systole, and is followed by a duplication. Or, instead of the rumbling, a series of one to four bruits is heard. When only one is superadded, the supplementary bruit may be

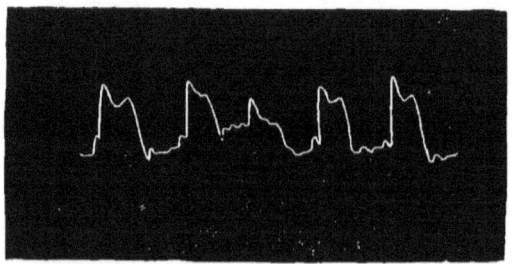

FIG. 46.—First Trace of Mitral Insufficiency in a Young Subject.

very near the first one and impart a galloping rhythm, or it is nearer to the second and gives the rhythm either of calling to arms or the "cry of the quail"[2] according as it is in greater or lesser proximity to the second bruit.

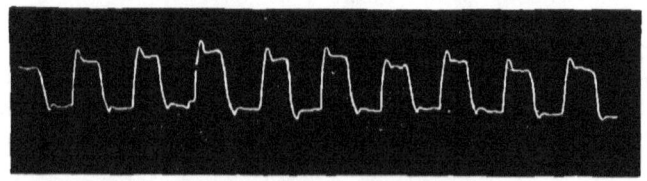

FIG. 47.—Second Trace, after Resting.

It is generally admitted in such cases that if the supplementary bruit is very near the first bruit, this is duplicated, and that if it is nearer to the second, the latter is duplicated.

---

[1] Rolling of a drum   [2] Bruit de caille.

Let us examine each of these two cases. In the case of duplication of the first sound, the explanation is easily given by the cardiograph. If the systole requires a certain time for its completion, and the impulse of the apex is delayed, two murmurs result, the first of which is the click of the tricuspid, the second that of the mitral valve. It can be understood very readily that a more or less indurated mitral valve will be delayed in arriving at complete tension.

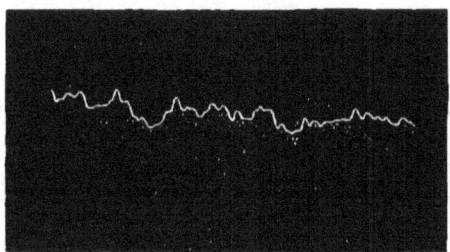

FIG. 48.—Trace of the Aorta in the Same Patient.

But in this case the mitral lesion plays only a minor part, and the method of contraction, or rather the energy of the left ventricular systole, plays the chief part, so that if one of these individuals is auscultated with patience, the duplications are found to appear and disappear under the ear. Such a variability of the murmurs cannot be due to a permanent lesion, but only to the mode of contraction, so much so that the delay in the impulse of the apex can be appreciated readily by the ear. If the delay is short, the pathological mitral murmur is distinctly preceded by the tricuspid click, but there is no period of silence between the two murmurs

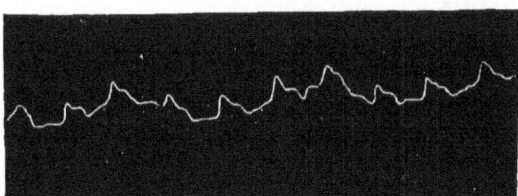

FIG. 49.—Trace of the Pulmonary Artery in the Same Individual.

and they are distinguished solely by a change of timbre. If the delay increases and the tricuspid bruit is heard first alone, and as it precedes the impulse of the apex, it is called presystolic. If the blowing murmur is short, the galloping bruit results; then if the mitral murmur is delayed and passes the half of the duration of the systole, the rolling rhythm results; if it is delayed still more and precedes the semilunar click by a short period, it produces the rhythm of the bruit de caille (quail).

In fact, even when the bruit de caille is produced, *i.e.*, when the supplementary murmur immediately precedes the semilunar click, it may not be concluded for this reason that the second sound is duplicated.

If the duplication belongs in reality to the second sound, when the stethoscope is applied to a point which is nearest to the aortic or pulmonary valves, the duplication of the second sound will be heard more clearly

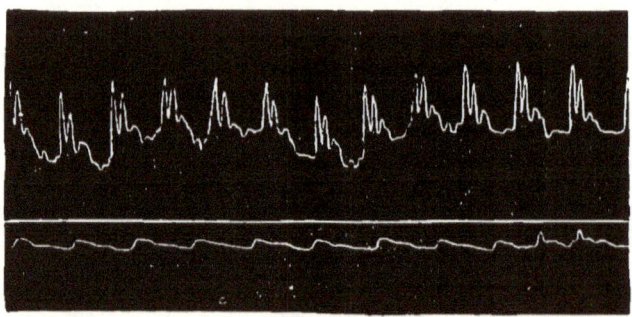

FIG. 50.

and loudly than at the apex, where it will be merely a transmitted murmur. But when the duplication is not heard more distinctly when the stethoscope is placed in the position mentioned, it is very evident that what appears to be a duplication of the second sound is not so in reality. What is it? Is it a duplication of the first sound with a corresponding delay. I have no longer any doubts on this question. Upon examining a certain number of cardiac traces I have found that in some cases the impulse of the apex is so retarded that it scarcely precedes the closure of the semilunar valves.

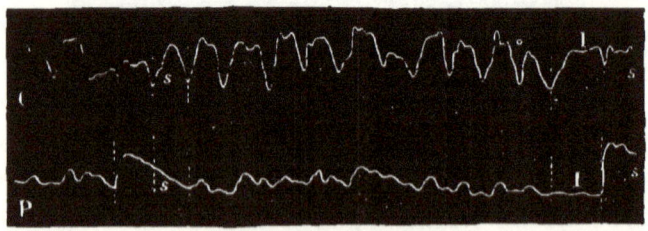

FIG. 51.—False Intermissions at S and I.

I have also learned to understand how, according to the brusqueness of the contraction of the ventricle, the rhythm of the heart was rapidly modified, and gave in succession the bruit de galop, de rappel (rolling of a drum) and de caille.

To recapitulate this long discussion upon the murmurs of mitral lesions, I will state :

The systolic blowing murmur is explained by the reflux of blood into the auricle during systole, it is a systolic murmur which is prolonged exceptionally and covers the greater period of silence. It is then converted apparently into a double murmur, to which I have applied the term paradox bruit.

The so-called presystolic murmur is in reality the sound caused by the closure of the tricuspid valve which begins the systole; it is followed by the delayed mitral murmur.

The duplication of the first sound is due to the same delay that the ventricular systole meets with.

The apparent duplication of the second sound may really be a duplication of this sound, and have its maximum at the arterial orifices, or it is nothing more than a more marked delay of the mitral bruit. In such an event the duplication does not extend to the arterial orifices.

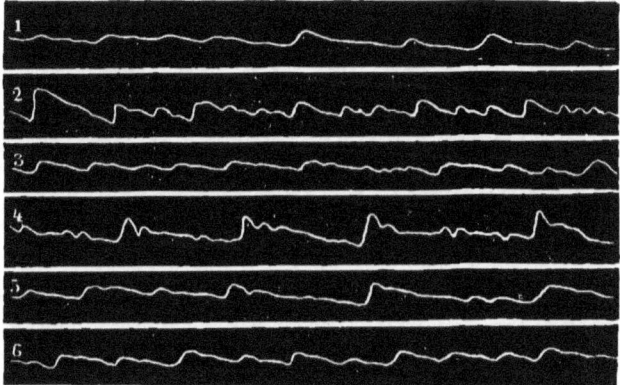

Fig. 52.—Radial Pulse in Mitral Insufficiency (Marey).

The duplication of the first sound is explained very readily by the delay in the impulse of the apex shown by the cardiograph. The duplication of the second sound is caused by the greater tension of the arterial system, which accelerates the occlusion of the aortic orifices.

THE PULSE IN MITRAL LESIONS.—In the earlier days of sphygmography it was believed that each diseased orifice would give a, so to speak, specific appearance to the pulse, that we would be able to recognize at first sight an aortic, a mitral pulse, etc. Experience has shown that the matter is not as simple as it seems.

In the first period of the disease, during the period of tolerance, if the patient is young and the peripheral circulatory apparatus still healthy, then if the myocardium is healthy and reacts vigorously against the obstacle created by the lesion of the orifice, the cardiac or arterial pulse is scarcely affected. The cardiac pulse may even present an absolutely normal appearance, as may be be seen in Fig. 47. It may even happen that the trace indicates nothing beyond nervous palpitation, as was shown in a vigorous girl of eighteen, who was suffering from a very evident mitral lesion.

At a later period, on the contrary, when the myocardium is altered and the other organs of circulation are enfeebled, the cardiac contractions grow weak, and at certain intervals cannot propel the arterial wave to the end of the arterial system; hence false intermissions (Fig. 51).

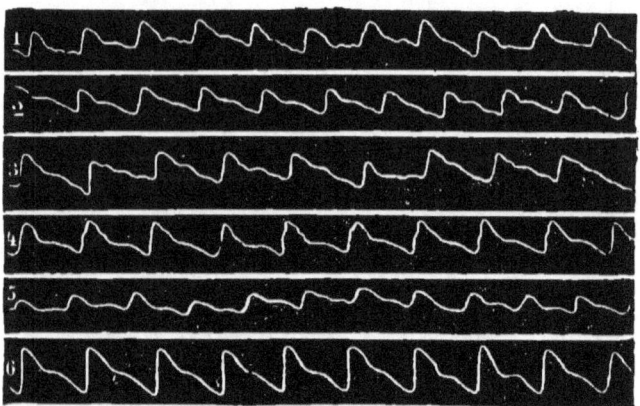

FIG. 53.—Radial Pulse in Mitral Stenosis (Marey).

Apart from these two extreme cases, we find that in mitral insufficiency, in proportion as the myocardium is changed, the pulse becomes irregular in its rhythm and in the amplitude of its pulsations. It seems as if the heart is obliged to rest from time to time and that, when it starts again, the first vigorous pulsation is followed by others whose energy constantly decreases, even despite the action of digitalis.

The pulsation is often dicrotic and the dicrotism corresponds to the abundance of the reflux.

It is evident from the traces in Fig. 52, which have been copied from Marey,[1] how variable is the form of the radial pulse in mitral insufficiency.

The characters of the pulse of mitral stenosis are usually negative. All that can be said is that frequently, though by no means always, the pulse of stenosis is less irregular than that of insufficiency (Fig. 53).

---

[1] Marey: La Circulation du Sang, p. 689.

CHAPTER XVI.

RÉSUMÉ OF THE SIGNS OF MITRAL INDURATION—COURSE AND PROGNOSIS OF THIS AFFECTION.

The necessary discussion of certain pathological bruits produced by induration of the mitral valve has drawn us away somewhat from a general view of these lesions. It seems to me to be necessary to recapitulate their characteristics and compare them with those of other affections, in a word to discuss their diagnosis.

In the first place, insufficiency of the mitral valve may be produced suddenly in the course of an endocarditis, either by the destruction of the attachment of a valve to its base, or by the rupture of the tendons of a papillary muscle, or by the adhesion of two leaflets; it may also be produced by ulceration or by an aneurism of the valves.

As the insufficiency is caused by vegetative endocarditis, and most frequently by sclerotic endocarditis, the induration generally follows gradually.

As a rule, sclerotic endocarditis causes induration of the mitral valves, the chordæ tendineæ and papillary muscles, and leads in the end to stenosis and insufficiency, not alone of the orifice, but of the edge of the valve and the fenestrated region formed by the tendons which constitute the real passage from the auricle into the ventricle.

I will first mention the signs of the mitral lesion, then the special signs which indicate whether the predominant lesion is insufficiency or stenosis.

The rigidity produced by sclerotic endocarditis renders the valve immobile, then leaves the orifice of communication open between the valve and the papillary muscles. On account of the induration of the valves narrowing occurs, since they are no longer pushed aside to allow the blood to enter the ventricle, and insufficiency results, because the immovable valve cannot be applied to its neighbor. Insufficiency alone does not exist except in cases of cardiac dilatation; stenosis never exists alone.

None the less certain differences arise when the canal formed by the indurated mitral leaves a wide or narrow communication. Not to break entirely with tradition, we will say that there is mitral induration with a large opening in which insufficiency predominates, and induration with a small opening in which stenosis predominates.

The general symptoms of mitral induration are at first cardiac dyspnœa or dyspnœa on exertion, then palpitation of the heart and a small pulse.

Inspection of the chest is negative unless considerable hypertrophy is present. Measurement of the heart soon shows depression of the apex in the sixth intercostal space, and a separation of the apex from the median line of 9 to 14 ctm., showing hypertrophy of the left ventricle. Palpation often shows a systolic thrill which lasts throughout the entire systole, and, as it precedes the impulse of the apex in certain cases, is considered presystolic; but I have shown that, in such instances, the impulse of the apex is delayed. This delay, which is indicated by the cardiograph, is more marked when stenosis predominates.

Auscultation reveals a blowing murmur whose site is always at the apex, and whose area may extend from this point in three directions, viz., toward the axilla, the median line, and upward. The maximum of intensity is heard almost always at the apex, exceptionally in the third, or even the second intercostal space.

With regard to the period of its occurrence, the bruit presents the following varieties:

1. When insufficiency predominates, the murmur begins with the systole. It may be short, and then does not cover the lesser period of silence; it merely prolongs the first sound, and partly covers the lesser period of silence, but leaves the latter and the semilunar click perceptible.

2. In other cases, the murmur begins with systole and lasts during its entire period, covers the lesser period of silence, and is terminated by the semilunar click.

3. At other times, finally, the murmur begins with systole and lasts during its entire period, but is not arrested by the semilunar click. It is prolonged into the greater period of silence, when it is always soft, and is reinforced a little before its termination by the closure of the semilunar valves, thus giving the appearance of two consecutive bruits. This is the bruit to which I have applied the term paradox.

The proof that all these murmurs are varieties of the systolic bruit is afforded by the fact that they are found to change into one another, according as the patient has rested for several days, or has been moving about.

When stenosis predominates, the following characteristics are noticed:

1. If the patient is thoroughly rested, under the influence of digitalis, and the pulse is slow, nothing is found but a rough systolic sound, very little prolonged, and without a murmur; this is followed by the semilunar sound which is often double. In such cases, there is a reduplication of the second sound which is heard more distinctly at the aortic orifice than at the apex. This bruit, which indicates increased aortic tension, is accompanied usually by a prolongation of the interval between the first and second sounds. It seems as if the ventricle must prolong its effort in order to force the wave of blood into the aorta.

8

2. The mitral sound may be retarded; if slightly, it begins before the tricuspid sound has been completed, and then the systolic bruit is prolonged; it has two timbres, one short, one long, without an interval between them. The first, corresponding to the closure of the tricuspid valve, has the character of marked brusqueness.

3. If the mitral sound is delayed still more, the tricuspid closure is heard first, then follows the mitral, leaving a short interval and coinciding with the impulse of the heart. This is the bicuspid bruit, improperly called tricuspid, then the delayed mitral murmur, which may be soft in quality. This results in a rhythm of three periods, composed of a click, a murmur, and another click, and according as the blowing murmur is more or less delayed, it is either followed or preceded by a short silence.

If the mitral sound is delayed but little and occurs during the first part of systole, it will be nearer to the first sound than to the second, and give rise to the rhythm du galop. If it is more delayed and occurs during the second half of the systole, it will be nearer to the second sound than to the first and give rise to the rhythm of the bruit de rappel (rolling), and finally, if delayed still more, it produces the rhythm of the bruit de la caille (cry of the quail). That the second sound of the heart has not been reduplicated is shown by the fact that the duplication is not heard at the origin of the aorta.

Finally, there are some cases in which both sounds are double, and four are heard in consequence. In such cases the duplication of the second sound is always heard at the aortic orifice.

4. The murmur, instead of being soft, may be replaced by a hard friction murmur which may begin with the systole or be delayed; the tricuspid sound is then heard first (so-called presystolic murmur), and after a longer or shorter period the click of the semilunar valves, which may be either reduplicated or single.

But it must not be forgotten that in the course of time the myocardium undergoes various degenerations, either hypertrophy with fatty degeneration, or cirrhosis; finally, asystoly occurs, and the pathological murmurs become weaker and weaker, and at length disappear.

In the production of these murmurs, the mode of contraction of the heart plays the principal part, so that they vary according as the organ is quiet or active.

It must be remarked, however, that they vary much more in cases of predominant stenosis than insufficiency. They vary, it is true, in the latter, but slowly, and according as the general condition of the patient is modified.

Another problem remains. Why are the murmurs of mitral induration heard usually at the apex, and exceptionally in the second intercostal space, although the mitral orifice is situated at the level of the third left cartilage near the sternum? This is due to the fact that the real site of the lesion is not at the orifice, but at the free border of the valve, and below it at the level of the chordæ tendineæ and papillary muscles, *i.e.*, very near the apex of the heart, which is an excellent conductor because the normal thickness of its walls is increased still further by the hypertrophy. It is only under exceptional circumstances that the murmur is produced at the level of the second intercostal space. Calcareous or other indurations are then found, which come in contact with the cartilages and constitute good conductors of sound.

Finally, the intensity of the murmur depends less upon the amount of the lesions than on their hardness and roughness. The pulse is generally small, frequent, unequal, more rarely irregular to the sphygmograph. At first, the irregularity is due to the respiratory movements; the rhythm becomes slower during inspiration, and is accelerated during expiration. At a later period, the irregularities are due to degeneration of the cardiac muscle.

Diagnosis of Mitral Induration.

The precision with which we have described the symptoms of mitral induration will enable us to differentiate it in a relatively easy manner from the diseases already described.

1. The differential diagnosis from anæmia is presented under two conditions. In the first case, blowing murmurs are present in the pulmonary artery and jugulars as well as at the apex. The question then arises, whether the anæmia is alone present, or whether it accompanies a mitral lesion. In some cases, the mitral lesion gives rise to anæmia. The diagnosis then remains suspended until the result of treatment shows whether the murmur at the apex persists for a longer period than the anæmia.

In mitral induration, accompanied by a bruit with its maximum at the second intercostal space, the general phenomena of the cardiac affection are very well marked and do not permit a mistake.

The anæmic murmur at the apex is never propagated toward the axilla, and it is never prolonged beyond the closure of the semilunar valves. It may be delayed sometimes, but it is never resolved into three or four bruits. Anæmia does not give rise to hypertrophy or arhythm. In cases of rheumatism, a guarded opinion should be given.

The differential diagnosis from pericarditis is usually very easy, and has been considered in the article on endocarditis. The diagnosis from simple hypertrophy will be discussed at a later period.

COURSE, DURATION, PROGNOSIS, TERMINATION.

It may be remarked at the outset that the intensity of the murmurs has no relation to the gravity of the disease.

If the lesion is circumscribed, the patient young, and the other organs of circulation healthy, life may be protracted for a long time, sometimes for a period of fifteen, twenty, thirty, even forty years.

There is in particular a form of this disease contracted during the course of an acute articular rheumatism in childhood or youth, which gives rise to slight hypertrophy with a faint murmur limited to the apex— a form which I have observed very frequently, and the long duration of its period of tolerance has induced me to call it stationary or latent mitral induration.

This affection may remain latent until the menopause in females, and somewhat later in males; but when the veins begin to lose their contractility, when the arteries lose their elasticity from atheroma and embonpoint, or abdominal plethora develops, etc., the obstacles are not overcome readily by the diseased heart, and the period of retrogression begins.

But if the lesion is grave enough to compromise seriously the functions of the organ, the compensation proves inadequate, and the organic disorders are not delayed in their development, as in the preceding case. A series of disturbances are then found to develop in succession.

The first is hypertrophy of the left ventricle, which is characterized by depression of the apex, its increased separation from the median line, and the obliquity of the lower border of the heart. Even the increased energy of the contractions often remains insufficient, and the heart supplements them by an increase in the number of beats, thus causing palpitation. This is not present constantly, but is caused by physical or moral excitement, the use of wine, tea, or coffee.

Irritability and weakness of the heart are two properties which always go together. Little by little, the excitable heart becomes exhausted, and its nutrition is affected either by extension of the sclerotic endocarditis from the valve to the papillary muscles, by the spread of the sclerosis into the myocardium, or by the degeneration of the muscular fibres. Little by little, also, the other organs of circulation take less part in the common action, and the insufficiency of cardiac activity becomes pronounced. The mitral induration, under these conditions, no longer permits a sufficient passage of blood from the central organ; hence pulmonary and bronchial congestion, which diminishes hæmatosis; hence dyspnœa, and increased respiratory efforts in the form of sighs, in order to introduce a larger amount of air; hence, also, a frothy expectoration, a sort of bronchial transudation, and, upon auscultation, at first, congestion of the lungs, and later, œdema at the base, with dulness, respiratory silence, and subcrepitant râles.

At other times hæmoptysis occurs. Then, the tension in the pulmonary artery increases, the venæ cavæ experience a greater resistance, and the right ventricle is dilated, rendering the valve insufficient.

At this period the tension increases in the portal system and the veins of the lower limbs. Œdema occurs at the malleoli, at first during the night, then during the day, and the dropsy gradually increases, filling the abdomen, pleura, and pericardium.

Then asphyxia develops, characterized by dyspnœa, congestion of the face and eyes, projection of the jugulars, cyanosis of the lips, nails, and knees, finally symptoms of intestinal congestion, frequently diarrhœa and renal congestion, with or without albuminuria. As the myocardium grows weaker, the pulse loses its fulness and increases in frequency. It gradually becomes so feeble that it cannot be counted, and then becomes imperceptible.

As less blood flows to the brain, the cerebral activity diminishes, the patient becomes somnolent; upon awaking he suffers from vertigo; hence diminished innervation of the heart, etc.

We shall return to a detailed consideration of these symptoms in the consideration of cardiac cachexia.

CHAPTER XVII.

AORTIC INSUFFICIENCY, OR CORRIGAN'S DISEASE.

Among a total of 1,046 cases of lesion of the cardiac orifices, the aortic orifice was found affected in the proportion of 36 to 100.

According to Ormerod, this proportion varies with age, as is shown in the following table:

From 0 to 30 years .... 21 cases.
"   30 to 50   "   .... 49 "
"   50 years upward....................... 38 "

Bamberger found that among 50 cases, 38 were males and 12 females.

Anatomical Data.—The aortic orifice is situated behind the sternum at the level of the insertion of the third costal cartilages; it is a little to the

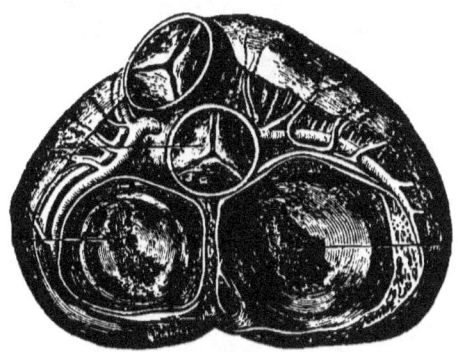

Fig. 54.

left of the median line. The orifice of the pulmonary artery is situated at the same level but more to the left.

Upon a horizontal section of a frozen body the aortic orifice is found between the two orifices of the right heart.

Anatomo-pathological Data.—The best means of recognizing aortic insufficiency in the cadaver consists in pouring water into the aorta from a sufficient height to produce tension of the valves. This water should be retained in the aorta, but we must be certain that it does not disappear through the cut coronary arteries.

Aortic insufficiency may be produced in various ways. It may arise suddenly, during exertion, from the rupture of a valve; this occurs usually at the free border, at other times at the insertion of a leaflet. In such cases we must admit the previous existence of a valvular lesion.

Apart from this accident, aortic insufficiency is the result of endocarditis; sclerotic endocarditis in particular hardens the edge of the valve and diminishes its perimeter in such a manner that complete closure is no longer possible. This is also true of fatty degeneration or calcareous incrustation of the valves. At other times the inflammation causes the adhesion of one of the leaflets to the wall of the vessel. More rarely a perforation has been found as the result of a local valvular aneurism. Perhaps the adhesion of two adjacent valves is found most frequently. The neighborhood of the mitral valve, which is attacked so often by endocarditis, in certain cases gives rise to an extension of the inflammation to the adjacent part of the aortic circumference and the corresponding semilunar valve.

At other times the lesion of the valves is merely an extension of the atheromatous degeneration of the aorta. Another rare lesion is a sort of atrophy giving rise to a reticulated condition of the valves.

Finally, insufficiency may be produced without valvular lesion, from mere dilatation of the aorta. This insufficiency is merely the result of the disease so well described by Hodgson, and which I have called "Hodgson's disease."

In youth, the semilunar valves have a greater area than is necessary to cover the section of the aorta, but as the aortic orifice enlarges after manhood, while the valves no longer increase in size, the latter finally become smaller than the area of the section of the aorta.

Beneke[1] has made a very large number of measurements which demonstrate this progressive dilatation of the aorta with age. It follows from his observations that the aortic orifice gradually increases in size from birth until the age of twenty-one years. The dimensions of the body and of the aorta then remain stationary until the age of forty to forty-five years, after which the relation changes, as the aorta continues to increase in size. In a word, it then begins to lose its contractility and elasticity, and yields gradually to the pressure of the blood.

The aortic orifice is closed by the three semilunar valves, which form a sort of pocket at the level of their insertion. The inferior, or adherent border of the valves, is curved, with the concavity upward. The superior, or free border, is almost horizontal when the valves are lowered, and is firmer than the pouch formed by each leaflet; at the middle of the free border is found a harder portion known as the nodule of Arantius.

---

[1] F. W. Beneke, Ueber das Volumen des Herzens, etc., Cassel, 1879.

Dimensions of the Circumference of the Aorta at its Origin, according to Age and Sex. Relations of this quantity to the length of the heart (the figure marked indicates the relation of this quantity to 100 centimetres in length of the body).

Age.	Male Sex.				Female Sex.			
	Number of cases observed.	Internal circumference of aorta in millimetres.	Relation to 100 centimetres of length.	Average length of the body in centimetres.	Number of cases observed.	Internal circumference of aorta in millimetres.	Relation to 100 centimetres of length.	Average length of the body in centimetres.
Fœtus ?............	1	9.5	30.6	31.0	..	....	....	....
Fœtus ?............	1	11.0	29.7	37.0	..	....	....	....
Fœtus of 6–7 months......	1	12.5	30.9	40.5	..	....	....	....
Fœtus of 7 months........	1	14.0	33.7	41.5	..	....	....	....
Fœtus of 7–8 months......	1	13.5	33.7	40.0	..	....	....	....
Fœtus of 7 months........	..	....	....	....	1	11.0	30.5	36.0
Still-born................	6	18.7	37.5	50.0	1	17.5	35.0	50.0
From 1–11 days...........	4	20.0	39.8	50.4	4	19.0	38.1	50.2
From 11 days–3 months....	17	23.7	44.2	53.7	14	23.5	42.4	55.4
From 3 months–1 year.....	10	36.0	48.6	74.0	14	28.3	45.4	62.3
From 1–2 years...........	11	33.9	46.9	72.7	9	33.2	44.1	75.4
From 2–3 years...........	12	40.0	48.8	82.0	9	34.6	41.5	83.5
From 3–4 years...........	4	39.3	42.5	93.1	2	39.3	42.5	93.1
From 4–5 years...........	1	38.0	38.0	100.0	4	40.1	41.3	97.0
From 5–6 years...........	5	40.3	38.9	103.7	3	40.0	36.4	110.6
From 6–7 years...........	6	43.0	37.0	116.1	2	39.5	40.0	104.5
From 7–9 years...........	6	46.6	38.3	121.4	3	43.0	35.9	119.5
From 10–11 years.........	8	47.3	38.8	122.4	2	44.0	35.1	125.5
From 11–13 years.........	5	50.8	36.9	137.4	4	48.5	34.8	139.5
From 13–14 years.........	4	46.2	32.3	143.5	5	49.6	36.3	136.9
From 14–15 years.........	7	49.0	33.9	144.3	3	49.6	33.7	147.3
From 15–16 years.........	9	51.9	33.2	157.0	3	56.3	36.9	153.6
From 16–17 years.........	3	55.1	35.3	156.6	5	49.6	32.5	152.5
From 17–18 years.........	7	53.5	33.1	161.3	5	55.4	34.9	159.0
From 18–19 years.........	5	57.8	34.6	166.8	3	53.3	32.3	164.6
From 19–20 years.........	11	57.8	34.3	168.6	4	53.5	34.1	156.7
From 20–21 years.........	3	62.8	36.0	172.0	5	56.2	36.7	156.7
From 21–22 years.........	14	60.9	35.9	171.3	..	....	....	....
From 22–23 years.........	6	60.8	35.7	170.1	6	60.0	37.8	153.6
From 23–24 years.........	12	59.0	34.7	170.3	..	....	....	....
From 24–25 years.........	9	59.8	34.9	171.5	12	56.0	34.8	153.6
From 26–27 years.........	14	60.7	35.5	171.6	11	57.6	37.6	154.1
From 27–28 years.........	..	....	....	....	..	....	....	....
From 28–29 years.........	15	63.6	37.3	170.2	9	67.1	38.2	159.9
From 30–34 years.........	22	63.4	37.5	169.2	17	59.4	38.4	154.7
From 35–40 years.........	24	67.2	39.4	170.6	18	59.4	38.4	154.7
From 40–45 years.........	24	71.4	42.6	167.9	9	64.8	41.1	157.6
From 45–50 years.........	29	72.6	42.9	169.5	7	67.8	43.1	157.5
From 50–55 years.........	22	75.0	43.9	171.1	15	77.0	48.3	153.4
From 55–60 years.........	21	75.8	44.6	169.7	11	72.2	45.6	158.3
From 60–65 years.........	18	81.4	47.5	173.8	11	75.6	46.9	158.7
From 65–70 years.........	11	79.8	47.1	169.3				
From 70–80 years.........	16	82.7	49.2	167.8	4	98.0	....	....

When the three valves are lowered, they touch upon their internal or ventricular surface, and upon the edges. The nodules of Arantius fill up the little triangular space which would otherwise remain between the three leaflets. As an anomalous condition, two valves are found instead of three, but in such cases they are generally larger, and suffice to close the orifice hermetically.

Symptoms and Diagnosis.

Corrigan's disease sometimes begins suddenly, usually during the course of an articular rheumatism. At other times it begins gradually, and is only noticed on account of the attacks of vertigo, pallor, and shortness of breath.

The diagnosis may often be made at first sight, because the attention is attracted by the abnormal pulsation in the arteries, especially in the primary carotids, the external carotids and subclavian, then in the vessels of the limbs.

Corrigan was the first to describe this phenomenon under the term visible pulse.

"The arterial trunks of the head, neck, and upper limbs, at once attract the eye by their peculiar pulsations. At each diastole, the subclavian, carotid, temporal, humeral, and sometimes even the palmar arteries, are projected forcibly from their bed, and bound under the skin. These pulsations are observed sometimes in healthy individuals, particularly after violent exercise, but in the affection at present under consideration, the movement of propulsion is excessive in amount."

The phenomenon of the visible pulse, which must not be mistaken for the bounding pulse, is not peculiar to Corrigan's disease. It merely indicates the loss of arterial contractility, and is also found in arterial atheroma.

In order to observe this phenomenon in the most satisfactory manner, the patient should bare one arm, and one side of the chest, particularly the right. The face is turned a little to the left, and the arm is separated from the body at an angle of 25° to 30°; the forearm is then semiflexed and supinated, so that the eye, at a single glance, may embrace the entire course of the arteries.

At each arterial systole, the vessels are then observed to increase considerably in volume. If the arm has been semiflexed, the curves of the arteries are found to be increased at each diastole to such an extent that true arterial locomotion occurs, with angular displacements, at times, of several centimetres. If the arm is extended, the movement is much less marked.

It must be remarked, also, that the pulse is sudden and bounding; it is elevated suddenly, and then falls immediately. This feature is well shown by the following traces:

The first is the trace of the apex of the heart, obtained by the cardiograph, upon a patient, aged seventeen years, suffering from aortic insufficiency.

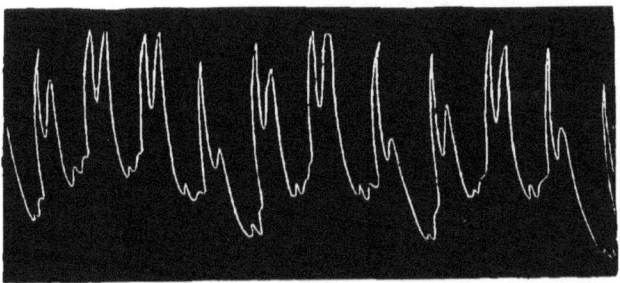

Fig. 55.

In older individuals, the ascent also occurs suddenly, but the oscillation of the needle is less marked.

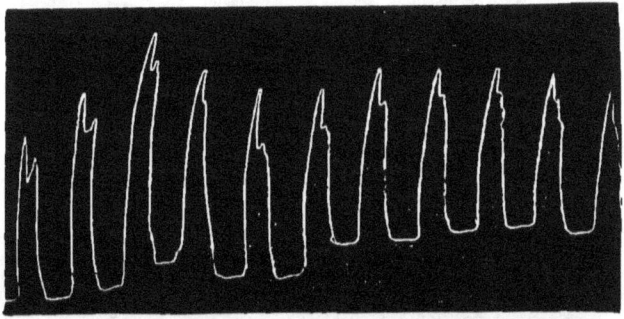

Fig. 56.

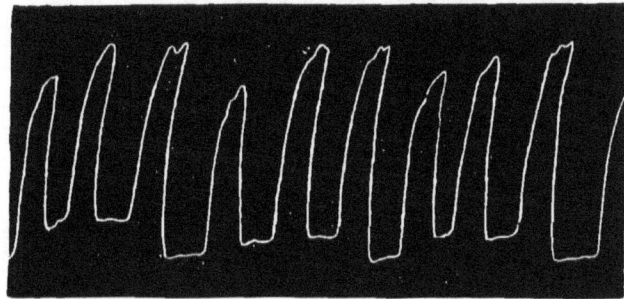

Fig. 57.

The same characteristic is found in the arteries in which the trace begins with a vertical line.

The following trace was taken upon the crural artery of the same individual.

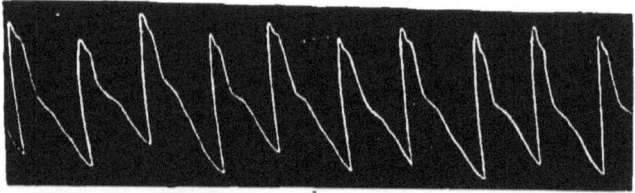

Fig. 58.

The trace of the radial pulse, upon the sphygmograph, also shows a vertical line of ascent.

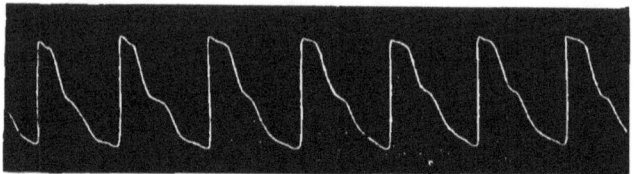

Fig. 59.

When the arteries also become atheromatous, the ascent becomes less brusque, as is shown upon the last two traces of Fig. 60, copied from Marey's work.

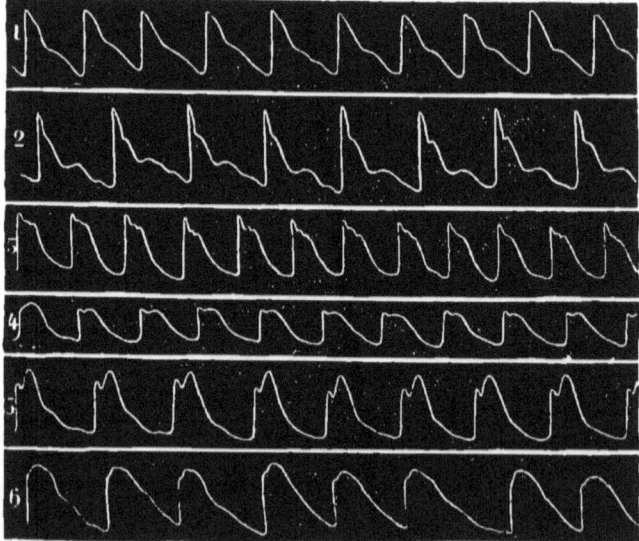

Fig. 60.

Corrigan gave the following excellent explanation of the visible pulse:
"When the semilunar valves are healthy, they are closed by the pressure of the blood, immediately after each ventricular contraction. When the occlusion is complete, the blood propelled from the ventricle is retained in the aorta, and the large vessels remain distended. These vessels, then, maintain almost the same calibre in systole as in diastole. But when the valves no longer close the aortic orifice, a certain amount of blood flows back into the ventricle after each systole; it follows, that the ascending aorta, and the arteries supplied by it, allow a certain amount of the blood contained in them to escape; they become flaccid after each ventricular contraction, and their diameter diminishes. At this moment, a fresh contraction of the ventricle rapidly forces into the vessels a quantity of blood which dilates them suddenly and forcibly. The arterial diastole is

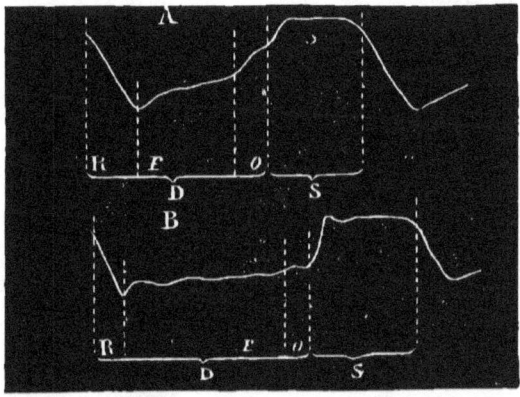

Fig. 61.

then marked by such a marked and sudden increase in the calibre of the vessel, that it produces a visible pulsation, constituting one of the signs of the disease."

Stokes has remarked, that the visible pulse may exist, even though the blood does not flow back into the ventricle, but into some other cavity, as, for example, when there is a communication between the aorta and the pulmonary artery, the aorta and infundibulum, the aorta and right ventricle. Stokes has also observed it in a case of aneurism of the ascending aorta.

We will also remark, that the filling of the heart occurs more rapidly during diastole.

In Fig. 61 the normal trace is below, that of aortic insufficiency, above.

At the beginning of the disease, there is no hypertrophy, and the symptoms are a certain pallor, which often suggests anæmia, and frequently accompanies it in young people. Auscultation alone furnishes any characteristic symptoms at the outset.

1. *Topography of the Bruit of Aortic Insufficiency.*—It begins at the level of the aortic orifice, *i.e.*, the third right costal cartilage, extends about one or two centimetres from the edge of the sternum on the right, throughout the entire width of the sternum on the left, ascending into the right second intercostal space, then descending along the sternum, which it occupies throughout almost its entire extent, and passing about one or two centimetres below the lower border of the heart. It inclines a little to the left in passing downward. But it is a remarkable fact, that the position of the murmur changes in time. As the secondary hypertrophy increases, the apex is lowered, and further removed from the median line; the reflux of blood is then less vertical, and becomes more and more oblique to the left, so that the area of the murmur then appears as in Fig. 63; Fig. 62 rep-

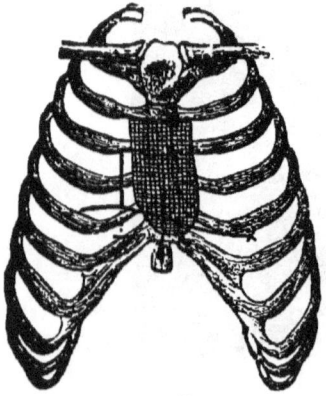

FIG. 62.

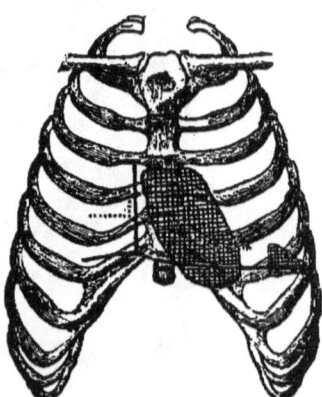

FIG. 63.

resenting its area at the beginning of the affection. The shaded portion, at the apex of the heart, in Fig. 63, represents a mitral systolic murmur, due to insufficiency of this valve from dilatation of the left ventricle.

The point of maximum intensity of the bruit corresponds generally to the xiphoid appendix.

Period of Occurrence.—The murmur of aortic insufficiency replaces the click of the semilunar valves and begins with diastole. But the duration of the systole is less than in the normal condition, as the systolic impulse is brusque and rapid, and its subsidence is also rapid. It often appears as if the lesser silence is very short, and the diastolic sound follows almost immediately upon the systolic sound. The murmur begins with diastole, and consequently covers a portion of the greater period of silence; it terminates with the diastole.

In cases of considerable hypertrophy of the heart, it is propagated to the apex, and covers the second sound and the greater period of silence;

sometimes it is prolonged and covers the following systolic sound. The rhythm heard at the apex then appears to be a double blowing murmur, although there is no mitral lesion. I will return to this point in the discussion of the differential diagnosis.

Timbre.—The murmur is soft and mellow; it is sometimes superficial, but usually deep. To hear it well, the patient should be placed in a sitting, or, better still, standing position.

Renaut has noticed a peculiarity in the pulse of patients suffering from extensive aortic insufficiency. He found that the sphygmographic trace of the carotid pulse sometimes presents a ridge or rather an elevation of the line corresponding to the systole of the auricle. This feature appears to indicate a large insufficiency and an hypertrophy of the auricle.

Finally, Tripier[1] has stated that in aortic insufficiency the retardation of the arterial pulse is greater than in the normal condition. F. Franck failed to discover this phenomenon by means of registering apparatus, and I have found as the result of sphygmographic tracings that the pulsation is somewhat ahead of time rather than delayed.

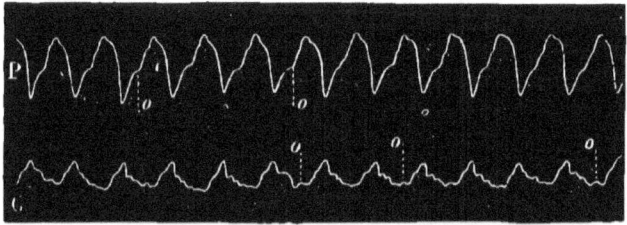

Fig. 64.—The Contraction of the Auricle is marked at O (Renaut).

Course of the Disease.—As soon as the disease begins it is evident that the heart is charged with a double labor. Hence results an hypertrophy which, at each cardiac revolution, propels a larger amount of blood than in the normal condition. Nevertheless, the arteries are not filled and relative anæmia occurs, characterized by pallor, vertigo, and often by other symptoms of relative cerebral anæmia.

If the patients are young the affection may remain latent for a long time, *i.e.*, no functional disorders are produced and the disease is manifested solely by physical phenomena.

The apex descends at an early period into the sixth intercostal space. Then it is gradually removed farther from the median line, sometimes even to a distance of 14 ctm.

The liver bends under the weight of the heart, and the right cardiac angle descends to the level of the insertion of the sixth and even the seventh cartilage.

---

[1] Tripier: Revue mensuelle de médecine et de chirurgie, janvier, 1877.

The vertical border of the heart does not move until a later period, when dilatation of the right auricle occurs from insufficiency of the tricuspid valve.

To recapitulate, aortic insufficiency is characterized by the visible pulse of Corrigan and the bounding pulse of Hope. The pulse is large, frequent, regular, intermittent only at rare intervals. Auscultation discloses a murmur, which, starting from the aortic orifice, descends along the sternum toward the right ventricle. The obliquity of the murmur turns more and more to the left in proportion as the heart increases in size. The murmur is heard with the greatest intensity from the third right cartilage to the xiphoid appendix.

The murmur is diastolic, and, in young subjects, no systolic murmur is produced. When an ascending systolic murmur is superadded, we must admit either an alteration of the semilunar valves or atheromatous and projecting deposits at the origin of the aorta.

The bruit begins with diastole and ends with diastole. Its timbre is soft, mellow, often distant, beginning with a certain intensity and gradually fading away.

The impulse of the apex is not always marked, though vigorous contractions are felt by the hand. This is due to the fact that the more the heart elongates the thicker is that portion of the lung which covers the apex and the more the heart tends to become parallel to the anterior wall of the chest.

An epigastric impulse, due to the depression of the heart, is often observed. In such cases, Potain and Rendu have observed that it precedes the impulse of the apex, and is due to the diastole.

During repose the patients are calm, but the slightest effort produces oppression, a feeling of heaviness in the epigastrium, dyspnœa, præcordial anxiety, palpitation.

From time to time a dull continuous sternal pain is noticed, usually at a late period; it is often absent.

The nervous symptoms may assume various forms, sometimes as attacks of angina pectoris, though this occurs rather in insufficiency, which is symptomatic of atheromatous dilatation of the aorta, than in Corrigan's disease pure and simple.

In truth, Potain and Rendu have justly remarked that aortic insufficiency is rather a disease of the aorta than of the heart, properly speaking.

Aortic insufficiency is readily distinguished from anæmia by the difference in the situation and time of occurrence of the bruit.

It must be distinguished also from aneurism. If the insufficiency is the sole lesion, it is easily recognized; but if it is accompanied by other grave lesions of the aorta, there is some risk of misinterpretation. The diagnosis then depends upon the care with which the examination is made.

The course of Corrigan's disease is slow when it develops in young subjects. It is not until the occurrence of hypertrophy and dilatation of the heart with degeneration of the muscular fibres and asystoly, that the phenomena of venous tension develop, with congestion, œdema around the veins of the lower limbs, the portal and pulmonary veins, and, finally, pulmonary apoplexy.

The termination may develop gradually, but death also occurs suddenly from angina pectoris or syncope.

CHAPTER XVIII.

SCLEROTIC ENDOCARDITIS OF THE AORTIC SEMILUNAR VALVES.

Stenosis, or Narrowing of the Aortic Orifice.

The lesions which produce stenosis of the aortic semilunar valves rarely leave them sufficient suppleness to enable them to close the orifice hermetically, so that it may be said generally that stenosis constitutes but a single period of the disease. Nevertheless, certain cases do occur in which stenosis continues without insufficiency until the death of the patient.

Anatomo-pathological Data.—Slight lesions are confined to a thickening of the nodules of Arantius and hypertrophy of the free edges ; sometimes to a fringe of vegetations upon the ventricular, more rarely the aortic surface of the valves.

At a more advanced stage the changes in the valves consist of thickening, atheromatous degeneration, and calcareous infiltration. The valves are usually rigid and adherent at the base, and then form a dome with the convexity turned toward the aorta and having at its apex a narrow opening which barely admits a quill, or is reduced even to a simple slit. In a still more advanced stage the stenosis affects the artery itself.

Another site of election of the stenosis, called subaortic by Vulpian, is situated at the level of the mitral orifice and is formed by a cicatrix of connective tissue, secondary to an abscess of the myocardium. Several cases of this kind have been described.

As a rule, when the aortic lesions are the result of an inflammatory condition, they are propagated by continuity from the lesions of the mitral valve.

Symptoms.—The disease often remains latent. It is manifested usually by dyspnœa on exertion and by oppression upon walking up-stairs.

The impulse of the apex is brusque, the ascent of the needle is rapid and accompanied by a marked purring thrill, which is often very strong at the level of the second right intercostal space, or a little lower if the heart is depressed on account of its increased weight.

Auscultation reveals a systolic blowing murmur, whose area forms an ascending column starting from the second intercostal space near the sternum and directed like the innominate artery into the carotids.

In the beginning its point of maximum intensity is situated at the level of the third intercostal space; as the heart hypertrophies the orifices are lowered and with them the valvular murmurs. I have seen a patient suffering from aortic stenosis in whom the maximum of intensity of the bruit was situated in the right fourth intercostal space.

OBSERVATION XXXI.—F. C——, sixty-six years of age, came to the infirmary June 14, 1872. Examination gave the following results: The patient suffers from dyspnœa and from pulmonary congestion, more extensive on the right side. The impulse of the apex of the heart is little felt; it beats in the fifth intercostal space, immediately below the nipple; the upper border of the liver corresponds to the insertion of the fifth costal cartilage.

The heart-sounds are feeble at the apex, stronger at the level of the tricuspid, but without murmur. To the right of the sternum is heard a very distinct, rough, systolic murmur, with its maximum of intensity in the right fourth intercostal space, near the sternum. There is slight reflux in the jugular veins. Some œdema of the lower limbs, slight œdema of the face. The pathological bruit corresponds to a murmur of aortic stenosis, but I could not explain why its maximum of intensity was situated so low.

The autopsy showed that the aortic valves were covered with rugosities and that the heart was lowered by its own weight.

The murmur of stenosis of the aortic orifice is situated, then, to the right of the sternum, generally has its maximum of intensity in the second intercostal space, ascending into the first space and the vessels of the neck.

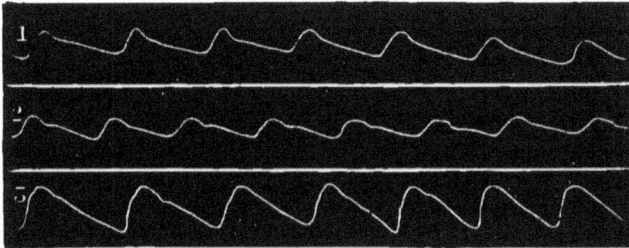

FIG. 65.—Trace of the Pulse in Aortic Stenosis, according to Marey.

Inferiorly, the murmur may extend behind the fourth cartilage and even the fourth intercostal space. It is vibrating, more or less rough, and accompanied frequently by a purring thrill. It is sometimes propagated to the apex. Its intensity bears less relation to the stenosis than to the roughness of the lesions.

The pulse is usually small, hard, and wiry; when the patient is in repose the systoles are long. Fraentzel[1] has remarked that the traces of stenosis and of cardiac weakness present slight elevation, but that in the former the tension remains great during the entire duration of the systole, while in the latter it falls immediately. Traube has noted retardation of

---

[1] Fraentzel: Berl. Klin. Wschr., pp. 49, 50, 1867.

the radial pulse, but this phenomenon is far from being certain, and, moreover, is observed also in mitral stenosis.

The diagnosis of aortic stenosis is readily made from the situation of the murmur, its time of occurrence, and its rough timbre with the purring thrill. It is also propagated into the vessels of the neck.

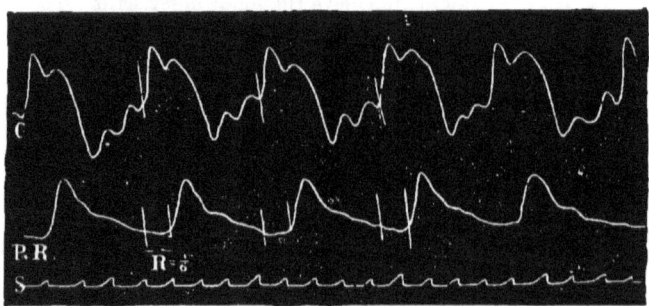

FIG. 66.—Retardation of the Pulse in Aortic Stenosis (Keyt and Marey).

This characteristic and the accompanying hypertrophy, which begins in the left heart and then extends to the other side, are readily recognized.

But there are cases in which a differential diagnosis must be made between slight aortic stenosis and anæmia, and this may prove embarrassing, as illustrated by the following case:

OBSERVATION XXXII. *Aortic Stenosis with Anæmia.*—J. P——, forty years of age, entered the hospital April 5, 1881, with all the signs of anæmia. Pallor, anorexia, blowing murmur on both sides of the sternum at the level of the second intercostal space.

Careful examination of the murmurs gives the following results: In the second left intercostal space, near the sternum, is found a systolic blowing murmur, followed by the sharp click of the semilunar valves. The murmur does not diminish in the sitting position. It is not limited to the second left space, but extends into the second right intercostal space. The murmur is synchronous with arterial diastole, and not with the venous depression. Consequently this murmur exists not alone in the pulmonary artery, but also in the aorta; it also extends into the first intercostal space, but less on the right side than on the left.

Apex-beat in the fourth intercostal space, 10 ctm. from the median line. The upper border of the liver corresponds to the insertion of the fourth costal cartilage; the apex is depressed 2½ ctm. The heart is therefore small. No murmur in the vessels of the neck and over the mitral and tricuspid valves. There is therefore a slight lesion of the aortic valves.

The progress of the disease is slow and its period of tolerance is of long duration. As a general thing, lesions of the aortic semilunar valves give rise to stenosis and insufficiency at the same time.

The disease causes hypertrophy of the heart, followed by dilatation; at the same time chronic aortitis is often produced, together with dilatation and thickening of the artery.

132   DISEASES OF THE HEART.

Aortic Stenosis and Insufficiency.

After the details given above the diagnosis of aortic stenosis and insufficiency, when existing in combination, becomes very easy, since the murmurs are double in number and direction and each of them retains its own characteristics.

Both start from the origin of the aorta at the right third cartilage, or a little lower. The murmur of stenosis ascends into the vessels of the neck, that of insufficiency descends along the sternum; the latter bruit is at first vertical, but becomes oblique to the left after hypertrophy of the left heart occurs. This is illustrated in the following figures:

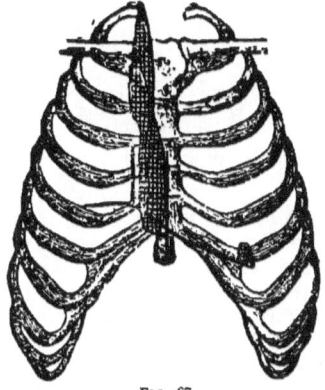   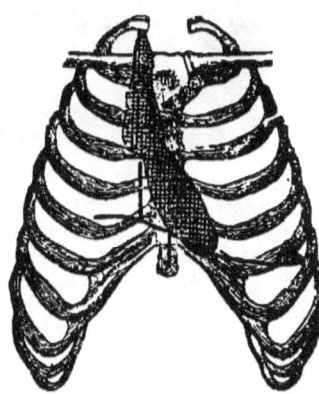

FIG. 67.   FIG. 68.

The maximum of intensity of the two sounds is not situated in the same place, that of the first being in the second right intercostal space, that of the second in the fourth intercostal space, at first on the right side, later on the left.

The disease soon gives rise to hypertrophy of the left side of the heart. The inferior border is very oblique, and it may be 15 to 15½ ctm. in length, instead of 11 ctm., the normal measurement.

Finally, I will state that a double blowing murmur is observed very frequently at the apex. This bruit is remarkable in the fact that it is double, confined to the apex, and not propagated into the axilla like mitral murmurs: this murmur is propagated from the base.

The pulse of the double aortic lesion has less amplitude than that of simple insufficiency, but it preserves the character of brusqueness (bounding pulse), and especially of exhaustion.

OBSERVATION XXXIII.—Joseph L——, thirty-six years of age, was taken sick October, 1875, with acute polyarticular rheumatism. He left the hospital two months later, suffering from palpitation and shortness of breath. Re-entered the hospital in the beginning of 1879.

ENDOCARDITIS OF THE AORTIC SEMILUNAR VALVES. 133

Examination gave the following results: the apex-beats in the sixth intercostal space, 12 ctm. from the median line; the upper border of the liver corresponds to the insertion of the sixth cartilage; vertical border of heart is 2½ ctm. from the median line. Pulmonary resonance ceases in the first and second right intercostal spaces, 3 ctm. from the sternum. Loud

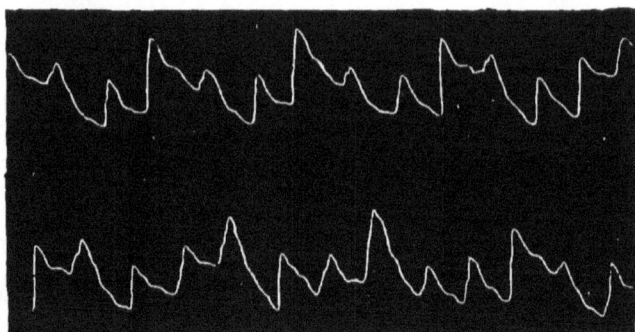

FIG. 69.—Stenosis of the Aorta with Beginning Insufficiency.

purring thrill in the right second intercostal space near the sternum. Auscultation shows a rough murmur to the right of the sternum, beginning in the third intercostal space, and continuing into the second and first; maximum of intensity in second space very near the sternum. This murmur is propagated into the brachio-cephalic trunk, the right, and even the left carotid. This murmur is systolic, begins with systole, covers the lesser silence and the second sound, but is not followed by the diastolic blowing murmur of insufficiency. Its timbre is very rough and vibrating.

At the apex is heard a distant murmur at the close of systole, and which is a propagated murmur; it is also heard at the xiphoid appendix.

July 15th, the signs have changed somewhat. The systolic bruit persists, but is followed by a soft blowing murmur, which descends as far as the fourth rib. Pulse always regular, slightly bounding. No stasis or œdema, merely anæmia. From that time until the present (August 16,

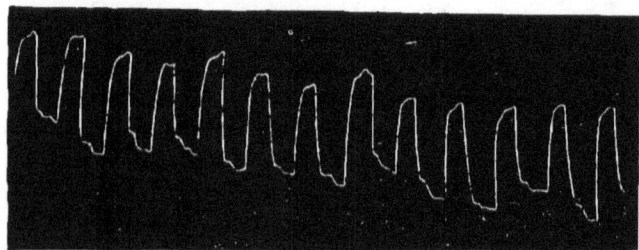

FIG. 70.—Stenosis of the Aorta with Beginning Insufficiency.

1881), the patient has remained about the same. When he feels well he returns to work; when he is again worn out he re-enters the hospital, is put on milk diet, and in a little while recovers satisfactory health. In 1882 he returned with more marked insufficiency.

The traces represented above were obtained upon this patient with the cardiograph.

CHAPTER XIX.

DISEASES OF THE ORIGIN OF THE AORTA—AORTITIS—ATHEROMA.

ANEURISMAL DILATATION, OR TRUE ANEURISM—HODGSON'S DISEASE.

Anatomical Data.—The aorta begins at the base of the left ventricle. Its origin is situated behind the sternum at the level of the third costal cartilages, a little to the left of the median line. It terminates at the level of the fourth lumbar vertebra.

The aorta is protected in a remarkable manner by the vertebral column, the sternum, and the ribs. The following is its calibre at its origin:

At birth	23 millimetres.
At 21 years	60 "
At 50 years	70 "
At 80 years	80 "

Its calibre diminishes constantly from its origin to its termination.

At its origin is a slight dilatation which tends to give it somewhat the shape of a clover leaf; the dilatation is formed by the three sinuses of Valsalva, which are spanned by the semilunar valves.

The aorta is composed of three layers—a gray inner tunic, a middle yellow tunic, and an external red tunic.

1. *Internal Tunic.*—The inner layer of this tunic is composed of epithelium which is destroyed very quickly by decomposition, and is never observed upon autopsy, except when animals are killed in experimentation. The cells composing it are flattened and polygonal in shape, and contain circular or elongated flat nuclei. The subepithelial layer is composed of stellate cells and a fibrillary substance having a longitudinal direction. The internal tunic contains no vessels.

2. *Middle, or Yellow Tunic.*—This is much thicker than the preceding layer: it is composed of elastic plates and fibres, which constitute by their anastomoses a continuous network. The interstices contain smooth muscular fibres which run in a transverse direction. The inner portion of the yellow tunic is formed by a thick and refractile elastic plate. The transition from the middle to the external tunic is less abrupt. The former also contains no vessels.

3. *External Tunic.*—This is composed mainly of arteries, veins, and lymphatics bound together by connective tissue.

The aorta is divided into three parts—the arch, the thoracic aorta, and the abdominal aorta.

The arch comprises that portion which is included between the origin and the passage of the left bronchus; the thoracic aorta extends from the left bronchus to the diaphragm; the abdominal aorta extends from the diaphragm to the fourth lumbar vertebra.

The arch is divided into three parts—the ascending or intrapericardial portion, the transverse portion, and the descending portion.

The intrapericardial portion is subdivided into two parts; the first, which extends from the origin to the second intercostal space; the second, or great sinus, which extends from the second intercostal space to the first part of the sternum. This subdivision is unnecessary from a pathological point of view.

The ascending aorta is covered at its origin by the infundibulum of the right ventricle, by the pericardium and sternum; it is directed obliquely upward, and to the right toward the second intercostal space, where it comes in contact with the superior vena cava. It is important to note that the great sinus, in the normal condition, does not pass beyond the right border of the sternum, but as soon as the vessel dilates it passes beyond the bone into the second intercostal space. In the normal condition, then, the external border of the aorta follows the right border of the sternum along the right first and second intercostal spaces.

Aortitis.

The aorta is yellow in the normal condition; the redness found after death is due to post-mortem imbibition.

In inflammation of the aorta the inner surface is pale, because the layers, in swelling, are removed farther from the vascular network. Inflammation is recognized better by the thickening of the coats of the vessel. The inner tunic presents circular swollen patches, either isolated or confluent; these have a soft gelatinous consistence. They are opalescent and rosy in color. Upon section of these patches they are found to show very distinctly the separation of the internal from the middle tunics. The patches contain spherical or oval cells, the nuclei of which swell up under the action of acetic acid; in addition, the large cells with multiple prolongations forming the deeper layer of the internal tunic.

The remaining lesions are observed especially in the external tunic and constitute periarteritis; the middle tunic is little affected.

Acute arteritis may be due to rheumatism, the puerperal condition and alcoholism, but it is very rare.

The clinical history is unknown ; it is accompanied usually by other grave lesions which mask its symptoms.

Chronic Aortitis and Atheroma.—Fatty degeneration is the ordinary senility of the aorta and this senility begins at an early period. After the age of forty to forty-five years this degeneration is marked by enlargement of the aorta, which yields to the pressure of the blood. The primary site of this degeneration is the part which supports the greatest force of the blood-pressure, viz. : the origin of the aorta and the great sinus.

This fatty degeneration is evidenced to the naked eye by opaque and striated white spots on the inner surface of the vessel. Upon making a longitudinal section fatty granules are found occupying the deeper parts of the internal tunic and the outer part of the middle tunic.

In the middle tunic the fatty drops have taken the place of the muscular fibres and are interposed between the more resisting elastic fibres.

The fatty degeneration always leads to chronic arteritis, but when the alteration has begun with arteritis the gelatiniform plaques of the inner tunic undergo fatty degeneration and become opaque.

The fatty foci are at first microscopic, but after coalescing they become visible to the naked eye, forming foyers situated in the deepest part of the internal tunic. Arteritis occurs at their edges with swelling of the inner tunic. This swelling around the plaque gives to it an umbilicated appearance. Upon incision it discharges a whitish mass composed of free fat granules, granular bodies, cholesterine, and crystals of fatty acids. If the foyer opens during life a small cavity is formed, with loss of substance of the inner tunic.

At a later period, if the degeneration proceeds slowly, calcareous infiltration occurs, at first isolated, then confluent. These patches lose their elasticity and often curl up in their inner surface, causing the projection of one edge. This projection may penetrate the inner layer and become covered with fibrine, thus forming the starting-place for emboli.

When these patches are numerous and the membrane has lost its elasticity, a cylindrical dilatation or true aneurism of the aorta occurs, the dilatation being most marked at the origin of the vessel.

Finally, the lesion may proceed deeper and destroy the middle tunic, but the external thick tunic still resists.

It is to this combination of lesions, beginning by fatty degeneration and terminating in true aneurism with secondary valvular insufficiency, hypertrophy and dilatation of the heart, that I have given the name Hodgson's [1] disease, in honor of him who first described it.

Indeed, Hodgson has so well described atheromatous degeneration of the arteries that the description sounds as if written to-day.

---

[1] Hodgson : Traité des maladies des artères et des veines, translated by Breschet, p. 19. Paris, 1819.

"The inner surface of the arteries is converted very frequently into a soft, pulpy substance. Sometimes it presents the appearance of small flattened tubercles, sometimes its surface is irregular and brawny. This is found very often in individuals suffering from aneurism.

"One of the most frequent affections of the arteries is the accumulation of atheromatous or purulent matter between the internal and middle coats of the vessels. The diseased part is yellow, opaque, and usually a little elevated. At times these elevations are considerable and very extensive, while in other cases they are circumscribed and present a tubercular or pustular appearance. Upon cutting into them a substance may be expressed the consistence of which varies from that of cheese to ordinary pus.

"An ulceration occurs sometimes at the surface of these elevations and, penetrating the middle layer of the vessel, produces an aneurism. Calcareous matter is deposited frequently in the centre of these eminences."

And farther on (page 22) Hodgson says :

"In advanced age it is common to find the arteries studded with calcareous matter, and the appearance of the vessel depends entirely upon the extent of the deposit. Sometimes the internal layer presents numerous fine white spots, while in other cases the calcareous matter is mixed with cheesy deposits. More generally it forms a sort of crust which crackles under the finger like an egg-shell. These incrustations have no particular shape ; they may form eminences which diminish the calibre of the vessel."

Hodgson has also given a perfect description of aneurismal dilatation of the aorta and the resulting valvular insufficiency.

"The disease consists of a permanent enlargement of the lumen of an artery and is accompanied usually by some morbid change in its walls.

"Its usual site is the ascending portion and arch of the aorta, which is dilated sometimes to an extraordinary degree, the dilatation beginning immediately above the semilunar valves. The walls of the artery, thickened in a remarkable manner and encrusted with calcareous and atheromatous matter, constitute the wall of the sac. Smaller pouches often develop upon the sides of the large sac and are also encrusted with calcareous matter. At other times the dilated walls appear to have opened at some place and an aneurism is engrafted, so to speak, upon the dilated artery. This circumstance may cause the disease to be mistaken for an aneurism, but in this case the place at which the neck of the sac begins is very evident, and its wall does not present the morbid appearances peculiar to arterial walls.

"The dilatation often terminates abruptly at the arch of the aorta, but at other times it diminishes gradually. Sometimes the dilatation is partial and occupies only one side of the vessel, forming a pouch like an aneurism ; but such partial dilatations rarely contain laminated coagula.

"This morbid condition is found more frequently in the ascending aorta than in any other vessel."

Hodgson also reports several cases in which secondary valvular insufficiency developed, as is illustrated by the following:—

OBSERVATION XXXIV.—A vigorous man, aged sixty years, complained for a long time of difficulty of respiration, great oppression on the chest, and frequent palpitation of the heart. He was subject sometimes to syncope, and preferred the recumbent position. The pulse was very small, frequent, and intermittent. While walking in his room one day, he suddenly fell to the floor and expired instantly.

The right ventricle was found to be soft and dilated. The aorta, enlarged from its origin to the descending portion, formed a large sac which readily admitted five fingers. This dilatation terminated gradually; the walls were thickened and covered in several places by calcareous ridges and atheromatous deposits. *The semilunar valves were thickened and separated from one another.* The enlarged left ventricle seemed healthy.

OBSERVATION XXXV.—A man, aged sixty years, had long suffered from symptoms of pulmonary disease; had frequent attacks of syncope, and the pulse was feeble and irregular; sudden attacks of dyspnœa, simulating asthma. He expectorated large quantities of mucus, and finally died of dyspnœa.

The ascending aorta was extremely dilated, and formed a sac four inches in diameter. The dilatation began at the origin of the vessel and terminated suddenly at the middle of the arch. The internal layer was irregularly thickened and covered with calcareous plates; above the semilunar valves were found two small sacs which were lined with a thick crust of earthy matter. The semilunar valves were separated from one another on account of the dilatation of this part of the vessel, and they were thickened and retracted. The dilated aorta pressed upon the lungs, and the bifurcation of the trachea.

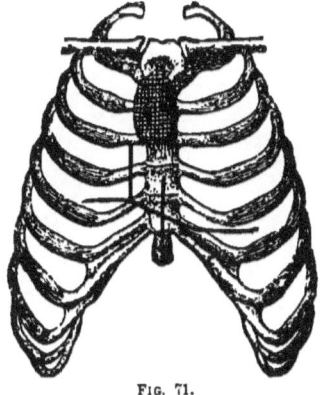

FIG. 71.

Diagnosis of Hodgson's Disease.—In the beginning, the disease is not well marked, and, if it occurs in young subjects, may pass unnoticed, or be regarded as anæmia, unless auscultation is carefully performed.

This mistake is so much the more possible, as the affection is accompanied by a certain pallor, and a knowledge of the exact situation of the anæmic murmur in the pulmonary artery is not very widespread.

OBSERVATION XXXVI.—For ten years I have had under my observation a woman who has suffered, for that period, from a slight pain behind the sternum, at the level of the second intercostal space. She was pale for some time, but no anæmic bruit was heard. The heart is healthy, but auscultation discloses a murmur corresponding exactly to the ascending aorta, from its origin to its exit from the pericardium.

The bruit is systolic, continuing throughout the entire period of systole; it is followed by the normal semilunar click. The murmur is very soft, but nevertheless very strong. Nothing is heard over the pulmonary artery; in addition, the bruit is not modified by change of position, arrest of respiration or exercise. Finally, it has the permanent character of organic murmurs. It is due to roughness of the origin of the aorta.

When thus limited, endaortitis in a young subject may last for a long time without bad effects. This is not true of Hodgson's disease, *i.e.*, fatty degeneration of the aorta, with its sequences. This will form the subject of the following chapter.

CHAPTER XX.

SYMPTOMS AND DIAGNOSIS OF HODGSON'S DISEASE, OR FATTY DEGENERATION PRODUCING TRUE ANEURISM OF THE ORIGIN OF THE AORTA.

BEFORE Hodgson's disease can be distinguished, it is usually characterized by the functional disturbances of cardiac disease. At first, dyspnœa on effort ; this is particularly well-marked on going up-stairs. The dyspnœa may be absent while the patient is walking on a level surface. Attacks of vertigo due to cerebral anæmia may also be present.

Apart from these phenomena, the attention is usually attracted at first by the condition of the arteries. The radial is found to be large, hard, destitute of elasticity, little dilated by the cardiac pulsation, and resisting the pressure of the fingers ; moreover, the artery is sinuous. If the neck and chest are bared it is found that the arteries are large, sinuous from elongation, so that they are curved even during extension. The elongation of the arteries can also be observed in the arch of the aorta.

In the normal condition the ascending aorta does not pass beyond the right border of the sternum, and is 15 to 20 mm. below the sternal notch. When the vessel dilates, it passes beyond the border of the sternum, and may be recognized in the second and first right intercostal spaces by palpation, the cardiograph, and percussion, and at the same time it may reach the sternal notch or even extend above it. This dilatation causes displacement of the principal arteries. I have seen a case in which the right subclavian artery no longer was given off from the innominate, but directly from the aorta. I have also seen the left subclavian pushed to the left, so that it formed a considerable curve upward before reaching the clavicle. The thyroid arteries may also be displaced.

The calibre of these vessels also enlarges, and the thyroids may enlarge to such an extent as to resemble their appearance in exophthalmic goitre.

The aorta and large vessels present a thrill which is observed not alone in the vessels, but also in the surrounding organs. Thus, it is felt sometimes over the middle and upper portions of the sternum, and more frequently still above the sternal notch.

Hope has observed that the purring thrill is often more marked in true than in sacculated aneurism, on account of the absence of stratified clots in the former.

HODGSON'S DISEASE. 141

As the dilatation is usually arrested before the end of the curve of the aorta, the arteries originating from the right side undergo dilatation, while the others do not, so that the pulse is unequal on the two sides. On the other hand, the irregular distribution of the atheromatous patches may cause obstruction of any vessel, so that the alteration of the pulse may be observed either on the same side as the true aneurism, or on the opposite side.

When the difference in the two pulses is not perceptible to the touch, it may be detected by the sphygmograph, as is shown by the following cases:

OBSERVATION XXXVII. *Hodgson's Disease.*—Sister P——, aged sixty-six years, entered the hospital October 20, 1872.

Examination of the heart in this patient, whose vessels generally were atheromatous, gave the following results:

Apex-beat in the fifth intercostal space, 10 ctm. from the median line; upper border of liver corresponds to the fifth cartilage; vertical border of heart is very near the sternum, 2 ctm. from the median line; the inferior border is slightly oblique.

Auscultation shows dull murmurs at the apex; in the aorta the murmurs are weak. Nothing found over the tricuspid valve or pulmonary artery. The left radial pulse is weaker than the right.

Percussion shows dulness behind the sternum, and in the first right intercostal space. Auscultation in these parts occasionally reveals considerable intensity of the murmurs. Here the dilatation is well shown by the arterial atheroma, the difference between the two pulses, and the dulness in the upper part of the sternum and the first intercostal space. The following traces represent the two pulses:

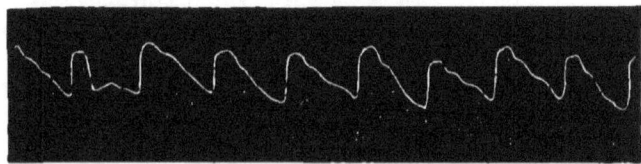

FIG. 72.—Right Radial Pulse.

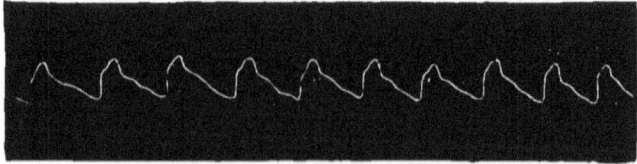

FIG. 73.

OBSERVATION XXXVIII. *Aneurismal Dilatation of the Aorta; Hodgson's Disease.*—B——, of a very vigorous constitution, enjoyed perfect health until the age of seventy, with the exception of occasional slight attacks of gout.

Last year he began to experience some difficulty of breathing; this year he presents marked cardiac dyspnœa, and œdema of the lower limbs.

Apex-beat in the sixth intercostal space, 12 ctm. from the median line; border of liver corresponds to fifth cartilage; vertical border of heart is 3 ctm. from the sternum; very great obliquity of lower border. Hypertrophy of the heart is therefore present, and is confined almost exclusively to the left side. Auscultation shows a soft systolic murmur at the apex.

Where the aorta extends to the right of the sternum, a loud murmur is heard and occupies the first three intercostal spaces. Its maximum of intensity is situated in the second intercostal space, in which it is heard 40

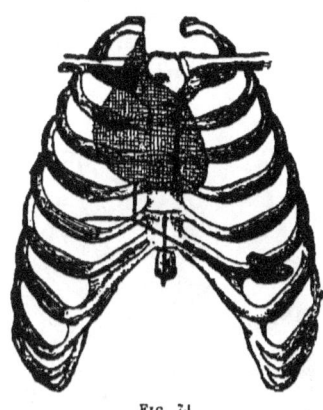

Fig. 74.

ctm. from the right border of the sternum; the bruit is propagated a little into the right carotid. The murmur is also intense to the left of the sternum, but nevertheless becomes weaker in the first three left intercostal spaces, and is heard feebly over the entire ventricle. At the apex is found a triangular focus, the apex of which is directed toward the axilla; this murmur is purely systolic, and is followed by a very distinct semilunar click. Nothing is heard over the tricuspid, the jugulars, or the pulmonary artery. Percussion shows dulness over the entire area of the murmur.

The right pulse is larger than the left. The diagnosis is easy; dilatation of the ascending part and of a portion of the transverse part of the aorta. The murmur at the apex is produced by the mitral, and is probably due to insufficiency caused by dilatation.

The patient died at the end of the year from cardiac cachexia.

As is evident from these cases, the disease is characterized by dulness over the upper part of the sternum and the right lung. This dulness is very characteristic when there is no pulmonary lesion in this locality.

We will now examine the character of the murmurs somewhat more carefully.

1. *Topography.*—The bruit extends in the beginning to the right border of the sternum, in the second intercostal space; as it becomes more extensive, it is heard more and more in the second intercostal space. It is remarkable that this bruit has a greater tendency to spread in width than in height. It is often found 4 ctm. from the right border of the sternum, although it does not extend into the first intercostal space, and still less into the right carotid.

Observation XXXIX. *Hodgson's Disease.*—S. D——, seventy-one years of age, entered the hospital October 12, 1872. The apex of the heart is in the sixth intercostal space, 9 ctm. from the median line. The border of the liver cannot be determined on account of a slight effusion in the right pleura; this is also true of the vertical border of the heart. The radial arteries are sinuous and atheromatous. Auscultation reveals slight prolongation of the first sound at the apex. A blowing murmur is heard, particularly in the second right intercostal space, is prolonged to the right, and is accompanied by dulness in this region. The murmur is systolic and slightly rough. (Fig. 75.)

OBSERVATION XL.—J. G——, seventy-one years of age, entered the hospital March 14, 1872. Apex-beat in the fourth intercostal space, 12 ctm. from the median line. The upper border of the liver corresponds to right fifth cartilage, lower border of heart is horizontal. The arteries are atheromatous. Dulness is present in the right second intercostal space, and a murmur which extends to the right of the sternum in the second intercostal space; it begins 1 ctm. from the right edge of the sternum and ends at 5¼ ctm. from this point. It is prolonged a little into the first intercostal space, but not at all into the vessels of the neck (vide Fig. 76).

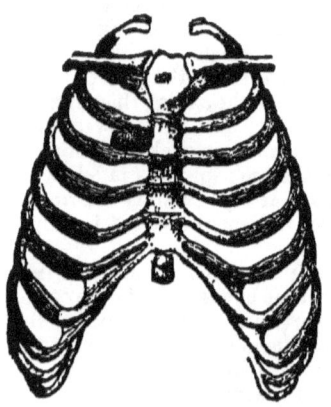
FIG. 75.

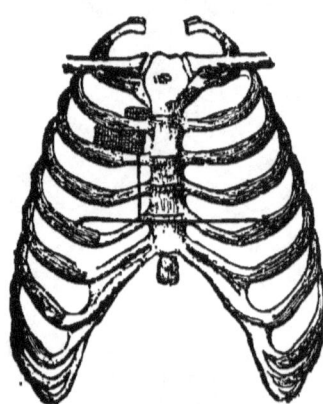
FIG. 76.

OBSERVATION XLI. *General Arterial Sclerosis; Aortic Dilatation, with Roughness of the Aortic Valves, and the Origin of the Aorta; Considerable Hypertrophy; Mitral Insufficiency from Dilatation.*—P. L——, sixty-two years of age, entered the hospital January 18, 1879. The affection began in February, 1877. The patient states that he was then seized suddenly with dyspnœa, and was confined to his room for a week. He then attempted to return to work, but was only able to work occasionally until January 2, 1879, when he was compelled to give up entirely.

Upon admission to the hospital, the patient was suffering from dyspnœa, with cyanosis and œdema of the lower limbs.

Apex-beat of the heart in the sixth intercostal space, 16 ctm. from the median line; upper border of liver in normal situation; the apex is 6 ctm. below the hepatic border. The vertical border of the heart corresponding to the right auricle is 1 ctm. from the right border of the sternum.

Hence there is chiefly ventricular hypertrophy, particularly of the left ventricle.

Auscultation shows a blowing murmur at the apex, which occupies this region and extends several centimetres toward the axilla. The murmur is systolic, begins with the impulse of the apex, lasts throughout the systole, covers the second sound, and is heard into the greater period of silence (paradox systolic blowing murmur). The murmur is soft; pulse 88, and regular. Nothing heard over the other orifices.

The lungs are œdematous in the lower part of the left, and whole of the right lung (fine crepitant râles).

Under the influence of digitalis the urine increased in quantity, and the dropsy and dyspnœa diminished to a certain extent.

A second examination (January 30th), gave very different results; the pulse is regular and has become larger; the arteries are large, sinuous, and present the phenomenon of Corrigan's visible pulse. Examination of the aorta at its origin reveals a blowing murmur situated in the right second intercostal space, and extending 5 ctm. from the right border of the sternum. The bruit is rough, systolic, and ceases at the closure of the semilunar valves.

The diagnosis is now complete, viz.: general arterial sclerosis, dilatation of the origin of the aorta; roughness of the inner wall, beginning at the orifice, without insufficiency of the valves; mitral insufficiency.

The condition remains the same during the month of February. In March the dyspnœa became more severe, but again improved under digitalis, and on April 2d the patient left the hospital.

March 3d he returned, suffering from considerable dyspnœa and intense pulmonary congestion. The heart beats feebly, and the asystole appears to be the cause of the pulmonary congestion. The patient continued to fail, and died on November 13, 1879.

The autopsy showed considerable hypertrophy of the heart (weight 845 grammes); the walls are considerably thickened, particularly those of the left ventricle.

The aorta is dilated from its origin; the dimensions are normal, immediately above the semilunar valves, but they soon increase. The brachiocephalic trunk is dilated, its circumference at its origin being 50 mm.

The aorta is still elastic but contains rough, projecting patches.

The semilunar valves are thickened, particularly the nodules of Arantius. There is undoubtedly slight insufficiency, and this explains the fact that toward the last we thought that we heard a systolic descending murmur.

The cavity of the left ventricle is considerably enlarged and there is mitral insufficiency from dilatation.

The right heart is enlarged, and its walls hypertrophic; the valves of the right side are normal. The liver and kidneys are congested.

In other cases the dilatation is more marked in the transverse portion of the aorta than its ascending portion, and then the aorta is often perceptible above the sternal notch. The roughness of the calcareous patches then gives rise to a transverse bruit situated at the upper part of the sternum; though often simple, this murmur is sometimes double.

Finally, when the dilatation is considerable, the bruit may extend beyond the left border of the sternum.

It only remains to say that the systolic aortic murmur is propagated very little toward the apex, where it produces merely a slight prolongation of the first sound.

2. *Period of Occurrence.*—The bruit begins with the ventricular systole, lasts during the entire systole, and is arrested at the moment of the semilunar closure.

According to Noël Guéneau de Mussy,[1] the second sound is even more intense than that which I have described as occurring in the pulmonary artery when an anæmic blowing murmur is produced. He states that it may be compared to the normal second sound as the amphoric souffle to the bronchial souffle, and that the word tympanitic appears to apply better than any other to the sensation which this murmur may produce. I have not observed such an intensity of the sound as that described by this writer.

Usually the second sound persists and is not replaced by a murmur. It may disappear under two conditions:

In the first, it may simply be masked by a blowing murmur produced by the sharp projection of a calcareous plate, which gives an ascending murmur during systole, and then a sometimes loud murmur when the blood flows back to close the semilunar valves. At least I have seen a case in which a projection of this character produced a double blowing murmur in the aorta without aortic insufficiency. In the second, the aortic dilatation produces insufficiency of the valves from the simple enlargement of the vessel.

3. *Timbre.*—In the beginning, when the artery simply yields and there are no projecting atheromatous patches, the dilatation is indicated rather by dulness on percussion than by the presence of murmurs, which are dull and without a blowing character.

As the aorta becomes more and more calcified, more or less rough blowing murmurs are gradually produced. The bruit due to insufficiency presents its ordinary characteristics of softness and depth.

Little by little the hypertrophy of the heart, especially of the left side, increases; the apex is felt in the sixth intercostal space and is removed farther from the median line.

The right heart is but little affected, as is shown by the fact that the right cardiac angle is not lowered and the vertical border is brought nearer to the sternum because the heart is drawn to the left. This displacement is shown by the schematic figure.

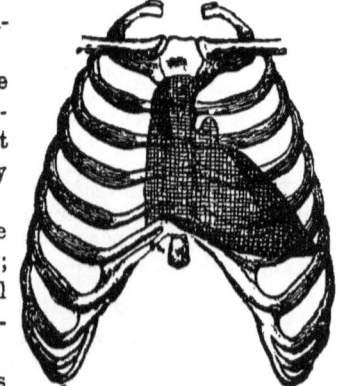

FIG. 77.

It is seen that the heart has turned, so to speak, around the attachment of the inferior vena cava, the most fixed point in the cardiac region.

The rotation around this point occurs from above downward, as we

---

[1] Noël Guéneau de Mussy: Sur la dilatation cylindrique de l'aorte ascendante et sur le caractère tympanique du second bruit cardiaque dans cette affection (France Médicale, Novembre 8, 1876).

have seen it occur transversely from before backward in emphysema, from behind forward in effusions into the left pleura.

As the hypertrophy of the heart in these cases is attended with dilatation of the cavities, the time comes when the mitral valve, although healthy, becomes insufficient from dilatation, and a systolic blowing murmur appears at the apex with all the qualities of a local bruit.

I will revert at a later period, *apropos* of hypertrophy with dilatation, to the characteristics of mitral insufficiency without lesion as the result of dilatation.

The pulse in Hodgson's disease presents the characters of atheroma (*vide* Figs. 72 and 73). Furthermore, the two radial pulses are usually unequal (*vide* Figs. 78 and 79). The characteristics of insufficiency are less marked than in Corrigan's disease.

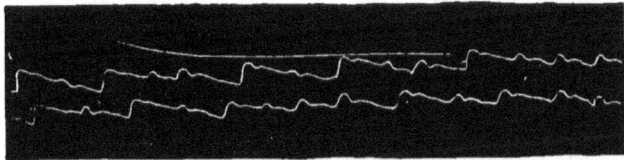

Fig. 78.—Right Radial Pulse.

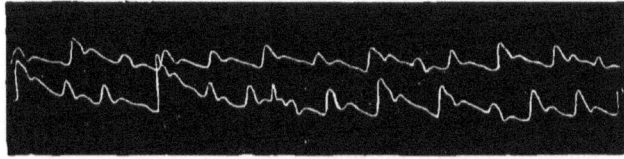

Fig. 79.—Left Radial Pulse.

CHAPTER XXI.

HODGSON'S DISEASE.—(*Continued.*)

COMPRESSION OF ADJACENT ORGANS — COURSE — DURATION — TERMINATION — ETIOLOGY.

THE superior vena cava is the first organ to be displaced by the aorta during the development of true aneurism. In fact, this vein is in immediate relationship with the great sinus of the aorta at points which are the first to be attacked by Hodgson's disease. This compression of the vein, or rather its being pushed away, is not attended generally with any great disturbance, as the anatomical conditions of the superior vena cava give it a certain facility of displacement which enables it to escape compression. It does not become inflamed from the contact of the true aneurism, while the contact of a false aneurism may cause adhesive phlebitis and obliteration of the superior vena cava.

This does not hold good with regard to the trachea. Monneret and Delabeye[1] state that compression of the trachea by a dilated aorta has given rise on several occasions to an affection of this canal, and has rendered tracheotomy necessary in order to relieve attacks of suffocation.

Compression of the trachea is, in fact, one of the complications of aortic dilatation, as the spinal column does not permit the trachea to recede to any great extent from the dilated vessel.

At first the opposition of the trachea against the vertebral column does not give rise to any respiratory disturbance, although it is characterized by a remarkable stethoscopic phenomenon, viz., the transmission of the tracheal sounds to the spine. This results in a tracheo-bronchial blowing murmur, presenting the rhythm of respiration and limited to the first few dorsal vertebræ, particularly the fourth.

I have not found this important symptom mentioned in any treatise on auscultation, but I have observed it on several occasions. The following case will serve as an illustration :

OBSERVATION XLII.—William J——, aged seventy years, entered the hospital June 2, 1872. He complained of difficulty of breathing, and the examination of the heart gave the following results :

---

[1] Compendium de Médecine, t. i., p 182.

Apex-beat in the fifth intercostal space, 105 mm. from the median line. The upper border of the liver corresponds to the insertion of the fifth cartilage; the vertical border of the heart is 2 ctm. from the right border of the sternum. Hypertrophy of the left heart with dilatation of the right is therefore present.

Percussion shows dulness extending into the first and second intercostal spaces. Auscultation of the aorta in this situation reveals a rough systolic murmur, most marked in the second space. Nothing abnormal over the pulmonary artery or mitral valve.

At the xiphoid appendix is found a systolic blowing murmur running along the length of the heart at this level, very loud at the cartilage, and accompanied at times by a reduplication of the second sound. The dilated cervical veins pulsate. These are the symptoms of dilatation of the ascending aorta (dulness and a rough murmur with the first sound in the right second intercostal space), with left hypertrophy and dilatation of the right heart with tricuspid insufficiency from dilatation.

The dyspnœa was not very intense. Nevertheless, although no pulmonary lesions were present, I found posteriorly, at the level of the upper portion of the dorsal spine, a tracheo-bronchial blowing murmur limited to the vertebral column and always presenting the respiratory rhythm. There was also inequality of the two pulses.

It will be found that the latter symptom also occurs in false, sacculated aneurism of the aorta.

In October, 1868, I had occasion to observe a patient at the Charité under similar conditions. The dilatation affected the transverse portion of the arch and was characterized by a transverse double murmur in the sternal notch; a tracheal murmur was heard over the first dorsal vertebræ. In this patient there was considerable compression of the trachea and the respiratory nerves. Prof. G. Sée, under whose care the patient fell in a few days, thought it necessary to perform tracheotomy, but the patient died during the operation. At the autopsy it was found that the compression was produced by dilatation of the transverse portion of the arch.

Course, Duration, Termination.—Hodgson's disease almost always begins gradually. Even when the dilatation is perceptible and demonstrable by characteristic signs, slight dyspnœa alone is produced, and the patients act like all others suffering from atheroma of the vessels.

They may remain in this stage for a long time, when cardiac cachexia develops at a certain period and progresses very rapidly. This aggravation is almost always due to albuminuria from interstitial nephritis.

Even sudden death may occur, as in primary aortic insufficiency (Corrigan's disease).

The disease usually is due to the same causes as atheroma, viz., age, gouty constitution, and alcoholism; to these must be added the use of tobacco.

The youngest of the patients under my observation was sixty years old. I have observed a number of cases in which gout was the etiological factor. Among the poor and in hospital practice, alcoholism is the principal cause of the disease. I have also observed a case in which the disease was traced to the abuse of tobacco, which so often gives rise to angina pectoris.

OBSERVATION XLIII. *Hodgson's Disease produced by Tobacco.*—V—— consulted me in 1874 concerning vertigo and a feeling of oppression. I found Hodgson's disease, dilatation of the ascending aorta, and hypertrophy of the heart without valvular lesion. Thinking that the cause lay in his abuse of tobacco, I restricted him to two cigars daily.

Three years later the patient again consulted me. I found a large, hard, bounding pulse with arterial locomotion, and the visible pulse of Corrigan. The heart was hypertrophied without valvular lesion. Auscultation of the aorta showed a hard, rough blowing murmur with its maximum of intensity in the second intercostal space along the right border of the sternum. In this space the murmur extends from the median line to 4 ctm. beyond the right border of the sternum. In the third space the murmur is also marked, but only near the sternum. It is not heard in the first intercostal space.

The patient suffers from vertigo, oppression, flatulent dyspnœa, and some failure of sight; the legs are swollen at night. Nevertheless the patient has improved greatly since he has reduced his allowance of tobacco. He can now attend to his business, while this was impossible when he first consulted me.

Syphilis constitutes the fifth cause of Hodgson's disease. Lancereaux thinks that these syphilitic vascular lesions may be distinguished from ordinary atheroma by the following characteristics: 1, the subject is syphilitic; 2, the lesions attack by preference the cerebral arteries, while ordinary atheroma affects the large vessels—the aorta, splenic, mesenteric, and renal arteries; 3, the subject is still young, while atheroma generally attacks individuals who have passed the age of sixty; 4, there is often a certain symmetry in the lesions. I think that I have observed two cases in which the dilatation was due to syphilis. The following is the history of one of these cases:

OBSERVATION XLIV. *True Aneurism of the Aorta, probably of Syphilitic Origin, very much Improved under Treatment.*—D——, aged sixty-one years. In 1846 he contracted syphilis. When the patient saw me in October, 1877, he complained that for three months he had suffered from a fixed pain to the right of the sternum and accompanied by palpitation of the heart. Upon examination a slight projection was found at the level of the right third cartilage; a marked impulse to the right of the sternum.

The apex-beat was situated in the fourth intercostal space (the thorax is very long), 12½ ctm. from the median line; the liver corresponds to the insertion of the fifth cartilage. No cardiac or aortic murmur. Antisyphilitic treatment. In three weeks the pain and palpitation had disappeared; slight difficulty of breathing alone remained. Upon examination the prominence appeared to be increased; it seems to elevate the second, third, and fourth right cartilages. The tumefaction is 6 ctm. long, 6 ctm. wide, and projects 1 ctm.

This region is dull on percussion and is the site of evident expansile movements, though much less than formerly; they are most marked in the first intercostal space; no murmur heard upon auscultation.

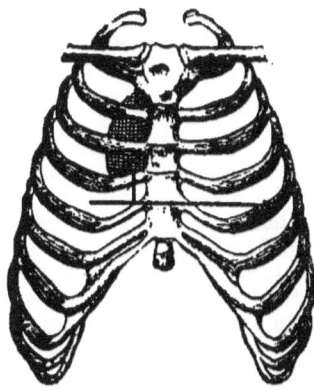

FIG. 80.

When the patient lies down he suffers from dyspnœa and violent cough. Upon auscultation feeble respiration is heard under the right clavicle. A tracheal blowing murmur is heard upon the vertebral column, over the first nine dorsal vertebræ.

Deglutition is difficult. Finally, the pulse is large and bounding, stronger on the left side than on the right.

Antisyphilitic treatment was continued. The patient returned three weeks later (June 13th), considerably improved; the projection of the ribs has disappeared. Slight oppression is still present in the recumbent posture, but deglutition is easy.

A month later the patient is merely annoyed by the cough; he can walk up a flight of stairs with ease.

October 1st, the patient is perfectly satisfied with regard to his condition. No pain or palpitation. Slight pulsation still felt along the arch of the aorta; this is felt less upon application of the hand than of the head in auscultation; no murmur. Very little difficulty of respiration, except in dorsal decubitus; frequent cough with expectoration. Over the left upper lobe posteriorly is heard a blowing murmur in inspiration, having the timbre produced by compressed bronchi; no dulness.

The patient then passed out of my observation.

CHAPTER XXII.

LESION OF THE DESCENDING PORTION OF THE THORACIC AORTA.

In proportion as the distance from the orifice of the aorta increases, the lesions produced by atheroma diminish in extent and intensity. The diseases produced by these lesions are, therefore, much rarer, and the greatest reserve must be exercised in making a diagnosis. But that this is possible is illustrated by the following case:

OBSERVATION XLV.—M——, aged fifty years, entered the hospital April 19, 1877. The patient has a strong constitution; has been a varnisher for the past twelve years; for five years presented symptoms of alcoholism. Marked tremor of upper limbs and head; occasional headache at night.

The affection on account of which the patient entered the hospital began suddenly four months ago. She complained of pain in the chest, slight oppression, and a moderate cough. A painful point appeared gradually to the left of the sternum a little above the breast. These symptoms increased until the patient entered the hospital.

Physical examination on admission gives the following results:

Marked development of the subcutaneous veins at the level of the right and left second ribs, especially on the left side; slight projection of the left second and third costal cartilages. Pulse 80 per minute and regular; it is hard, from the condition of the arterial walls.

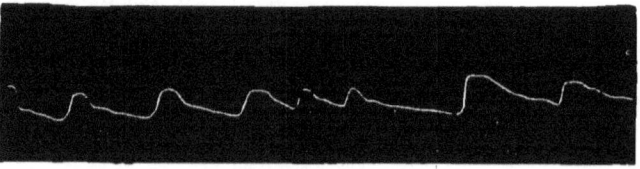

FIG. 81.—Right Radial Pulse (Atheromatous).

Apex beat of heart in fifth intercostal space, 10 ctm. from the median line; upper border of liver corresponds to origin of fifth costal cartilage; pulmonary resonance extends to right border of sternum. The apex is much lower than the upper border of the liver (on account of the great obliquity of the ribs), and the obliquity of the inferior border of the heart indicates hypertrophy confined chiefly to the left heart.

Auscultation at the apex shows two dull sounds at almost equal intervals, without any blowing murmur.

The situation of the enlarged veins and the elevation of the cartilages would lead us to believe that the lesion is situated at the origin of the great vessels.

Auscultation over the right second intercostal space discloses no bruit. In the vessels of the neck on the right side is heard a slightly rough, continuous bruit.

Auscultation of the second and particularly the third left intercostal spaces reveals a strong blowing murmur corresponding to the second sound; its maximum of intensity is found at the sternal insertion of the left fourth cartilage. Careful attention shows that the murmur is double; the valvular click is not heard.

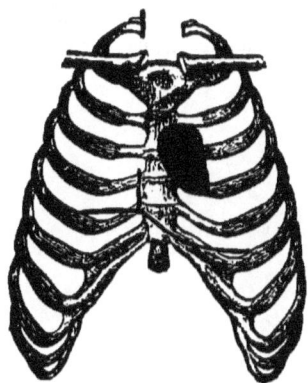

Fig. 82.

The differential diagnosis lies between a lesion of the pericardium, the aortic orifice, the arch of the aorta, and the pulmonary artery.

As the murmur is double, superficial, and slightly grating, we might think of a lesion of the pericardium. But this is improbable for the following reasons: 1, we must then admit that the pericarditis was chronic from the beginning; 2, the pulse is not frequent; 3, there is no thrill; 4, the murmur is propagated along the sternum and ascends toward the left carotid, while it decreases markedly in the lower part of the sternum, the usual site for the maximum of intensity of pericardial murmurs.

The diagnosis of aortic insufficiency is negatived for the following reasons: the murmurs in this affection are not situated usually to the left of the sternum at the origin of the fourth cartilage; in addition, the sphygmographic trace is that of arterial atheroma, not that of insufficiency.

Let us consider the third theory. We have to deal not with an affection of the convex surface of the arch but of the left or concave surface. Now, a number of cases have been reported, particularly in the treatise of Duroziez, in which a lesion of this locality (to the left of the sternum, in the third intercostal space) has given rise to exactly similar phenomena. This hypothesis would explain very well the elevation of the ribs and the development of the varices.

The theory of a lesion of the pulmonary artery, characterized by insufficiency alone, need scarcely be discussed. This affection is extremely rare, and has never been met with alone.

In conclusion, the diagnosis appears to be a lesion of the arch of the aorta, situated probably on the left side of the vessel.

The autopsy, made shortly afterward, confirmed the diagnosis. The cardiac orifices were entirely intact. The lesion consisted of atheromatous degeneration of the arch of the aorta, its maximum being found at the level of the descending aorta beyond the origin of the left subclavian. At this point the plaques are rough and calcified, including the entire circumference of the artery, and forming here a rigid tube with a rough surface. There were one or two very smooth patches in the ascending aorta.

To recapitulate : careful examination enabled us to determine that hypertrophy of the heart, accompanied by a double blowing murmur at the level of the third left costal cartilage, belonged in reality to an affection of the descending aorta, and this diagnosis was substantiated by the autopsy.

CHAPTER XXIII.

AFFECTIONS OF THE RIGHT HEART.

While congenital or rather intrauterine lesions are relatively common in the right heart, they are rare in the left. After birth, organic lesions in the right heart become very rare; the majority of writers state that they constitute about two per cent. of cardiac lesions. A distinction must be made in this respect between endocarditis of the right heart producing tri-

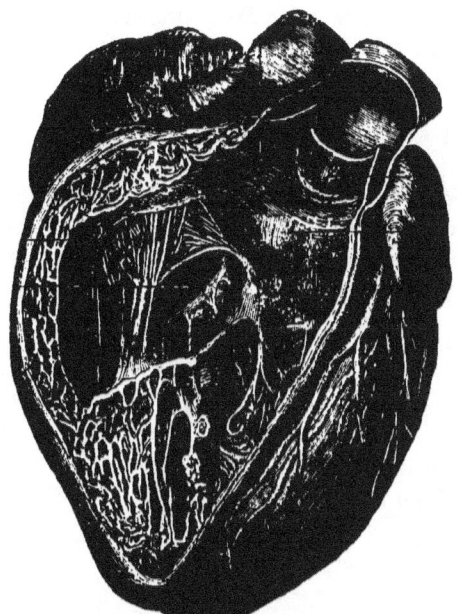

Fig. 83.—The Right Ventricle.

cuspid insufficiency, and dilatations of this side leading to relative insufficiency. If the latter is included among the lesions of the heart, the proportion becomes much greater, since every chronic cardiac disease gives rise, at a certain period, to dilatation of the right heart and relative tricuspid insufficiency.

Anatomical Data.—The right ventricle is situated transversely in front of the left ventricle, and, with the right auricle, constitutes the anterior surface of the heart, with the exception of the apex.

The right ventricle forms a triangular pyramid, which overlaps, as it were, the left heart. The latter projects on the left side at the apex; the right auricle projects on the right side.

The ventricle presents three surfaces and three borders. The anterior surface is vertical and constitutes the anterior surface of the heart, and, at the same time, the infundibulum of the pulmonary artery. The inferior or diaphragmatic surface of the heart is divided into two equal parts by the interventricular groove, of which the right ventricle occupies the anterior half. The posterior surface is formed by the groove.

The cavity of the ventricle does not correspond to its external shape. The apex is entirely filled by a network of columnæ carneæ of the second order, and if the anterior columna of the first order is lifted up, the cavity is found to be ovoid, or rather pyriform in shape, the apex corresponding to the infundibulum of the pulmonary canal, the base to that portion of the ventricle situated in front of the tricuspid valve. When the anterior wall is removed, care being taken not to disturb the columna of the first order which is inserted into it, a vertical membrane of a fan shape is found to start from the almost vertical line which forms the anterior portion of the auriculo-ventricular orifice.

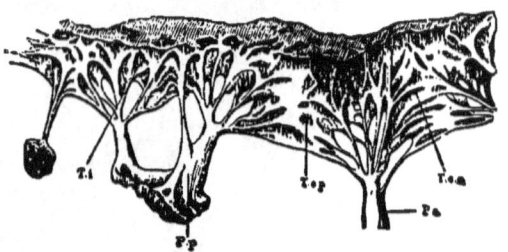

FIG. 84.—Tricuspid Valve Separated (M. Sée).

This membrane is the anterior tricuspid valve, and fulfils almost entirely the functions of this valve. To the left or free border of this valve are attached the chordæ tendinæ, the majority of which pass downward and to the left to the columna carneæ of the first order. This columna is inserted into the anterior wall below, near the inferior border of the heart, and near the apex, or rather into the columnæ which occupy this region (Fig. 83). The other papillary muscles are very short, and vary in number and shape.

The posterior valve adheres on one side to the posterior part of the auriculo-ventricular orifice, and upon the other border it receives the tendons of the posterior papillary muscles, and some chordæ which are inserted directly into the auriculo-ventricular groove. In addition, it receives two or three tendons of the anterior principal columna.

If the tricuspid valve is spread out, it is found to be short and composed in great part of intersecting tendons (Fig. 84). The principal leaflet divides the cavity of the ventricle into two parts, an anterior or ventricular cavity and a posterior or auricular cavity.

The anterior cavity is pear-shaped, and its small extremity corresponds to the infundibulum of the pulmonary artery. The large extremity is found in the middle portion of the ventricle, in front of the anterior valve and anterior papillary muscle. It seems that the cavity of the right ven-. tricle, instead of being parallel to the left, has undergone a rotation upon itself, so that the apex, instead of corresponding to the apex of the heart, has been directed toward the pulmonary artery (Fig. 83).

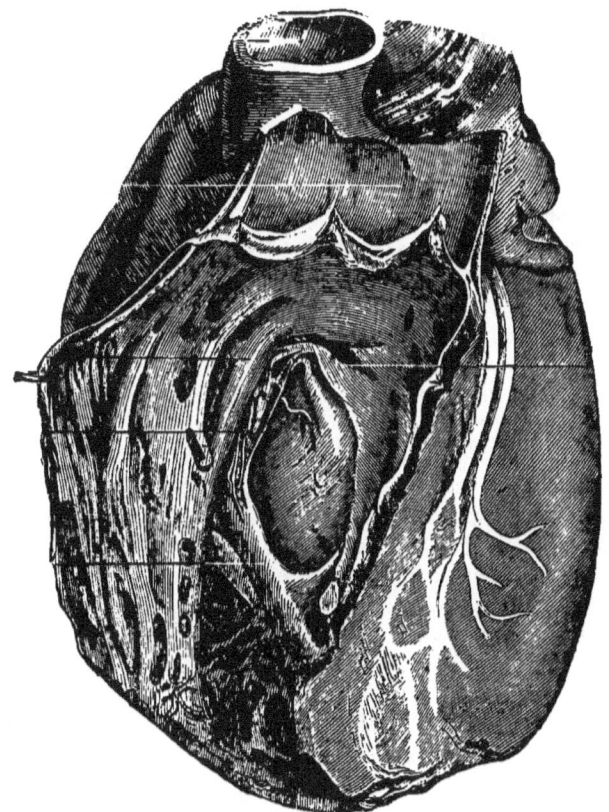

Fig. 85.—The Pulmonary Infundibulum, Conus, or Canal (M. Sée).

The cavity is bounded at the apex by columnæ carneæ of the second order, and by a muscular bundle which starts from the base of the anterior papillary muscle and ascends vertically upon the interventricular groove.

On the side of the base of the cavity is also found a considerable muscular bundle, called by Marc Sée the compressor muscle of the tricuspid valve.

As in the left ventricle the posterior portion is concealed by the columnæ and the valve, so in the right ventricle the auricular portion situated behind the valve only exists toward the base ; to the right of the projecting portion of the groove behind the anterior valve it is filled by the two complementary valves and columnæ carneæ of the first order.

If the heart is opened carefully, the clots in the right ventricle are always found separated into two layers : that of the ventricular portion, which is pear-shaped, the small extremity being engaged in the pulmonary canal ; then below the valve a black clot, which occupies the right auricle, the right auriculo-ventricular orifice, and the auricular portion of the right ventricle.

The play of the valves is thus clearly indicated. The venous blood, entering the right auricle from the superior and inferior venæ cavæ, takes a horizontal direction, and, at the movement of diastole, passes through the auricular orifice into the ventricle, raising the anterior valve, which it separates from the septum. Then during systole this anterior valve tends to become applied against the septum, the auricular part is concealed by the approximation of the columnæ carneæ, and the blood takes the direction of the pulmonary artery, being thrown in this direction by the contraction of the muscles which I have described above. These views have been fully confirmed by the investigations of Marc Sée.

In conclusion, the orifice of the inferior vena cava corresponds to the insertion of the right fifth cartilage into the sternum ; the external border of the right auricle forms the vertical border of the cardiac triangle, and is situated $1\frac{1}{2}$ ctm. from the right edge of the sternum ; the auriculo-ventricular orifice corresponds to an oblique line from below upward and from right to left, starting from the insertion of the right fifth cartilage into the sternum, and terminating near the sternal insertion of the third left costal cartilage ; the most projecting portion of the heart corresponds to the left border of the sternum, from the third to fifth cartilages.

ENDOCARDITIS OF THE RIGHT HEART—INSUFFICIENCY AND STENOSIS OF THE RIGHT AURICULO-VENTRICULAR ORIFICE.

Endocarditis of the right heart is rare, especially when unaccompanied by that of the left heart.

Endocarditis vegetans attacks the right heart in the same manner that it does the left, first upon the edge of the valve, then upon its auricular surface.

Sclerotic endocarditis is observed much more frequently ; the valve may be thickened and no longer closes the orifice completely. Bouyer has given the following description of this lesion as observed at the autopsy :

"Upon its inner surface, between its insertion and fringed border, the tricuspid valve presented an annular vegetation of fibro-cartilaginous hardness and white color, two lines in height, directed toward the auricle and tearing readily under the pressure of the finger. By this arrangement the valve formed a circular orifice two or three lines in diameter, constantly open, and which, in addition to obstructing the flow of blood from the auricle into the ventricle, also permitted the reflux of blood into the auricle."

Ulcerative endocarditis may also occur in the right heart as it does in the left.

The general symptoms of tricuspid lesions are similar to those of mitral disease, with some differences resulting from the difference of function.

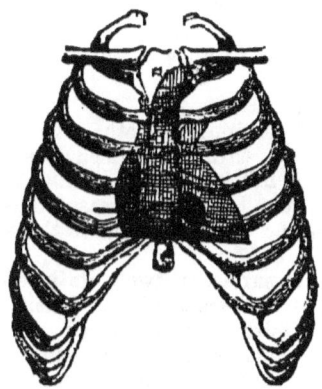

Fig. 86.—Topography of the Tricuspid limit. Depression and Separation of the Right Angle of the Cardiac Triangle.

The symptoms consist of dyspnœa on effort and palpitation of the heart. The pulse differs from the mitral pulse inasmuch as it remains regular and full. Upon examining the heart the apex is not found to be lowered, but the hepatic angle is depressed, so that the obliquity of the lower border of the heart diminishes. At the same time the vertical border seems to be separated from the edge of the sternum. The purring thrill is not felt on palpation if the tricuspid valve is alone affected.

The following are the characteristics of the murmur as heard upon auscultation. The blowing murmur produced by endocarditis of the tricuspid valve is heard upon the lower part of the sternum to the left, under the left border of this bone 1 ctm. from the median line, and in the inner part of the left fifth intercostal space. Its area is usually very limited, being scarcely larger than a five-franc piece. It is not propagated to the apex.

The murmur begins with systole, covers the first sound and almost the entire lesser period of silence, but leaves the second sound intact; it is usually rough in character.

The situation of the murmur in the region of the tricuspid, and its absence at the apex, suffice for its recognition. It is also intermittent, *i.e.*, it is present on some days, absent on others.

The following case will serve as an illustration:

OBSERVATION XLVI.—A laundress, aged twenty-four years, has complained for some time of palpitation of the heart and marked diminution of the menses; the latter finally ceased entirely. Five months after the appearance of the first symptoms the abdomen began to enlarge.

Present condition (August 24th): face pale and œdematous; general anasarca; difficult respiration; cough with mucous expectoration mingled with bloody streaks. No pulmonary expansion on the left side inferiorly;

dulness in this situation changes with position, showing the presence of fluid in the left pleural cavity. The increased dulness in the præcordial region, the distant heart sounds and enfeebled impulse of the apex, indicate the presence of fluid in the pericardium.

Auscultation reveals a very marked blowing murmur in the middle of the cardiac region, and which is sometimes so strong as to resemble a purring bruit. Pulse small, hard, 120 per minute. Abdomen enlarged; fluctuation present; notable diminution of the quantity of urine excreted. Digitalis administered in increasing doses.

September 3d.—The blowing murmur hardly audible; pulse 92 per minute.

September 6th.—The bruit reappears; pulse accelerated to 112.

September 7th.—Almost complete absence of the murmur; pulse 96. Urine begins to increase in quantity.

September 17th.—Bronchial breathing, crepitant râles, obscure sound in the left infraspinous fossa; irregular pulse.

September 18th.—Respiration easier, pulse hard.

During the night, increased difficulty of respiration, death. At the autopsy serous effusion was found in the three cavities in which it was diagnosed during life; a large portion of the left lung is firm and infiltrated with bloody serum. The volume of the heart is increased one-third; its cavities are very large, the walls of the ventricles equal in thickness.

The tricuspid valve presents upon its inner surface, between its insertion and free border, an annular vegetation of fibro-cartilaginous density, of a white color, and 2 lines in height. The valve forms a circular opening, 2 or 3 lines in diameter, which is constantly open. Some similar vegetations, as large as hemp seeds, are found upon the mitral valve, but they are too small to interfere with circulation to any considerable extent. (Martin-Solon, *Journ. hebdomadaire*, t. ix., 1832.)

The following case, in the service of Dr. Mesnet, has been reported by Deschamps, one of my former internes:

OBSERVATION XLVII.—L——, aged thirty-four years, entered the hospital with the general symptoms of heart disease. Examination gave the following results: the apex beats in the fifth left intercostal space, 11 ctm. from the median line. The upper border of the liver corresponds to the lower border of the right fifth cartilage. The vertical border of the heart is near the sternum, but it is difficult to determine this exactly, because there is a more marked projection of the chest on the right side than the left. Auscultation reveals a blowing murmur, with its maximum at the level of the xiphoid cartilage, and propagated toward the apex without reaching it; the murmur ceases 4 to 5 ctm. before reaching the apex. The murmur is systolic, covering the entire first sound and the lesser period of silence; it is rough, loud, like a jet of steam.

Subcrepitant râles are found over both lungs, particularly at the base. The sputa are puriform, and have a fœtid odor.

Very marked enlargement of the liver, slight ascites, and considerable œdema of the lower limbs are observed. No albumen in the urine.

The diagnosis was emphysema, catarrh, and tuberculosis at the apex, organic affection of the heart, situated probably at the tricuspid orifice, with insufficiency and secondary venous disturbances, pulmonary œdema, hepatic congestion, ascites, œdema of the lower limbs. The patient died a few days after admission to the hospital.

The autopsy disclosed fibrous induration of the upper lobes of the lungs, dilatation of almost all the bronchi of the lower lobes, with thinning of the walls and disappearance of the pulmonary tissue.

The heart presented a sclerotic endocarditis of the tricuspid valves with insufficiency.

In addition to the anatomical disturbances, functional disorders are also produced.

The first phenomenon is jugular congestion and venous reflux in the neck. This consists at first of simple stasis in the jugular and thyroid veins. In dorsal decubitus this stasis increases to such an extent as to form venous sinuses. In certain cases in which the patient lies upon the side of the face, the sternomastoid muscle divides the vein into two parts, and if the patient has marked dyspnœa, the muscle is seen to rise and fall, and form a sort of intermittent bar during the respiratory efforts. As the vein is lowered when the muscular contraction ceases, this may produce the appearance of true reflux. At a later period the venous pulse with reflux develops. In such cases, if the vein is emptied by ascending pressure, and the entrance of blood from the peripheral parts is prevented, an ascending venous pulse with undulations will be observed.

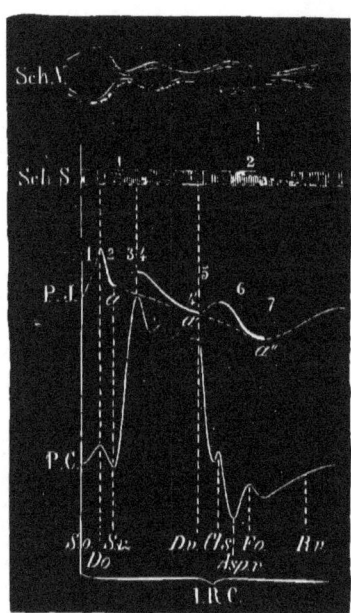

Fig. 87.—Normal Jugular Venous Pulse (F. Franck).

What is the cause of this reflux? It is not the action of the auricle, since, as a general thing, its contraction does not produce a venous pulse visible to the eye; the weight of the column of blood, aided by the aspiration of diastole, suffices to counterbalance the action of the auricle.

These details are shown much better upon the traces of Marey's polygraph. I have already (Figs. 23 and 24) given Potain's traces, and F. Franck[1] has since furnished more detailed traces with the aid of perfected instruments.

---

[1] F. Franck: Mouvements des veines du cou en rapport avec l'action de la respiration et du cœur, Gaz. hebdom. de Méd. et de Chir., Mars, Avril, 1882.

The normal trace of the jugular pulse furnished by Franck clearly shows the following facts:

1. The normal jugular pulse presents brusque elevation and depression at the beginning of the entire curve; these are related to the systole and diastole of the auricle.

The first elevation connected with the auricular systole extends from 1 to 2 in the figure. The first depression corresponds to auricular diastole and ventricular systole, and extends from 2 to 5.

2. A second elevation and depression occur at the close of the ventricular systole. The second elevation (5 to 6) corresponds to the closure of the semilunar valves, and the second depression (7) to the ventricular diastole Dv.

Now, as the first elevation in the vein corresponds to the auricular systole, is it produced by a simple retardation in the venous current, or is there in addition a reflux of blood from the auricle into the vena cava?

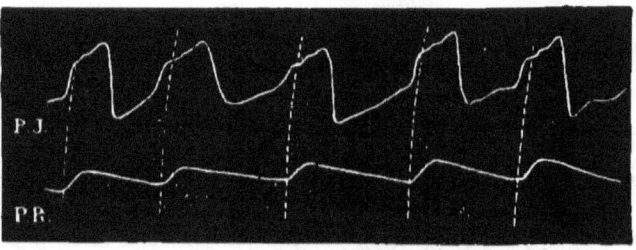

FIG. 88.—The Jugular Pulse with Reflux compared with the Cardiac Pulse (Marey).

Franck and Arloing have shown by experiments on horses that a very short and slight reflux occurs in the vena cava, but as this is not propagated into the jugulars, the elevation in the latter must be due solely to the increase of pressure from the arrest of the current of blood.

If a trace is taken of a jugular pulse with reflux, it will be seen that at the moment when the radial pulse is in diastole, the vein is elevated (*vide* Fig. 88). If the traces of the cardiac and jugular pulsations are taken simultaneously, it will be found that at each ventricular systole, the vein, instead of being depressed as in the normal condition, gives an ascending curve proportionate to the reflux. This is marked in Fig. 89, taken by Franck and myself upon a patient suffering from mitral induration with insufficiency, and complicated by tricuspid insufficiency from dilatation.

I have not spoken of the influence of respiration upon the venous pulse, in order not to complicate the question.

The reflux, then, only exists in insufficiency of the tricuspid valve, which permits the blood propelled by the ventricle to flow back into the vein.

Friedreich believes that venous pulsation in the neck is far from being characteristic. He thinks that a true reflux may occur when the tricuspid valve closes normally, and also that tricuspid insufficiency may exist without the venous pulse. He admits, however, that if the pulsation of the veins of the neck is very marked, the existence of tricuspid insufficiency is very probable.

I believe that Friedreich has confounded the true venous pulse with the undulations of which I have spoken above. The absence of reflux in cases of asystole is due to the fact that the ventricular contraction is unable to support the column of blood. An obstruction to the reflux by the closure of the valves of the cervical veins is not impossible, since the internal jugular possesses two valves at its entrance into the brachio-cephalic trunk. But the valves of the external jugular occlude the vessel only in

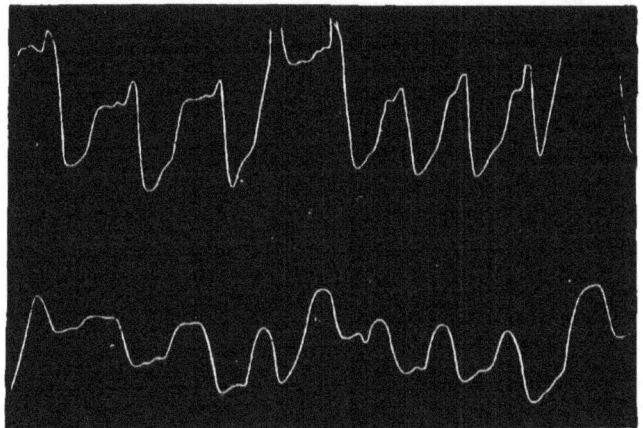

Fig. 59.—Jugular Venous Pulse with Reflux below, Cardiac Pulsation above.

part, and do not appear to me to be able to arrest the flow of blood propelled by the right ventricle unless, as I have said, asystole is present.

Another symptom produced by tricuspid insufficiency is the hepatic venous pulse. If the patient lies upon the back with the head inclined slightly to the side, the thighs flexed and carried to the outside, a slight pulsation is observed in the epigastrium and right hypochondrium, more marked over the left lobe of the liver. This pulsation is sometimes perceptible upon the side and even posteriorly.

The pulsation is slow and progressive, and is confined to the liver. If the abdominal walls are relaxed and the anterior border of the liver can be grasped by the hand, a movement of expansion is felt rather than a transmitted elevation. The impulse follows immediately after the impulse of the apex; it precedes the radial pulse by a very short interval. According to Mahot, it is sometimes dicrotic, but then the first pulsation always precedes the impulse of the apex. This venous hepatic pulse is synchronous with the cervical venous pulse.

Furthermore, on examining the traces furnished by the cardiograph, it is found that the line rises during inspiration and falls during expiration. This modification of the trace is the same as that presented by the cardiographic traces taken upon the chest. Here the ascent of the line is produced by the pressure of the diaphragm upon the liver during inspiration, while the reverse takes place during expiration.

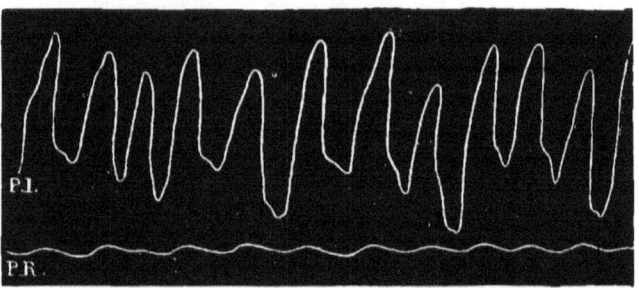

FIG. 90.—Hepatic Venous Pulse compared with the Radial Pulse.

Sometimes flexions in the trace are observed in the ascent and descent. Friedreich has applied to the former the term anadicrotism; it indicates the succession of the auricular and ventricular systoles. This distinct indication of the two systoles appears less and less in proportion as the auricle is weakened.

According to Friedreich, the flexion during the descent of the trace, which he calls catadicrotism, is due to the fact that, at the beginning of diastole a flow of venous blood arrives in the auricle and delays for a moment the termination of the trace. But asystole is the cause of the general diminution of the trace.

The venous hepatic pulse is of rare occurrence, and I have only observed it upon two occasions.

OBSERVATION XLVIII. *Mitral and Tricuspid Insufficiency; Hepatic Venous Pulse.*—M——, aged thirty-five years, entered my service on June 23, 1877, for slight ascites with shortness of breath and palpitations. In the past eleven years, he has had five attacks of articular rheumatism. Present condition: face slightly jaundiced; when excited, it becomes congested and the veins of the neck are agitated by tumultuous movements. The abdomen appears to be increased in size, particularly in the median line, the hypogastrium and epigastrium. At the level of the umbilicus the abdominal walls look as if strangulated by a circular cord. In the epigastrium palpitation gives the sensation of a hard, smooth body occupying the entire epigastrium. Percussion gives incomplete but very clearly defined dulness, and the body is evidently connected with the liver. It appears to be formed at the expense of the left lobe and extends into the left hypochondrium, more than 10 ctm. from the median line. It extends inferiorly to the umbilicus; on the right side, it ascends to the natural limit of the liver, so that while percussion denotes an enormous enlargement of the liver in the xiphoid line, this hypertrophy is scarcely perceptible in the axillary line. The upper border of the liver corresponds to the lower border of the sixth cartilage. The softness of the abdominal walls enables the anterior border of the liver to be grasped in the hand, and it is found that the hepatic pulse gives rather the sensation of expansion than of elevation.

Examination of the heart gives the following results : The apex beats in the eighth intercostal space, 10 ctm. from the median line ; the impulse is irregular, very feeble, without thrill. A second centre of impulse is found at the level of the xiphoid appendix. The enlarged liver is elevated by the pulsations of the heart. At the apex the first sound is replaced by a murmur. This blowing murmur, when examined carefully, presents two foci, one at the apex, the other at the xiphoid cartilage. At the apex is heard a soft, prolonged systolic blowing murmur with a rolling bruit. The topography of this bruit gives a triangle. The lower border of the triangle is situated 1 ctm. below the inferior border of the heart ; it extends 1½ ctm. to the left of the apex ; its right angle is 5 ctm. from the apex ; it is about 5 ctm in height.

The second focus corresponds to the xiphoid appendix ; it has an ovoid shape, 7 ctm. in the vertical direction, and 5 ctm. in the transverse direction. This focus is situated a little to the left of the median line. This bruit is also systolic and very strong ; it is not a prolongation of the first bruit. The clearness of the murmurs, the movements in the jugulars, the hepatic pulsations, and the enormous congestion of the liver, enable us to affirm that we have to deal with a tricuspid insufficiency more considerable than that ordinarily seen. The pulse is small and irregular, and has the qualities of a mitral pulse.

The patient left the hospital after a certain amount of improvement.

The explanation of this form of pulsation is apparently easy.

The vena cava and its ramifications contain no valves until the primary iliacs. Furthermore, the hepatic veins have no valves and are always patulous. It is admitted that the venous tension begins in the iliac veins, then extends to the renal, then to the mesenteric, then to the hepatic veins, and that, consequently, the reflux will make itself evident in the hepatic veins, provided that the others have undergone their maximum tension. But a doubt is cast upon this explanation by the fact that the phenomenon is so exceptional. The reasons mentioned above would be valid if the venous hepatic pulse were a common and almost constant result of tricuspid insufficiency.

On the other hand, if the theory of ascending tension were true, albuminuria should always be found before the hepatic pulsation, but this is not constant.

It should be noted that in almost all the cases of venous hepatic pulse, there is not alone tricuspid insufficiency, but also a lesion of the mitral valve ; furthermore, the hepatic pulsations may disappear at any moment.

Upon post-mortem examination of the liver in these cases, Mahot has found marked dilatation of the venous trunks, extending even to the small veins. In addition, he has observed atrophy of the liver with induration of the connective tissue.

To recapitulate the characteristics of tricuspid insufficiency from endocarditis, we notice at first the general phenomena of cardiac disease, viz., dyspnœa on effort and palpitation. The pulse is regular, quite full, and not mitral, *i.e.*, not irregular or excessively compressible.

Examination of the heart shows lowering of the hepatic angle, making the inferior border of the heart horizontal, and dilatation of the right auricle, which perceptibly separates the vertical border of the heart from the edge of the sternum.

Auscultation reveals a distinct, superficial blowing murmur, confined to the lower part of the sternum on the left border. This murmur is systolic, rarely covers the lesser period of silence, and, unlike the mitral murmur, does not last until the second sound. It is sometimes rough, sometimes smooth.

Another remarkable fact is that the murmur is intermittent; it is present on certain days, absent on others. When it is feeble, it is only heard during deep inspiration and disappears during expiration.

The dominant feature is the venous tension in all the organs and the pulsation of reflux. Venous stasis is produced in the liver, where it causes increased volume, lowering of the organ, a slight icteric discoloration, and finally the hepatic venous pulse.

The stasis in the mesenteric veins produces ascites; in the renal veins, albuminuria and anasarca; in the veins of the limbs, œdema; in the skin, cyanosis; in the lungs, œdema, hydrothorax, and asphyxia; in the gastric mucous membrane, a sort of catarrh which is aggravated by the anæmia, then intestinal hemorrhages.

Friedreich[1] states that he has found a blowing murmur in the crural vein; it is sometimes single, sometimes double. When single, the murmur is systolic; when double, one is presystolic the other systolic. According to him, they are due to the tension of the valves of the crural vein near Poupart's ligament and, when the valves are absent, to the tension of the walls of the veins. All this is pure imagination. The murmurs are undoubtedly produced in the crural artery by the pressure of the stethoscope.

Finally, the head symptoms observed are congestion and uræmic changes in the retina, cerebral, cerebellar, and bulbar congestion, *i.e.*, asphyxia. The course of the disease, accordingly, is a venous dyscrasia due to paralysis of the veins, aggravated by the distention of the lymphatics on account of the stasis produced in the thoracic canal, and which terminates in uræmia.

---

[1] Friedreich: Ueber Doppelton an der Cruralarterie, sowie ueber Tonbildung an der Cruralvenen, Deutsch. Arch. f. Klin. Med., 1878.

SCLEROTIC ENDOCARDITIS OF THE RIGHT HEART WITH STENOSIS OF THE ORIFICE.

There is no necessity for the separate consideration of sclerotic endocarditis of the tricuspid with stenosis. It is the same disease as that which accompanies insufficiency. It is even rarer than tricuspid induration with a large opening, but Duroziez has collected the reports of ten cases.[1] It should be noted that tricuspid stenosis never occurs alone, but is always associated with lesions of the mitral and the aortic valves.

As in the previous case, a xiphoid systolic murmur of a limited area is heard. It is a remarkable fact that this murmur is not diastolic or presystolic, thus confirming the theory which I have advocated concerning the so-called presystolic bruit. For if this bruit were due to the passage of blood through the narrowed orifice at the moment of the auricular systole, then it should be presystolic.

---

[1] Duroziez: Du retrecissement de la tricuspide, Gaz. des Hôpit., 1868.

CHAPTER XXIV.

OBLITERATION OF THE SUPERIOR VENA CAVA.

LESIONS of the superior vena cava are rare and are always secondary to lesions of neighboring parts. When thrombosis of the vein develops, it is almost always due to an extension of cancerous thrombosis of adjacent veins. At other times it is the result of compression produced by a lesion in the vicinity,[1] such as aortic aneurisms, mediastinal tumors, tumors of the bronchial glands, thyroid gland, lungs or sternum.

OBSERVATION XLIX. *Obliteration of the Superior Vena Cava; Aneurism of the Arch of the Aorta; Chronic Inflammatory Induration of the Apex of the Right Lung.*—J. M——, aged sixty-two years, entered the hospital July 2, 1875. Six weeks ago he noticed slight redness and swelling of the infra-orbital regions, spreading gradually to the eyelids. At the same time an oppression upon the chest, which had existed at intervals for some time, increased markedly. At the present time this oppression is only violent upon exertion. Complete apyrexia; pulse small. Dulness at the apex of the right lung; almost complete absence of respiratory sounds in upper part of both lungs.

It is found now that the œdema extends over the entire face and neck, the upper limbs and the anterior surface of the thorax, ceasing at its base. Varicose dilatation of the subcutaneous veins arranged in the form of a bluish girdle is observed at the base of the thorax on both sides. On the left side, this girdle runs along the fifth and sixth ribs; on the right side, a little lower. It terminates abruptly upon the sides. Anteriorly the halves unite in front of the sternum, along which they run in a vertical direction and form a band 8 to 10 ctm. in width along the median line.

At the base of the neck the right external jugular vein elevates the skin of this region, despite the œdema, and presents nodosities in which the tension is much greater than normal. The venous circulation also appears to be impeded to an extreme degree at the root of the upper limbs.

---

[1] Oulmont: Des oblitérations de la veine cave sup., Mémoires de la Soc. Méd. d'Observation, t. iii. Follin: Oblitération de la veine cave sup. et de ses branches par une tumeur anévrysmale, Soc. Anat., t. xxii., p. 365. Reynaud: Oblitération presque complète de la veine cave sup. par suite de la présence d'un sac anévrysmal, Journ. Hebd., t. ii., p. 112. Bailly: Oblitération de la veine cave sup. par une tumeur squirheuse, Bull. de la Soc. Anat., t. xxix, p. 61. Barth: Coagulation spontanée du sang, de la veine cave sup. chez une cancéreuse, Bull. de la Soc. Anat., t. xxviii., p. 4. Dujardin-Beaumetz: Arrêt brusque de la circulation de la veine cave sup. chez un malade atteint d'anévrysme de l'aorte, (Soc. Méd. des Hôpitaux, 14 Mars, 1879).

The apex of the heart is situated in the left fifth intercostal space, 10 ctm. from the median line. The upper border of the liver corresponds to the insertion of the sixth cartilage. The vertical border of the heart is 1 ctm. from the right edge of the sternum. Hence the heart does not seem to be hypertrophied, while the part which corresponds to the right auricle has markedly lowered the right cardiac angle. A slight, soft blowing murmur is heard at the apex and replaces the first sound ; second sound normal. An unusual silence is noticed at the level of the origin of the arteries ; nevertheless, when the patient stops breathing, the sound is heard in the pulmonary artery but not in the aorta.

The venous circulation appears to be immovable in the external jugular, the pulsations of which appear to be transmitted from the arterial pulsations. There is no pulmonary œdema, showing that the lesser venous circulation is intact. The diagnosis made was : obliteration of the superior vena cava and aneurism of the arch of the aorta.

July 10th.—The patient feels somewhat better. Auscultation shows greater intensity of the cardiac sounds over the left third intercostal space and also the presence of aortic bruits, which are very dull and difficult to detect. The murmur at the apex has diminished in intensity.

The improvement continued for two months, and at times the varicosities of the thorax disappeared almost entirely. After the beginning of September the patient gradually grew worse, and died in collapse November 15, 1875.

Autopsy.—Brain and *meninges* passively congested.

Lungs : Upper third of right lung indurated, of a uniform grayish color ; thickened fibrous bands observed in the parenchyma. Slight congestion of the remaining parts of the lungs.

Heart : Left ventricle slightly dilated and hypertrophied ; right auricle and ventricle dilated. No tricuspid or mitral insufficiency.

Vessels : Aortic aneurism as large as an orange, beginning 1 ctm. above the semilunar valves and terminating at the highest part of the arch. The artery is 7 to 8 mm. thick and compresses the adjacent vessels, including the superior vena cava, to which it is intimately adherent. The semilunar valves are insufficient.

Superior vena cava diminished in calibre and partly hidden by the vessel. In its lumen is found a brown, solid, cylindrical clot, moulded upon the walls of the vein and adherent throughout its entire extent by means of small fibres which are easily torn, showing the reddened, granular wall of the vein ; the vessel is completely obstructed.

The restriction of the œdema to the upper limbs, and the distention of the veins emptying into the vena cava constituted the first phenomena of the disease. An important diagnostic symptom is the enormous dilatation of the intercostal veins, indicating the tendency to supplementary circulation through the venæ azygos ; this sometimes extends to the mammary, lumbar, and even epigastric veins. This is also true of the spinal veins.

An interesting feature in the above case was the fact that the supplementary circulation proved almost entirely sufficient for a few weeks and caused almost complete disappearance of the œdema.

The course of the disease is shortened at times by the formation of an embolus which may obstruct one of the larger branches of the pulmonary artery.

I will here add the history of a case observed by my friend Dr. Dujardin-Beaumetz:

OBSERVATION L.—P. G——, aged forty-two years, entered the hospital July 29, 1878; contracted syphilis while a young man.

The disease began seven weeks ago with a stitch in the right side under the clavicle, but especially on the antero-lateral portion of the chest, in the fourth, fifth, and sixth intercostal spaces; occasionally the patient expectorated bloody sputa.

July 29th.—Sudden cyanosis and oppression on the chest.

July 30th.—Present condition: the trunk, head, and upper limbs are bluish; the abdomen and lower limbs have the normal color. The cyanosed parts are swollen and œdematous; the veins of the neck are enlarged. In the sixth, seventh, and eighth intercostal spaces on the left side are marked capillary varices, evidently of long standing. On the right side, over the upper part of the abdomen are large varicose veins.

Thorax: Frequent respirations; no râles or dulness on the right side; dulness over the right infraspinous region and inferiorly, over which is heard a very loud murmur which obscures the respiratory murmur almost entirely. Similar dulness anteriorly on the right side, extending from the clavicle to the fourth intercostal space; murmur also heard in this situation. Lower down pulmonary resonance is heard, showing that the right lung is pushed downward.

The murmur is rough and systolic. It is heard slightly at the apex, very clearly at the base along the course of the aorta. Apex-beat under the fifth rib, 13 ctm. from the median line; vertical border of the heart 2 ctm. from the sternum.

Aorta: A rough systolic murmur in the right second intercostal space, extending 4 ctm. from the sternum; the same bruit is also heard in the first, third, fourth, and even the fifth intercostal spaces on the right side.

July 30th, evening.—The patient complains of pains in the arms, which are enlarged and of a bluish color, but do not pit on pressure.

July 31st.—Another attack of suffocation with increased cyanosis.

August 3d.—The patient is in a more satisfactory condition; the cyanosis has diminished considerably, though the upper part of the body is still discolored. The abdominal subcutaneous circulation is marked by two vertical, parallel veins; no venous pulse in the jugulars. The arms and forearms are hard, as in sclerema; the superficial veins are moderately distended; the pulse is felt with difficulty. Temperature: right axilla, 35.2°; left axilla, 34.6°; rectum, 37°.

August 8th.—A very curious sign is observed on the right side, near the spine, in the eleventh and twelfth intercostal spaces; it consists of a continuous, distinct bruit, which is reinforced at the moment of cardiac systole. It does not extend to the left of the spine, but is heard more feebly in the right ninth and tenth intercostal spaces.

August 11th.—The oppression increased. The system of the inferior vena cava remains intact; no œdema of the lower limbs. The previous symptoms in the domain of the superior vena cava increase in intensity.

August 12th.—The patient is somnolent; the œdema of the trunk has increased, particularly upon the back. Death occurred at two o'clock the following morning.

This patient was seized suddenly (July 29th) with the symptoms of arrest of circulation in the superior vena cava. He had experienced some pain in the right fifth and sixth intercostal spaces anteriorly, but these symptoms were so slight that he had not given up work.

When the patient was examined for the first time, he presented all the classical symptoms of compression of the superior vena cava.

To what was this compression due? Examination of the thorax revealed the signs of aortic aneurism ; a systolic blowing murmur in the right second intercostal space, extending into the adjacent spaces ; pulsation and expansile movements in this situation, dulness on percussion. All these signs enabled us to affirm the existence of an aneurism of the ascending portion of the aorta.

It now remained to explain the sudden arrest of circulation in the vena cava. Careful examination had enabled us to establish two phases of the disease ; in addition to the complete cyanosis observed in our patient, he presented, in the dorsal region and the lateral portion of the thorax, numerous varicose patches, which must have existed a long time before the appearance of the cyanosis and œdema. It could then be affirmed that this patient had been suffering for a long time from aortic aneurism, which had not manifested its presence in the last few weeks, except by intercostal neuralgia, but had long presented the symptoms of compression of the superior vena cava, in the midst of which had occurred the sudden arrest of circulation.

With regard to this point, two opinions were entertained by my colleagues in the hospital. Some explained the sudden arrest by the rupture of the aneurismal sac into the vena cava ; others (and I was of this number) thought that thrombosis had occurred in the vein as the result of compression by the aneurism. The following lesions were found upon autopsy.

Autopsy.—Upon opening the chest, a tumor as large as a fist was found above and to the right of the heart, pushing the latter to the left and downward, and situated in great part within the pericardium. The tumor is situated upon the right auricle, which it occludes entirely. The superior vena cava is situated behind it, pushed to the right of the thorax, and adherent to the sac. The left wall of the ascending aorta appears entirely intact ; the sac is formed at the expense of its anterior, right, and posterior walls. Above, the right brachio-cephalic venous trunk has almost its normal position, but the left trunk makes a very sharp bend, having been pushed to the right side.

Heart : The heart is very slightly hypertrophied ; its long axis is almost transverse, the base of the organ being pushed downward by the aneurism. *The right auricle* has disappeared, as it were ; its walls are flattened and about as large as the little finger ; the upper part has coalesced with the walls of the aneurism ; this is also true of the pulmonary artery, where it passes behind the aorta.

Superior vena cava : This vessel is 6 ctm. long, 2 ctm. in circumference; its walls are extremely thin, and on the left side and posteriorly they have coalesced with the aneurism; the walls are pressed against one another so that it is difficult to pass a fine sound through the canal, through which very little, if any, blood could possibly pass.

Vena azygos: This is 2 ctm. in circumference in its upper portion and evidently replaces the superior vena cava. It appears to be continuous with the brachio-cephalic trunks, to judge from the direction of these trunks and the modifications which their calibre undergoes, in proportion as they approach the azygos. Its dimensions are as great as those of the vena cava, but at the level of the dorsal vertebræ its calibre has diminished to one-quarter of that noticed above. There is no direct communication between the azygos and inferior vena cava. About 4 or 5 ctm. from the bend which the vena azygos makes above the right hip is found a clot which fills the vessel for a distance of 5 or 6 ctm.

Aneurism.—This measures about 10 or 11 ctm. in its three principal diameters. It begins immediately above the semilunar valves, occupies the entire ascending portion of the arch, and encroaches upon the transverse portion. The brachio-cephalic trunk and left primary carotid arise from the sac, which is directed toward the right clavicle.

Liver and *kidneys* congested; *spleen* weighs 400 grammes.

Brain: meninges normal, veins enlarged, bloody suffusion in posterior part of membranes.

Lungs markedly congested; œdema, bronchitis, emphysema.

The autopsy justified our diagnosis in great part. It disclosed an aneurism at the origin of the aorta, in the posterior wall of which was the markedly compressed vena cava. The thrombus was not found in this vessel, however, but in the dilated vena azygos, although the result was the same as if the clot had been situated in the superior vena cava.

CHAPTER XXV.

DISEASES OF THE PULMONARY ARTERY.

THESE diseases are of two kinds, those contracted during intra-uterine life (congenital) and those contracted after birth; the latter are much rarer than the former. We will now discuss those alone which are acquired after birth.

Anatomical Data.—The pulmonary artery arises from the infundibulum of the right ventricle, and is directed upward and to the left in front of the aorta, which it crosses at an acute angle. After having encircled the concavity of the aorta it divides into two branches, one passing to the right lung, the other to the left.

The origin of the pulmonary artery corresponds to the inner part of the second left intercostal space and extends 3 ctm. outside of the sternum. There are slight differences, however, in the position of the orifice.

The course of the artery is very short, and it terminates about at the level of the junction of the second rib with its cartilage. For this reason the murmur of stenosis of the pulmonary artery is scarcely audible in the first intercostal space.

The artery increases in size with age; as a rule, it is much larger than the aorta until old age supervenes.

NARROWING OF THE ORIFICE OF THE PULMONARY ARTERY.

Lesions of the right heart, particularly sclerotic endocarditis of the pulmonary orifice, had not been observed except upon the cadaver, so that its description was entirely wanting. One reason which prevented its ready recognition was the fact that it was supposed that cyanosis would occur in such cases; so that in 1869, when I had occasion to diagnose a remarkable case of this affection, I could find scarcely anything on the subject in medical literature.

Cruveilhier's Atlas contains a description of this lesion,[1] and I have collected a few others in the treatise[2] which I published in 1871.

---

[1] Cruveilhier: Anat. pathol., liv. xxviii., p. 4.

[2] C. Paul: Du rétrécissement de l'artère pulmonaire contracté après la naissance, de ses symptômes, de ses complications, et particulièrement, de la phthisie pulmonaire consécutive, Soc. Méd. des Hôpitaux, 11 Août, 1871.

The following is the history of my case:

OBSERVATION LI. *Stenosis of the Orifice of the Pulmonary Artery ; Hypertrophy of the Right Ventricle ; Closed Foramen of Botalli ; Secondary Cheesy Pneumonia.*—H. A——, thirty-six years of age, entered the hospital July 16, 1869. He had an attack of acute articular rheumatism ten years ago ; did not suffer from cardiac symptoms at that time ; has had a cough for two years, and for three months a quotidian intermittent fever with nocturnal sweats. July 12th had an attack of hæmoptysis.

Present condition : notable emaciation, slight œdema of the face. The apices of both lungs present diminished resonance, feeble respiratory murmur, and moist subcrepitant râles ; these signs are much more extensive on the right side.

The heart is found increased in volume, particularly transversely. A loud blowing murmur, which replaces the first sound, is heard over the entire heart ; it covers the lesser period of silence and the second sound. The maximum of intensity is situated in the left second intercostal space, from 2 to 4 ctm. from the edge of the sternum.

The bruit is propagated in an almost vertical direction toward the left clavicle, but diminishes considerably a little before it reaches that bone. It is very feeble in the carotids and is not heard posteriorly over the descending aorta.

The second sound, masked by the first, is not heard distinctly except in parts which are remote from the pulmonary orifice—for example, at the base of the sternum.

The pulse is feeble but regular, and the sphygmograph shows no intelligible trace.

The hæmoptysis reappeared on July 27th, and did not cease until death (August 2d).

The diagnosis was not doubtful : an organic affection of the heart complicated by pulmonary phthisis. But what was the nature of the cardiac affection ?

The point of departure for the diagnosis was furnished by auscultation. The blowing murmur, which replaced the first sound and covered the heart, had its maximum of intensity toward the base, but did not present the characteristics of an aortic bruit.

In the first place the second sound was distinct, so that there could be no question of aortic insufficiency but only of stenosis. I was surprised to find, however, that the symptoms did not correspond to aortic stenosis.

The murmur was not rough, was unaccompanied by thrill, had its maximum of intensity in the second left intercostal space, and was propagated very feebly into the vessels of the neck. Moreover, the pulse, instead of being hard as in aortic stenosis, was full, soft, and compressible, the arterial tension was feeble, and the sphygmographic trace was not characteristic of aortic stenosis.

I was led to believe, accordingly, that the lesion was stenosis of the pulmonary artery, particularly on account of the situation of the maximum of intensity of the murmur over the pulmonary orifice, its propagation along the pulmonary artery, and its almost complete disappearance at the level of the passage of the arch of the aorta.

Upon post-mortem examination the walls of the right ventricle were found as thick as those of the left ventricle, and the cavity of the former was as large as that of the latter ; the interventricular septum projected into the left ventricle. The semilunar valves of the pulmonary artery were adherent over that portion of their border nearest to their base, and the narrowed orifice only permitted the introduction of the little finger. At the adherent portions the valves were thickened and indurated, and had lost their transparency and elasticity. However, the valves retained a certain degree of suppleness, and their free edges came in apposition in such a manner as to prevent insufficiency. The pulmonary artery itself was thinned and dilated, its circumference near the bifurcation measuring 16 ctm. The inter-auricular septum was normal.

The foramen of Botalli was completely closed. Taking this latter fact into consideration, there can be no doubt of the extra-uterine production of the stenosis.

The cardiac muscle presented a certain degree of fatty degeneration. The upper lobe of the right lung was in a condition of cheesy pneumonia; similar condition of the left apex.

In 1882 Vimont had collected in his thesis the reports of 52 cases of acquired affections of the pulmonary artery, including stenosis and insufficiency.

It follows from these observations that sclerotic endocarditis may attack the semilunar valves of the pulmonary artery, as it does those of the aorta, though more rarely. The absence of the ordinary lesions of diseases which have developed during intra-uterine life enables us to state that we have to deal with acquired lesions.

But in order to remove all hesitation in this regard, I will quote a case observed by Mayer,[1] in which the endocarditis assumed an ulcerative form and the patient succumbed during the acute stage.

OBSERVATION LII. *Stenosis of the Pulmonary Orifice Secondary to an Acute Endocarditis of the Semilunar Valves.*—A woman, aged twenty-six years, first complained in the spring of 1877 of violent palpitations which made their appearance mainly on exertion; also complained of painful twinges in arms and legs; gradually grew weaker and entered the Freiburg Hospital.

The first examination (August, 1877) showed a soft systolic blowing murmur at the apex, which increases in intensity at the base and has its maximum in the second intercostal space. It can be followed to the left into the axillary region and even into the corresponding supra-clavicular space. It is also heard, but feebly, to the right.

The second pulmonary sound was less distinct than the second aortic sound. Posteriorly, to the left and upward, could be heard equally the systolic murmur of the heart, and to the right as far as the spine of the scapula.

A diagnosis was made of stenosis of the left branch of the pulmonary artery, undoubtedly caused by some fibrous band.

September 4th the patient sank into a typhoid condition and died on October 17th.

Autopsy.—The heart is hypertrophied; right ventricular wall 1 ctm.

---

[1] Mayer: Ueber ein Fall von Stenosirung der Pulmonal-arterie in Folge von acuter Endocarditis der semilunar Klappen (Deutsch. Arch. f. Klin. Med., Bd. XXIV., H. IV., V., p. 435, 1879).

thick; pulmonary orifice is found obstructed by soft clots, vegetations, and excrescences. The ventricular septum is perforated above the conus of the pulmonary artery. Two small vegetations are found on the tricuspid valve. The right auricle is small and its walls thin. The foramen ovale is entirely obliterated. Nothing especial in the other organs. This is evidently a case of acquired lesion.

Allied to these lesions are the ones in which the stenosis is situated not alone at the level of the orifice but also of the infundibulum, or solely at the latter, and is due not to endocarditis but to myocarditis. To this lesion I have applied the term prearterial pulmonary stenosis.[1]

The stenosis leads to hypertrophy of the right ventricle, which acquires the capacity and thickness of the wall of the left ventricle and gives to the heart the shape of a sac formed of two equal ventricles. Indeed, in certain cases the right ventricle becomes larger than the left and the convexity of the septum projects into the left ventricle.

The hypertrophy extends to the columnæ corneæ, especially those of the first order, so that despite the increased volume of the right ventricle, its capacity does not exceed that of the left ventricle. The tricuspid valve is sometimes affected at the same time.

The pulmonary artery is usually dilated and its walls thinned. This is explained by the stasis of the venous blood, which is no longer thrown into the artery with sufficient force and induces a loss of contractility and elasticity in the vessel.

Finally, at a later period the hypertrophy leads to fatty degeneration of the muscular fibres.

Diagnosis.

The diagnosis of stenosis of the orifice of the pulmonary artery depends upon the following signs:

Measurement of the heart shows depression of the hepatic angle of the cardiac triangle, and consequently a more horizontal position of the lower border. In certain cases palpation may reveal a purring thrill, which is more marked in the left second intercostal space.

Auscultation discloses a peculiar blowing murmur. This murmur occupies particularly the left second intercostal space, from the sternum to 4 or 5 ctm. from the edge of the bone. If the murmur is intense it is transmitted readily by the thickened wall of the ventricle and then covers the entire anterior surface of the heart; but the maximum of intensity remains very distinctly in the second intercostal space. It is not conveyed to the first intercostal space unless hepatized pulmonary tissue acts as a conductor.

---

[1] Observations VII., VIII, IX., and X. of my treatise.

The murmur begins with ventricular systole, covers the first sound and the lesser period of silence, and terminates with the closure of the semilunar valves. It is usually rough ; moreover it is superficial.

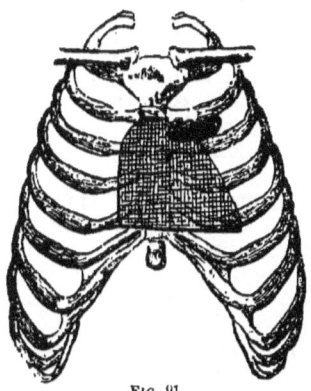

Fig. 91.

The murmur has its maximum of intensity when the patient is ausculted in the recumbent posture ; it diminishes considerably and sometimes disappears entirely in the sitting position. This is due to the fact that the force of gravity is almost sufficient to counterbalance the propulsive force of the right ventricle in the pulmonary artery.

If respiration is stopped the murmur is at first more marked during the first cardiac revolutions, but it gradually grows weaker and weaker. Then if the patient is asked to make a vigorous expiratory effort while the nares are closed, the murmur diminishes until it disappears almost entirely. Finally, when the patient is again allowed to draw breath he makes deep inspirations, and the pulmonary circulation becomes freer and freer, and the murmur returns in proportion as the waves of blood increase. These qualities of pulmonary bruits, which I have observed so often in anæmic murmurs of the pulmonary artery, are shown very clearly in pulmonary stenosis.

OBSERVATION LIII. *Stenosis of the Pulmonary Artery* ; *Secondary Phthisis.*—The patient was a railroad porter, who continued at work until last May. At that time he was suddenly seized with a chill and tremor, and fell asleep on a bench. Upon awaking he felt bad and could not return to work. A physician who was called diagnosed pulmonary congestion. The patient remained a month in bed and has since been unable to work. Entered the hospital November 9, 1878. In January, 1879, I found him in the following condition : complains of weakness ; coughs, and appears to be suffering from pulmonary disease. The chest is markedly emaciated, and the ribs project strongly ; right lung entirely healthy. On the left side dulness below the clavicle and in the first intercostal space ; feeble respiration in this locality, with a large number of moist crepitant râles. Inferiorly the respiration is normal, but there is considerable arterial impulse with a blowing murmur. Posteriorly, diminished dulness in the infraspinous fossa with numerous subcrepitant râles.

Apex-beat of heart behind the fifth rib, 10 ctm. from median line ; upper border of liver at level of fifth right cartilage ; vertical border of heart $1\frac{1}{2}$ ctm. from right edge of sternum. The hypertrophy, therefore, affects chiefly the right heart.

No murmur is heard over the aortic, mitral, and tricuspid orifices. A pulmonary blowing murmur is heard, with the following characteristics :

Situation : In the left second intercostal space, beginning at the left border of the sternum and extending 8 ctm. toward the axilla ; it is also heard in the third intercostal space over an extent of 1 to 2 ctm. It is heard to the right of the sternum, but there sounds distant. Its maximum is situated 3 ctm. from the left border of the sternum.

Period : The bruit is systolic ; begins with ventricular systole, develops during the duration of the systole, and ends a little before the sharp closure of the semilunar valves. But this bruit presents the special characteristics which I have already recognized as belonging to anæmic murmurs in the pulmonary artery : there is considerable difference in its intensity according as the patient is standing or lying down. When he is lying down it attains its maximum of intensity and is accompanied by a purring thrill ; when standing, it diminishes considerably.

The second characteristic of pulmonary bruits presented by this patient is the following : If the nose and mouth are closed while the patient is making a vigorous expiratory effort, the murmur diminishes in duration and finally disappears almost entirely. When the patient is again allowed to take full inspirations the murmur gradually resumes its former intensity and even exceeds it for a moment.

All these phenomena are evident signs of an alteration of the pulmonary artery, situated at its orifice and diminishing its calibre. It is also found that at the xiphoid appendix the cardiac impulse has a remarkable intensity, and that at the left border of the sternum, at this level, the heart sounds tend to assume the pendulum rhythm, *i.e.*, the lesser and greater periods of silence tend to become equal, a phenomenon belonging to hypertrophy of the right heart. A slight musical murmur in the veins of the neck.

The murmur is evidently not situated at the aortic orifice. We must also discard the hypothesis of an aneurism of the aorta giving a blowing murmur in the second left intercostal space. Such aneurisms occur and are situated in the descending part of the arch of the aorta, immediately below the emergence of the subclavian ; but they give rise to an affection of the recurrent nerve and to aphonia, which are not present in this case ; there are no expansile movements.

Finally, we may state that the pulse is regular ; there is no cyanosis, pulmonary œdema, congestion of the liver, or œdema of the lower limbs.

The persistence of the lesions for a number of months will not permit us to diagnose simple anæmia, but rather an acquired stenosis of the pulmonary artery with secondary phthisis.

But these symptoms are not constant. When the heart is markedly hypertrophied and the impulse of the right heart approaches in energy that of the left heart, the force of gravity no longer causes the murmur to disappear. Arrested respiration renders the murmur more distinct, and exertion diminishes it but does not lead to its disappearance. These phenomena are well marked in a young patient under my observation at the present time.

The lesion which produces stenosis of the origin of the pulmonary artery is not always situated at the level of the valves ; it may be below them and affect both the valves and a part of the infundibulum. Such, for example, are the cases reported by Philhouze,[1] Bock, Duguet, and

---

[1] Observations II. and VIII. in my treatise on Stenosis of the Pulmonary Artery.

Landouzy,[1] Havage,[2] and Dittrich. In these observations the endocarditis appears to have been complicated with myocarditis, and in Dittrich's case it is probable that the lesion was traumatic in its character, and caused by the blow upon the sternum from a horse.

In other cases, the valves of the pulmonary artery are normal and the stenosis affects only the infundibulum. It is to this form of stenosis that I have applied the term præarterial. In all these cases the lesion results from cicatrization of a myocarditis.

These cases differ also with respect to the symptoms. The blowing murmur is indeed situated to the left of the sternum, but the maximum of intensity, instead of being found in the second intercostal space, is in the third space, and the murmur is also present in the first and second intercostal spaces (vide Fig. 92).

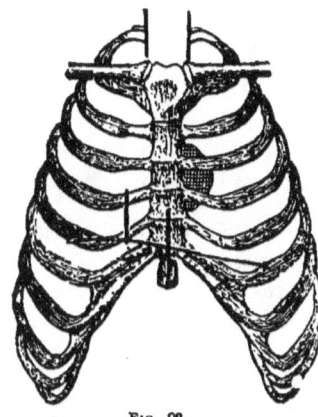
Fig. 92.

The other symptoms are the same as those of valvular stenosis, provided the patients are observed at a late stage, when the lesions have become fixed. On the other hand, if the patients are observed during the existence of the myocarditis, the symptoms of this affection will complicate the diagnosis.

Let us resume the diagnosis of stenosis of the pulmonary orifice.

One of the most important functional signs is the fact that none of the patients, who present stenosis of the pulmonary artery uncomplicated, suffer from cyanosis. Apart from this, we are astonished not to find more venous disturbances. As a rule there is no congestion of the liver, no ascites, and very little œdema of the legs, despite an advanced stage of the malady.

There is no disturbance on the part of the arterial circulation; the pulse is regular and presents no abnormal features. According to Potain and Rendu the patients complain of a feeling of peripheral coldness despite the integrity of the arterial circulation. For my part I have not observed this symptom which, on the contrary, is a constant feature in cases of cyanosis from congenital lesion.

The patients are less subject, perhaps, than those suffering from other cardiac diseases, to dyspnœa on exertion, at least if the rhythm of the heart and the myocardium are intact. In a word, if hæmatosis is diminished by stenosis of the pulmonary artery, it may result in insufficient oxygenation

---

[1] Duguet and Landouzy : Soc. Méd. des Hôpit., November 22d, 1878.
[2] Havage : Soc. Anatom , 1879, p. 562.

of the blood, and consequently in deterioration of the organism advancing to phthisis; but this deterioration does not assume the form of venous stasis with asphyxia, dropsy and their sequelæ.

Some tendency to hemorrhages, particularly to hæmoptysis, has been observed and must be regarded as a prodrome or symptom of the secondary pulmonary phthisis. It is a very peculiar feature of this affection that it leads to phthisis in the same manner as does congenital stenosis. The phthisis does not supervene immediately, though it may occur at an early period. In the last case under my observation, in which phthisis was undoubtedly present, the disease had lasted scarcely a year. In the first case, it had lasted seven years.

The patients rarely resist the phthisis and succumb to asystole, in which event cyanosis may appear at the termination of the disease, as in every other organic disease of the heart.

Stenosis of the Trunk and Branches of the Pulmonary Artery.

Cases of this kind are rare. In my treatise on Stenosis of the Orifice of the Pulmonary Artery, I have reported an observation by Willigk, in which the stenosis affected the right branch of the vessel. Its diameter was reduced to 2 mm., though the diameter of its origin was 16 mm., the same as that of the normal left branch. The trunk of the vessel was thickened and presented evidences of endarteritis. The patient, aged forty-nine years, died of tubercular phthisis of the lungs, larynx, and intestines.

In another case, observed by Karl Bettelheim, and published by Vimont, the stenosis affected the right branch of the pulmonary artery. The patient died of phthisis with tubercles in both lungs.

Finally, in a third case, the lesion affected both branches of the pulmonary artery (Observation XII. in my treatise). The patient observed by Tomassi Crudeli died suddenly from thrombosis of the artery.

CHAPTER XXVI.

DISEASES OF THE PULMONARY ARTERY.—(*Continued.*)

INSUFFICIENCY OF THE VALVES OF THE PULMONARY ORIFICE.

INSUFFICIENCY of the valves of the pulmonary artery, contracted after birth, is undoubtedly the rarest of all valvular lesions. Cases have been observed by Benedict,[1] Frerichs,[2] Kolisko,[3] Von Wahl,[4] Klob,[5] Roeber,[6] Weiss,[7] Morison,[8] Whitley,[9] Decornière,[10] Vast,[11] Budin,[12] and Vimont.

It follows, from these observations, that insufficiency of the pulmonary valves may be the result either of endocarditis, or of the atheromatous process, which thickens and shortens the free border of the valves. In Von Wahl's case, the endocarditis destroyed the valves almost entirely. In another case, one of the valves has been found perforated at the base, as the result of myocarditis. In addition, the pathological process which has changed the valves to such an extent as to render them insufficient, has often produced a certain degree of stenosis.

It is said that dilatation and hypertrophy of the right ventricle and auricle have been observed as the result of this lesion ; also a certain amount of dilatation of the pulmonary artery, with secondary hypertrophy of the

---

[1] Benedict: Fall von Insufficienz der Valv. semilunar. arter. pulm., Wien. Wochenschr., 1854, 35.

[2] Frerichs: Ibid. 1853.

[3] Kolisko: Fall von Insufficienz der Pulmonal arterien Klappe, Zeitschr. d. Wiener Aerzte, 1859.

[4] Von Wahl: Acute Endocarditis der Pulmonal Klappen, Petersb. Med. Zeitschr., 1861.

[5] Klob: Beitraege z. Pathol. d. Pulmonal arterien Klappen, Zeitschr. d. Wiener Aerzte, 1861.

[6] Roeber: Ein Fall von Insufficienz d. Pulmonal Klappen, Berl. Klin. Wochenschr., 1870.

[7] Weiss: Ein Fall v. Insufficienz d. Pulmonal Klappen, Wien. Med. Presse, 1876, 1.

[8] Morison: A Case of Disease of the Pulmonary and Tricuspid Valves, Trans. of the Path. Soc., 1876, xxvii., p. 83.

[9] Whitley: Guy's Hospital Rep., 1858, third series, vol. v., p. 252.

[10] Decornière: Essai sur l'Endocardite puerpérale, Thèse de Paris, 1869.

[11] Vast: Endocardite Ulcéreuse, Thèse de Paris, 1864, p. 53.

[12] Budin: Thèse d'agrégation de Porak, 1880.

left ventricle and auricle. Lobular pneumonia and infarctions have also been found upon autopsy.

Examination of the heart has shown enlargement of the right side, and a diastolic murmur having its maximum of intensity at the left border of the sternum, in the second intercostal space. This murmur has a high pitch, and extends, to a certain extent, over the lower part of the sternum along the right ventricle, in the direction followed by the blood which flows back into the ventricle.

According to Rosenstein, Weill has confirmed the statement of Dusch that this murmur may be propagated in the direction of the vessels of the neck. A systolic blowing murmur has been heard in certain cases, but this is explained by the coexistence of stenosis and insufficiency.

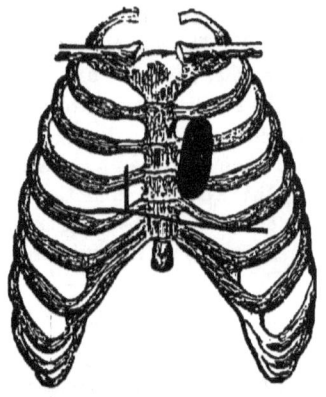

Fig. 93.

Among the functional disturbances have been noticed a certain amount of dyspnœa and palpitation of the heart. Compensatory hypertrophy of the right ventricle develops slowly. In Morison's case the patient complained of acute pains in the region of the heart, radiating into the right shoulder and hand. The pulse is normal, or approximately so.

After the period of compensation and tolerance has terminated, the patients suffer from venous stasis and dropsy. Death is the inevitable result, but life may be prolonged for a considerable period.

I have not observed any case of insufficiency of the pulmonary valves in which the diagnosis has been confirmed by post-mortem examination.

At the present time I have a patient under observation, who is suffering from acute articular rheumatism, and presents in the clearest manner the symptoms of insufficiency of the pulmonary artery.

OBSERVATION LIV. *Acute Rheumatism; Pulmonary Insufficiency.*—L. P——, twenty-six years of age, entered the hospital July 7, 1882. Present sickness has lasted a week; high fever, profuse sweats, pain and swelling in ankles and knees. In the second left intercostal space the first sound is heard normally, then a blowing murmur is heard, which is propagated from above downward along the sternum, and is heard better on the left side than on the right. Nothing heard at the apex.

Apex-beat in fifth intercostal space, 12 ctm. from the median line; vertical border of heart 4 ctm. from edge of sternum; upper border of liver at level of fifth cartilage.

July 13th, the murmur has increased considerably in intensity. Its area forms an elongated ellipse beginning in the second left intercostal space, 3 ctm. from the median line, and directed toward the fourth intercostal space, at its junction with the sternum; the ellipse is 8 ctm. long, and 3½ wide. It corresponds, therefore, to the pulmonary artery and its infundibulum. The first sound is rapidly followed by a blowing murmur, which is prolonged into diastole. At the aortic orifice, on the contrary, the semilunar click is heard distinctly. The murmur is produced very probably by insufficiency of the pulmonary valves.

ENDARTERITIS OF THE PULMONARY ARTERY.

Inflammation of the pulmonary artery, though much rarer even than that of the aorta, has nevertheless been observed, and even in the acute form.

It has not been found isolated, but is mentioned in certain observations of acute endocarditis with vegetations upon the pulmonary semilunar valves. In one case, it is said that endarteritis was present with rugosities upon the surface, and adhesion of a clot which divided into two and entered the branches of the vessel. There is no description of the disease, as the lesions have been found merely on autopsy.

Chronic endarteritis has also been observed. In a man of nineteen years, Fenger[1] found, in addition to vegetations upon the pulmonary valves, three vegetations as large as hazel-nuts, and sixty small ones, which were situated at the level of the bifurcation of the artery, and were prolonged into the two branches.

O. Weber[2] gives the following appearances in a case in which the lesion was attributed to syphilis.

OBSERVATION LV.—The individual presented syphilitic lesions in the vertebral column and liver. The right branch of the pulmonary artery contained patches 2 to 3 mm. long, and 5 mm. thick, which projected into the cavity of the vessel so that the lumen was reduced to a mere slit. The patches were covered by the inner tunic of the artery; they were soft and yellow, composed of granulation tissue which had developed at the expense of the middle tunic. The external tunic was still elastic, but very much thickened. Infarctions in both lungs. Death occurred from hæmoptysis.

ATHEROMA OF THE PULMONARY ARTERY.

Atheroma of the pulmonary artery is sometimes met with, although much more rarely than in the aorta, but the appearances are the same in both. The patches are situated mainly at the bifurcation of the artery, and near the semilunar valves.

One of the first results of the lesion is hypertrophy of the right heart. This has been noticed not only in France, but also by Klob[3] in Germany,

---

[1] Fenger: Nord. med. Ack. V., 1873, 4. Jahresber., II., 140, quoted by Quincke, Krankh. d. Gefaesse, p. 368.

[2] O. Weber: Syphilit. Neubildung in d. Wand. d. arter. pulmon. (Med. Central Ztg., 1862, No. 52), and Schmidt's Jahrb., Bd. 123, p. 174.

[3] Klob: Wochenblatt d. Ges. d. Aerzte, 1865.

who found hypertrophy of the right heart in an autopsy upon a man who presented dilatation of the pulmonary artery.

It is difficult to determine the importance of the lesions, since atheroma of the pulmonary artery is associated almost always with other diseases of the heart and vessels.

The lesion is rarely suspected during life. Nevertheless J. Hope [1] has recorded an observation in which a characteristic bruit was heard on auscultation.

OBSERVATION LVI.—A dilatation of the pulmonary artery, although not large, produced a perceptible thrill and impulse in the second left intercostal space. A small tumor could be mapped out between the ribs. Auscultation revealed a very loud, superficial, dry murmur, like the creaking of a saw, above the clavicles and over the entire præcordial region, but most markedly in the second intercostal space ; hypertrophy and dilatation of the heart.

Atheroma of the pulmonary artery is said to be produced by old age, alcoholism, rheumatism, gout, and syphilis.

ANEURISM OF THE PULMONARY ARTERY.

Aneurisms of the pulmonary artery are very rare. Among 950 cases of aneurism, Crisp collected only 4 in the pulmonary artery. We here refer only to aneurisms of the trunk of the vessel, not to those of the small branches, and which are found in phthisis.

Both forms of aneurism have been observed, the true or fusiform, and the false or sacculated.

The symptoms have been a prominence in the left second intercostal space near the sternum ; here auscultation reveals very loud murmurs, sometimes systolic, sometimes diastolic.

OBSERVATION LVII. *Aneurism of the Trunk and acquired Stenosis of both Branches of the Pulmonary Artery.*—A sailor, aged forty-two years, who had suffered several times from muscular rheumatism and bronchial catarrh, was admitted to the clinic of Prof. Carlo Maggiorani, December 26, 1867. Since September, 1866, he had had attacks of dyspnœa and palpitation of the heart ; severe bronchial catarrh developed in the last two months. Pulse 88, area of cardiac dulness increased. Between the two heart sounds was heard a rough murmur, which was very perceptible at the base. A soft blowing murmur was heard instead of the second sound, more marked over the right than over the left ventricle ; the two periods of silence were shortened. In addition, signs of bronchial catarrh, pulmonary œdema, and hepatic congestion. Urine, sp. gr. 1.022, certain amount of albumen. Lower extremities slightly œdematous.

January 7, 1868 : the murmur heard between the first and second

---

[1] Compendium de Médecine, article Aorte, t. I., pp. 181-186.

sounds became rough, and the first sound less clear; face slightly cyanosed, pulmonary râles more numerous, pleuritic exudation on both sides.

January 16th, condition grew worse; dyspnœa and cough more distressing, pulse 102. January 17th, under the influence of veratrum viride, pulse fell to 52 and became intermittent. January 19th, patient died suddenly while eating.

The autopsy showed œdema of the brain, cerebral sinuses and meningeal veins filled with blood, and moderate effusion in the ventricles. Small amount of fluid in pleuræ and pericardium; lower lobe of right lung œdematous. Arteries and veins of right lung contained little blood, those of left lung almost entirely empty. Nutmeg liver. Enlarged and softened spleen; kidneys congested and showed beginning parenchymatous nephritis.

Heart hypertrophied, especially the right auricle and ventricle, which were filled with blood; left cavities entirely empty. Both venæ cavæ dilated. Transverse diameter of heart measured 12 ctm. at the base, longitudinal diameter 13 ctm. from the apex of the right ventricle to the base. The right ventricle and auricle were doubled in volume; the endocardium of right side had lost its transparency, and the columnæ carneæ were hypertrophied. Wall of right ventricle 7½ mm. thick; right auricle 6 mm. thick in parts; circumference of right auriculo-ventricular orifice measured 120 mm. (normally 103 to 104 mm.); circumference of pulmonary artery 95 mm. (normally 72 mm.); semilunar valves enlarged and thickened; trunk of pulmonary artery dilated like an aneurism, and its transverse diameter measured 50 mm. Above the semilunar valves were two small diverticula, the walls of which were ½ mm. thick, while the remainder of aneurismal wall was 1½ to 2 mm. thick. No trace of endarteritis in aneurism, but hyperplasia of the adventitia. Between the orifices of the two branches of the vessel the arterial wall was formed of sclerotic connective tissue, hard and partly calcified, so as to form a ridge in the cavity of the aneurism. Farther on, the right branch dilated; in the left branch the walls were thickened, reducing the lumen of the vessel from 13 to 2 mm. At this point the lumen was obliterated by a recent clot (probable cause of the sudden death); a little farther on the lumen gradually enlarged.

The trabeculæ and papillary muscles of the left heart were atrophied; the ventricular walls were from 7 to 4 mm. thick (normally 15 to 16 mm.); the muscular tissue was soft and friable, while that of the right ventricle was firm and brownish red in color. Under the microscope the primary fibrillæ of the left heart were found to be half as thick as those of the right side; the former contained fatty and pigmentary granules, the latter were normal. The left auriculo-ventricular orifice was normal; the aortic orifice had a circumference of 64 mm. (normally 66½ mm.). The semilunar valves were shortened and insufficient. The aorta contained several patches of chronic endarteritis (Corrado Tomassi Crudeli de Palerme, Riv. clin. vii., 2, p. 37, 1868, in Schmidt's Jahrb., 8, p. 169, 1870).

THROMBOSIS AND EMBOLISM OF THE PULMONARY ARTERY.

There is nothing more common in post-mortem examinations than to find clots in the pulmonary artery. These are usually formed either after death or in the last stage of the agony; they are recognizable particularly by their stratification. Seen in the ordinary position of the corpse, the upper part of the clot is composed almost exclusively of fibrin, containing a few globules, but the majority of these have accumulated in the lower parts. Furthermore, the clots are not adherent to the walls, and are held in position simply by prolongations which they send into the branches within the lungs.

The clot formed during life is not stratified, and is darker in color the more recent its formation; it is firmer, as if condensed by the expulsion of the serous contents. Furthermore, it adheres to the walls. This adhesion may be due merely to a plastic fluid which is readily detached; but at a later period the clot is joined to the wall and its detachment leaves a roughened, unpolished surface. Indeed the adhesion may be of such a nature that it becomes difficult to determine accurately the wall of the vessel. Certain clots are composed entirely of leucocytes, others of red globules.

Emboli differ from thrombi in this respect, that instead of finding a more or less homogeneous tissue, the interior of the relatively recent clot contains a mass the color and structure of which are different. This is sometimes an old, more or less degenerated clot; at other times a still recent clot, upon which may be recognized the impression of the valves of the veins. Or the nucleus of the embolus consists of the débris of cancer, particles of the valves, or pus. The differential qualities of thrombi and emboli are presented in the following table:

Thrombi.	Emboli.
Begin by small ramifications; clot arborescent.	Are arrested upon the projections and valves, and at the bifurcations.
Structure homogeneous throughout the entire clot.	A whitish mass surrounding soft coagula.
Regular shape.	Irregular shape.
Cardiac extremity rounded.	
Texture stratified.	Texture irregular.
Do not entirely fill the calibre of the vessel.	Distend the vessel without adhering to its walls.

The most striking phenomenon is the sudden death during a cachectic condition. The onset of the symptoms is as sudden in thrombosis as in embolism.[1] Sometimes the sudden symptoms have been preceded by dyspnœa for a couple of weeks.[2]

Thus in cachectic individuals or in the puerperal condition the presence of thrombi in the lower limbs, the increasing dyspnœa, the sudden onset of excessive respiratory anxiety, cyanosis, a feeling of coldness, ex-

---

[1] Homolle: Soc. Anat., p. 103. 1874.
[2] Bucquoy and Hanot, quoted by Favre: Étude sur la Thrombose de l'artère pulmon. Thèse de Paris, 412. 1875.

pectoration of blood—these are the signs which, combined with rapid death, indicate the occurrence of thrombosis or embolism of the pulmonary artery.

Death is inevitable unless a small embolus is present and does not obliterate the entire circulation of the pulmonary trouble.

Finally, clots may be found in the pulmonary artery, produced by pathological causes situated outside of the vessel. Oulmont[1] has reported a case of sacculated aneurism which had produced compression of the left branch of the pulmonary artery and chronic interstitial pneumonia. At other times the compression has been produced by bronchial or mediastinal glands,[2] finally by retraction of the lungs as the result of interstitial pneumonia.

---

[1] P. Oulmont: Prog. Méd., January 1, 1881.
[2] Barety: Thèse, Obs. LXXXIII.

CHAPTER XXVII.

DISEASES OF THE MYOCARDIUM.

THE myocardium is a peculiar tissue. Although composed of muscles which are not subject to the will, it is allied to the voluntary muscles by its connective-tissue framework, the terminal tendons of its fibres, red color, and finally by the striation of the fibres. It differs from the voluntary muscles in the following respects: the size of the fibres is smaller, the diameter of the latter being six- to eight-hundredths of a millimetre, that of the former being only two-hundredths of a millimetre.

The fibres of the voluntary muscles are grouped in secondary, then tertiary bundles, isolated from one another by connective tissue forming a sheath; those of the heart are not grouped in parallel, gradually increasing bundles.

Each fibre of voluntary muscle is surrounded by a sheath of sarcolemma; the heart fibres are not provided with a sarcolemma. The muscular fibres of the heart, by their incessant divisions and continual anastomoses, form an inextricable network which compels them to contract at the same time.

Although examination of the myocardium differs from that of the voluntary muscles, the traces furnished by the myograph are very much alike.

While the curve made by voluntary muscles is formed by shortening along its length, that made by the cardiac muscle is furnished by increase in width at the moment of contraction. But these two phenomena are so related that the curves obtained are identical. Thus, the form of the systole of the cardiac muscle is that of a muscular contraction. The ascending period, which corresponds to the shortening of the muscle, is shorter than the descending period, *i.e.*, than the return of the muscle to its former length.

Fatigue modifies the systole of the heart and the contraction of a muscle in the same manner; it causes diminution of the amplitude and increase in the duration of the movement. Both forms of contraction are also affected in the same manner by heat and cold.

The systole does not present an equal duration in the different parts of the heart. The auricle effects its systole much more rapidly than the ventricle.

At the beginning of each systole an electrical variation is produced, and is manifested in the galvanoscope.

When the heart contracts, all the fibres do not enter into contraction simultaneously; the systolic movement is propagated from the base to the apex, *i.e.*, from the auricles to the ventricles, and runs along the length of the heart by a sort of peristaltic movement. According to Engelmann,[1] the propagation in the muscular tissue occurs from cell to cell, independently of all nervous action.

The series of cardiac movements being equivalent to a series of successive contractions, it follows that these movements will be more and more extensive in proportion as the time occupied in their accomplishment is greater. Acceleration of the cardiac rhythm, therefore, will render the movements less extensive.

The force expended by the myocardium in its contraction is conveyed to the heart; furthermore, the empty heart ceases to contract as soon as it exhausts the blood contained in its vessels. Ligature of the coronary arteries in a mammal arrests the action of the heart in about two minutes. A drop of blood allowed to fall upon a frog's heart almost immediately reawakens the movements which had been extinguished.

The myocardium is excitable to traumatic stimuli, but these irritations are confined to the parts which have been stimulated. This excitability is purely reflex, since the myocardium is insensible, as was shown in two cases in which the heart was laid bare and could be directly handled.

Another question remains for discussion. It is known that the rhythm of the heart is not produced either by the pneumogastric or the sympathetic. The former merely moderates or suspends the rhythm, the latter accelerates it. When the heart is cut below the auricular groove, its rhythm is still retained. Ranvier has shown, by a series of experiments, that the rhythm appertains to the cardiac muscle itself, and not to the small ganglionic centres. He re-established the rhythm by the action of an induced current upon the separated apex, which does not contain any nerve-cells.[2] Marey had shown, in 1876, that the excitability of the myocardium to electrical currents exists only during the period of repose, and that they are ineffective during the systolo-diastolic period of activity. Dastre and Marcocci have confirmed this experiment, not alone upon the heart as a whole, but also upon the apex, which is destitute of ganglia (*Gaz. des Hôpit.*, March 28, 1882).

Laborde has found the cardiac rhythm established in the embryo of the chick before the existence of the developed nervous system. This law proves that nervous disorders may indeed change the frequency of the

---

[1] Engelmann, quoted by Marey: La circulation du sang à l'état physiologique et dans les maladies, p. 32. 1881.

[2] Ranvier: Cours du Collège de France, xxvi° leçon, Prog. Méd , August 16, 1879.

pulsations of the heart, but that alterations of the myocardium are the cause of alteration of rhythm.

Myocarditis.

Myocarditis, or inflammation of the myocardium, may be acute or chronic, general or partial; the latter form may be purulent or fibrous.

Myocarditis is often an extension of endocarditis or pericarditis. The left ventricle is the site of election during extrauterine life; prior to birth, on the contrary, myocarditis of the right ventricle is the usual cause of congenital heart disease. The auricles are rarely affected alone.

ACUTE MYOCARDITIS.

Myocarditis is rarely ever primary, perhaps only in some cases of traumatism. Secondary myocarditis ordinarily follows endocarditis or pericarditis; it is rarely produced by disease of neighboring organs (pneumonia or pleurisy). It is only in rare cases that ulcerations of the lung have given rise to thrombosis of the pulmonary veins, and these to emboli which have been arrested in the coronary arteries, producing circumscribed purulent myocarditis.

Diffuse parenchymatous myocarditis has been observed in the course of general diseases, in rheumatism, and particularly in infectious diseases, such as typhoid fever. Purulent myocarditis may occur in purulent infection and the puerperal condition.

When the lesion is extensive the pericardium is less transparent, and contains ecchymotic spots. The cardiac muscle is dark or discolored; its consistence is less elastic, rarely increased, generally diminished, and very friable. The softening is visible to the naked eye.

The microscope shows granulo-fatty and vitreous degeneration, disseminated irregularly among the different fibres. The muscular fibres are swollen and cloudy, the striæ have disappeared; they contain fatty granules, and the nuclei have undergone proliferation. A large number of new-formed cellular elements are found in the interstitial tissue. At a more advanced stage the fatty degeneration is complete. Finally, spots of softening, true abscesses, are formed.

The most frequently affected part is the apex of the left ventricle, then the base, posteriorly, near the aortic valves, then the septum near the base, and more rarely the papillary muscles and fibres of the right ventricle.

INTERSTITIAL MYOCARDITIS.

This is the purulent form. The pus is collected between the fibres in small deposits, from the size of the head of a pin to that of a bean, or in longer or shorter streaks.

O. Weber[1] believes that the pus is formed at the expense of the muscular fibres. Demme[2] thinks it is formed from the nuclei of the vascular walls. In certain cases of pyæmia there are a large number of small foci. Rindfleisch has found them filled with vibriones.

ABSCESSES OF THE MYOCARDIUM.

When the inflammatory process in endocarditis terminates in an abscess, the following characteristics are observed: the abscess is large and projects beneath the pericardium or endocardium, which has the appearance of an ecchymosis at this spot. The abscess may open into the pericardium and produce purulent pericarditis, or it may open into the endocardium. In the latter event the pus is carried into the circulation, and gives rise to metastatic abscesses in the spleen, kidneys, brain, etc. If the perforation occurs into the right ventricle, the metastatic process develops in the lungs.

The process may terminate in rupture of the heart or one of the papillary muscles, or finally by the formation of a partial aneurism of the heart. On the other hand, the recovery of such an abscess is possible, but it then leaves behind a connective tissue cicatrix. Finally, abscesses may become encysted and undergo calcareous incrustation.

When the abscess is situated in the upper part of the septum there is no tissue present in which it may spread, and it may then open into both ventricles, thus establishing a communication between the two cavities. Finally, these abscesses of the septum may extend to a semilunar or an auriculo-ventricular valve.

A certain amount of dilatation is produced in myocarditis as the result of the loss of tonicity. Hence follows a diminished energy of circulation and a tendency to stagnation of the blood, and to the formation of clots in the heart. The enfeebled circulation also gives rise to venous stasis, pulmonary congestion, bronchial catarrh, pulmonary œdema, hemorrhagic infarctions, œdema of the meninges and brain, congestion of the liver and spleen.

Diagnosis.—The myocarditis which develops during the course of a grave affection like typhoid fever, variola, etc., can only be recognized

---

[1] O. Weber: Virch. Arch., Bd. xv.
[2] R. Demme: Beitr. z. Anat. u. Diagn. d. Myocarditis, Schweiz. Journ. f. Heilk., i., 79, 461.

by the careful daily examination of the heart and other organs of circulation, and also of the disturbances of the nervous system.

The disease begins with a certain excitability of the heart, and with dyspnœa which is not explained by the condition of the lungs.

Stokes has observed merely enfeeblement of the first sound, but a number of writers have noted the presence of a blowing murmur, due to atony or paresis of the myocardium. It will be recollected that I suggested this explanation for certain temporary blowing murmurs in rheumatic endocarditis. The murmur is heard at first near the apex, then it moves toward the sternum and ascends on the right side and toward the origin of the great vessels. The murmur is systolic; in exceptional cases, in which a semilunar valve becomes affected and rendered insufficient, the bruit occurs at the close of systole. Finally, the murmur is soft, diffuse, deep.

The disorder is first manifested by simple enfeeblement of the first sound, and the murmur does not develop until a later period. The murmur is only temporary; either the patient recovers, the heart resumes its vigor, and the bruit ceases, or he grows worse, asystole becomes marked, and the blowing murmur diminishes and finally disappears.

It does not persist unless the patient recovers with a cicatrix which impedes the play of a valve or causes stenosis of one of the canals.

Reduplication of one or the other heart sounds has also been noticed, and is explained less by the increase of certain resistances or tensions than by a loss of harmony of action in the ventricular systole.

The impulse of the heart diminishes in vigor and consists merely of a sort of tremor, its area becomes greater toward the end, and this phenomenon is justly attributed by Desnos and Huchard to cardiac dilatation. The pulse is first strong, but soon becomes soft, small, compressible, and intermittent on account of the insufficiency of the ventricular contractions (false intermissions). It should be noted that the loss of rhythm is the real sign of the myocarditis, the acceleration or retardation being due to the action of the sympathetic and pneumogastric nerves.

The weakness of the pulse indicates insufficiency of the arterial supply and obstruction to the venous circulation in the viscera. This condition in the brain is soon shown by the delirium; this is a mild delirium of depression with general or partial convulsions, or sometimes merely tremor of the limbs. It is followed by coma. The respirations become irregular, jerky, less frequent, while the movements of the heart become more and more frequent and feeble.

Passive congestion of the bronchi may be added to the other pulmonary lesions, such as infarctions, metastatic abscesses, etc.; then asphyxia develops, with cyanosis, clammy sweats, and coldness of the limbs.

When the myocarditis produces merely an enfeeblement of the first sound or a blowing murmur, it can be distinguished from the murmur of pericarditis by the distinct signs of the latter which I have previously discussed. But when the period of asystole has developed it is difficult to say whether we have to deal with myocarditis, pericarditis or hydropericardium. In such an event we should be guided by the general course of the disease.

Myocarditis usually runs its course in three to eight days; the duration is longest in the cases which recover. Demme has observed one case which lasted forty-three days. The usual termination is death; recovery is exceptional.

OBSERVATION LVIII. *Stenosis of the Infundibulum of the Pulmonary Artery; Old Myocarditis.*—A. K——, aged thirty years, entered the hospital March 6, 1851. A little while before, pains had appeared in the lower limbs and left half of the chest; palpitation of the heart at the same time.

At the moment of systole the entire cardiac region, from the third to sixth intercostal spaces, was slightly retracted, as was evident from the movements of the breast, and during diastole the chest resumed its shape, so that, upon superficial examination, this diastolic movement might be mistaken for the impulse of the apex, which, on the contrary, was not perceptible. The dulness on percussion extended from the third to the sixth ribs, and from the middle of the sternum to the mammary line; the upper boundary was almost as broad as the lower. Inspiration and expiration unchanged. Along the left border of the sternum was felt a strong thrill, filling the entire period of systole; it is prolonged more to the left than to the right, and has its maximum of intensity at the insertion of the left third cartilage into the sternum. The thrill is accompanied by a loud murmur, which was feebler over the aorta and apex than over the left border of the sternum; the second sound was audible though feeble. The pulsations of the heart were irregular; radial pulse very small and weak.

The murmur referred to above was heard on the left side of the thorax posteriorly near the spine, from the spine of the scapula to its apex. It also appeared to us that a diastolic murmur was heard at the middle of the sternum, but it was so faint that we could not be positive.

Diagnosis.—Adhesion of the pericardium to the heart, to the left side of the mediastinum and the costal pleura appeared to us to be established according to the teachings of Skoda and our own experience. With regard to the vessels we assumed stenosis of the aorta and perhaps of the pulmonary artery.

The patient became dropsical and then was seized with right apoplexy. He died July 16, 1854.

Autopsy.—The heart, throughout its entire extent, is adherent to the diaphragm, the costal pleura, and left side of the mediastinum. The muscular tissue is retracted. The endocardium near the aortic valves is opaque and looks like a cicatrix. The infundibulum of the pulmonary artery is retracted evidently below the valves by a cicatrix, and the opening will merely admit the passage of a pea. The posterior mediastinum is filled with cellular masses; deep abscesses are found, extending to the bones.

Liver, spleen, and kidneys indurated and contain little blood. The right ilium is smaller than the left; the pelvis is contracted; synchondrosis of the left side, and an abscess at the level of the crest of the ilium, and osteophytes upon its inner surface (Cejka, *Prager Vierteljahrsschr.*, xlvi., p. 128. 1865).

The autopsy, made by Professor Willigk, is completed by the following details:

The heart has a flattened, rounded shape, 11 ctm. broad, 9 ctm. long; its entire surface is covered by connective tissue 1 to 2 mm. thick; the layers of the pericardium are adherent to one another. The right ventricle is more dilated than the left. The infundibulum is 8 mm. in diameter, and in its inner layers, to a depth of 5 or 6 mm., it is transformed into a thin, brownish-white cicatricial tissue, which projects, like a little tongue, to below the valves, especially upon the posterior wall, and forms a septum with an opening 1 ctm. long and 5 mm. broad. The walls of the pulmonary artery are slightly thickened, and at the emergence of the left branch are atheromatous; the valves are very thin. The extremity of the tricuspid valve is retracted and thickened and, at the base, is adherent to the cicatrix mentioned above. This valve closes the right auricle, despite the dilatation and hypertrophy of the ventricle. (Willigk, "Sections ergebnisse an d. Prager path. anat. Anstalt,"—*Vierteljahrschr. f. d. prak. Heilkunde*, vol. 51, p. 22. 1856.)

CHRONIC MYOCARDITIS.

Chronic myocarditis may follow the acute condition, but it generally develops primarily; it is confined usually to the connective tissue.

It occurs sometimes during intrauterine life, sometimes after birth.

During the latter period it may be secondary to endocarditis or pericarditis, or is due to traumatism or the presence of foreign bodies. As a general thing, however, it is produced by arthritis (gout and rheumatism), alcoholism, and even syphilis.

Two forms of myocarditis must be recognized. In one form, atrophy of the myocardium occurs at the same time as the development of interstitial tissue; the other variety is one of the forms of hypertrophy.

Atrophic sclerotic myocarditis appears to be more frequent in males than in females; it occurs chiefly about the age of thirty and affects preferably the left ventricle.

The lesion consists of proliferation of the interstitial cellular tissue and atrophy of the muscular fibres; it is found especially in the left ventricle, particularly the septum and apex. In the ventricle the lesion involves the papillary muscles, particularly toward their upper extremity. In some of these cases almost all of the muscular tissue is destroyed.

OBSERVATION LIX. *Stenosis of the Infundibulum of the Pulmonary Artery.*
—L——, aged fifty-six years, entered the hospital May 13, 1856. She presented the ordinary symptoms of advanced organic disease of the heart. Twenty years ago she had an attack of acute articular rheumatism, after which she began to suffer from dyspnœa and other disturbances of circula-

tion. About a month before admission the dyspnœa became more severe and was accompanied by a distressing cough.

Present condition : pulse 90 to 100 per minute, weak, small, irregular. Apex beat in sixth intercostal space ; very strong impulse. A very loud, rasping blowing murmur is heard, accompanying and covering the normal first sound, with the maximum of intensity at the apex.

Very marked anasarca, chiefly in lower limbs ; congestion of lungs. No trace of cyanosis ; no pulsation in the jugulars. The pulmonary congestion and the anasarca increased gradually, and the patient died May 21st.

Autopsy.—Considerable cardiac hypertrophy, affecting mainly the right side ; no lesion except hypertrophy found on left side of heart. In the infundibulum of the right ventricle was found a fibrous ring, firm and resisting, and at least 2 or 3 mm. in thickness. This ring admitted the tip of the little finger ; it is situated at least 1 ctm. from the insertion of the pulmonary valves, which are perfectly healthy. The pulmonary orifice presents the ordinary dimensions. Above this point the artery has undergone a marked dilatation ; the other parts of the heart are normal.

Sclerosis with hypertrophy will be discussed at a later period.

Syphilitic lesions may appear under two forms, either as ordinary sclerotic myocarditis, accompanied by other syphilitic lesions, or in the shape of gummy tumors. Observations have been recorded by Ricord,[1] Virchow,[2] E. Wagner,[3] Morgan,[4] R. Fowler,[5] and Lancereaux.[6]

These gummata appear in the shape of yellowish, cheesy tumors, exactly like those met with in the liver ; the tumors are situated in the interstitial tissue, and give rise to atrophy of the muscular fibres. The endocardium and pericardium project above them, and at the same time are inflamed and thickened. Under the microscope these tumors are found to be composed at the periphery of firm connective tissue, and at the centre of small cells with a nucleus, which rapidly undergo fatty degeneration. The number of these gummata in one individual is sometimes very considerable.

The diagnosis of sclerotic myocarditis is difficult ; nevertheless it may be suspected from the weakness of the heart, and particularly from the change of rhythm, unless these phenomena are due to the want of occlusion of the valves. The diagnosis becomes more probable if the patient presents the signs of old syphilis, but if the myocarditis has produced val-

---

[1] Ricord : Traité complet des maladies vénériennes. Clinique icon. de l'Hôpital des Vénériens, pl. XXIX., p. 120. 1862.

[2] Virchow : Ueber d. Natur d. Const. Syphilis, Virch. Arch., 1858.

[3] E. Wagner : Das Syphilom d. Herzens u. d. Gefaesse, Arch. d. Heilk., 34, 1866.

[4] Morgan : Gummy Tumors in the Wall of the Left Ventricle, Med. Press and Circular, November 18, 1868.

[5] Fowler : Fibroid (probably Syphilitic) Degeneration of the Heart, Trans. of Path. Soc., xix., p. 108. 1869.

[6] Lancereaux : Traité de la syphilis, 2ᵉ édit, p. 295. 1873.

vular changes the diagnosis becomes more difficult. I will discuss this in detail in the chapter on cardiac hypertrophy.

I will add one word more. Pericardial adhesions usually give rise to hypertrophy of the heart; nevertheless I once found, in a case of old pericardial adhesions, an induration of the organ with considerable atrophy, which could scarcely be interpreted otherwise than as a sclerotic or interstitial myocarditis.

The disease has a slow course, but the termination may be very rapid. Ricord's patient, after having lost a little blood from an ulceration of the penis, died very suddenly with symptoms of cardiac distress.

CHAPTER XXVIII.

PARTIAL ANEURISMS OF THE HEART.

THESE may be divided into two categories, viz., acute partial aneurisms and chronic partial aneurisms.

FALSE ANEURISMS—ACUTE PARTIAL ANEURISMS.

One of these forms of aneurism has been described under the head of endocarditis. Pelvet has shown that when the acute endocarditis is situated on the valve, the multiplication of the cells, their embryonic condition, and the disappearance of the elastic fibres deprive the valve of the power of resistance, and cause it to yield under the pressure of the blood; if the course of the endocarditis is rapid, the valve ruptures; if its course is slower, the valve is distended and forms a little sac; and if the inner wall of the sac is ruptured, the sac is soon enlarged by the pressure of the blood.

These valvular aneurisms have only been found in the left heart. Upon the aortic valves they project into the ventricle; upon the mitral valves they project into the auricle.

The history of these aneurisms is confounded with that of acute endocarditis, in which they become one of the means for the production of valvular insufficiency.

Other small acute aneurisms form in the walls of the heart, always at the ends of the left ventricle, either at the apex or in the septum near the aortic orifice. These are due to acute partial myocarditis.

CHRONIC PARTIAL ANEURISMS.

These are usually the result of Pelvet's valvular aneurisms or of the endomyocarditis of which we have spoken. The place of election of these aneurisms is the apex of the left ventricle in the anterior wall. Among ninety-seven aneurisms of the heart Pelvet[1] found the lesion eighty-five times at the apex of the left ventricle, and only three were situated in the

---

[1] Pelvet: Des anévrysmes du cœur, Thèse de Paris, 1868.

right heart. This is due undoubtedly to the fact that the energy of systole is transmitted especially to both ends of the ventricle.

The dimensions of the sac may be quite large, they may even attain the size of a fist. The walls are usually so much the thinner [1] the larger the sac; they are protected against rupture by the deposit of stratified fibrin.

The largest aneurism is that described by Berthold.[2] This aneurism, which occupied the entire right auricle, had acquired the size of a man's head; it had eroded a large part of the sternum, from the second to the fourth ribs. It was covered only by the integument, and this was so thin in parts that the blood could be seen through.

These partial aneurisms occur much more frequently in males than in females. Apart from the aneurism, the myocardium has sometimes been found normal, sometimes hypertrophied.

The following case was reported by Potain:[3]

OBSERVATION LX. *Intracardiac Aneurism; Interventricular Perforation; Aberrant Tendon; Anomalous Bruits.*—The patient, aged fifty-five years, had been in perfect health until three months ago, when mild gastric symptoms developed. Two weeks before admission to the hospital he suffered from palpitation of the heart. Upon admission a systolic thrill was felt in the middle of the præcordial region; no enlargement detected on percussion. Auscultation revealed a double murmur—a loud, systolic one, with two maxima of intensity, one at the base of the heart, the other at the apex; the second murmur was diastolic, feeble, soft, lasting throughout the entire diastole. Pulse 112 to 116 per minute.

In view of these evidences of valvular lesions, together with the absence of cardiac hypertrophy, the diagnosis remained extremely obscure. The only interpretation possible was that, the lesions being recent, sufficient time had not elapsed for the production of hypertrophy. But, on the other hand, how explain such marked alteration of the orifices without morbid phenomena of great intensity? Another explanation was therefore forced upon us, and was sought either in an ulceration of the endocardium, in some vascular rupture, in valvular destruction, or in the existence of an aberrant tendinous cord.

The autopsy showed, at the base of the right ventricle, a secondary cavity large enough to contain a nut of ordinary size, with two small openings, one anteriorly, the other at the bottom and in front of the sinus of Valsalva. This cavity is an aneurism of the cardiac walls.

A certain number of such cases have been reported; they are situated usually at the apex or base of the heart in the vicinity of the arterial orifices, and particularly toward the septum.

In our patient the rupture occurred at the anterior and middle

---

[1] Die Defecte der Scheidewand des Herzens (Wien, Braumueller, 1875); quoted by Schroetter, Krankheiten d. Herzfleisches, p. 263.

[2] Berthold: Merkwuerdiger Fall eines von der rechten Vorkammer ausgehenden Herz-aneurysmas. Toeplitz. 1859.

[3] Gazette des Hôpitaux, 8 Août, 1882.

portion of the right semilunar valve through two openings, one into the ventricle, the other into the bottom and in front of the sinus of Valsalva.

It remains to establish the relations existing between this aneurism and the anomalous bruits heard on auscultation. These lesions are rarely isolated, but are associated almost always with changes at the orifices.

I have found reports of six cases which will serve for purposes of comparison. In the majority the existence of murmurs was noted together with thrill, but it must be added that the lesion was complicated with aortic stenosis and insufficiency. Gordon speaks of a "rolling" bruit in his case. In Todd's observation a soft systolic murmur is reported, and also a loud diastolic murmur at the base; but a channel of communication, large enough to admit the little finger, existed between the ventricle and aorta. In our patient, on the contrary, this channel is very narrow and barely admits the passage of an ordinary stylet.

We are, therefore, scarcely able to determine to what we should attribute the loud murmur heard by us. This is also true of the thrill, which is very common in aneurisms of the septum, while it has not been observed in cases similar to mine. Their interpretation must be sought, therefore, in some other lesion.

If we carefully examine the posterior part of the left ventricle, we notice a slender muscular tendon which, instead of being inserted into the lower part of the mitral valve, passes upward to the septum so that it is situated in the current of blood and makes it vibrate. We may presume that the presence of this aberrant tendon has modified the abnormal bruits produced by the aneurism, *i.e.*, the diastolic murmur. In the same manner the intensity of the systolic bruit may be explained by the vibrations of the column of blood under the influence of the same tendon.

Aneurism of the septum always terminates fatally, with more or less rapidity. Its course is acute if there has been an abscess of the ventricular walls; the rupture is followed immediately by accidents, either ulcerative endocarditis with the entrance of pus and detritus into the circulation, or acute asystole from a sudden alteration of the cardiac muscle. In the former event death occurs almost immediately; in the latter within eight to twelve days.

When the disease runs a chronic course it presents from time to time accidents resulting from a slight ulceration—phenomena of asystole which last for a few days and then subside. The disease may thus run along for a number of years.

In our case the lesion does not appear to have been of very long standing, although these aneurisms sometimes remain latent for a considerable period, until the first symptoms arise as the result of exertion or some cardiac excitement. In my case it seems very probable that the disease began in the aorta, which was very atheromatous. This patient died very quickly, in the course of six hours, although the autopsy revealed nothing which would account for such a sudden termination; the only explanation to be given is the development of asystoly or an interruption to the circulation.

The diagnosis of these aneurisms is extremely obscure until they have given rise to tumors which approach the thoracic walls. In one case Skoda observed enlargement of an intercostal space.

PARTIAL ANEURISMS OF THE HEART.

These aneurisms are usually mistaken for hypertrophy of the heart, and if they assume large dimensions it is difficult to differentiate them from aortic aneurisms. They almost always terminate by embolism or rupture of the heart.[1]

---

[1] For further details, *vide* J. Skoda and Klob: Fall von ausgebroiteter Schwielenbildung im Herzen, Wien. Med. Wochen., 1856. Skzeczka: Eigenthuemliche cavernoese Entartung der Muskelsubstanz des Herzens, Virch. Arch., xi., p. 181; 1857. Jasinsky: Zur Casuistik der Herzanevrysmen, Wien. Med. H. 2. Griesinger: Anevrysma der Ventrikelscheidewand, Arch. d. Heilk., 1864. Jaccoud: Un cas d'anóvrysme ventriculo-aortique, Union Méd., 1866. E. Kock: Anóvrysme partiel de la pointe du cœur, Presse Méd., xix., 9, p. 69; 1867. Spencer Watson: Small Partial Aneurism of the Apex of the Left Ventricle, Med. Times, July, 1867. T. L. Walford: Aneurism of the Heart, Brit. Med. Journal, July 3, 1869. Henry Arnott: Aneurism of Left Ventricle with partially Ossified Walls winding around the Root of the Aorta, Trans. of Path. Soc., xix., p. 149; 1869. Peacock: True Aneurism of Apex of Left Ventricle, Trans. of Path. Soc., xxi., p. 118; 1871. Simon: Zur Entstehung d. Herzaneurysma, Berl. Kl. Wochenschr., 45, 1872.

CHAPTER XXIX.

THE COURSE OF DISEASES OF THE HEART.

When a patient is affected with an acute disease of the heart, whether pericarditis, endocarditis, or myocarditis, he does not always succumb, but, on the contrary, usually recovers. But this first affection is often merely the first manifestation of a constitutional affection, the later attacks of which follow the ordinary law, *i.e.*, they undergo resolution less and less, and the return to the normal condition becomes more and more difficult.

The intervals of health between the successive manifestations of the diathesis may be very great and, indeed, a second attack may never occur. However this may be, a first affection of the heart or several successive attacks usually leave behind in the tissues either cicatrices or lesions which impede function.

If the individual is young and the lesion is not considerable, the other organs of circulation will lend their aid, and the patient will enjoy a satisfactory condition of health.

This period of tolerance of the organism to cardiac lesions varies according to several factors. It is greater in valvular insufficiency than in stenosis, and greater in lesions of the aortic orifice than in those of the mitral and pulmonary orifices. It is greater in circumscribed lesions of the pericardium than in those of the endocardium, and especially of the myocardium. It is also greater in young people and adults than in old people.

The period of tolerance may last for a very long time. Individuals suffering from an affection of the left heart, either Corrigan's disease or mitral induration, have been known to present a period of tolerance of ten, fifteen, twenty years, or even longer, before the secondary cardiac lesions manifested themselves with an intensity which threatened life.

Accordingly, the period which intervenes between the beginning of a cardiac affection and death, either from syncope or asystoly, is extremely variable—from a few days to twenty-five years.

The secondary results of heart disease differ according to the primary lesion, and while Corrigan's disease leads to sudden death, stenosis of the pulmonary artery leads to phthisis. Sclerosis of the mitral valve produces passive congestion, dropsy, and asystoly.

But it must be remembered that, whatever the lesion may be, the heart must summon the aid of the other organs of the circulation; and, at a later period, each of these organs being involved in its turn, certain secondary affections result which are common to all cardiac lesions. It has appeared to me more natural to discuss separately each of these complications.

Dyspnœa on Exertion.

When a patient has been affected with facial paralysis with deflection of the face, and he is on the way to recovery, a moment arrives at which the tonicity of the diseased muscle is restored, so that there is no deflection of the face during repose. But if the patient speaks or laughs the muscles of the healthy side distort the features, as the energy of contraction of the paralyzed muscles has not been completely restored. Then the affected muscles gradually regain their power until complete recovery occurs.

A patient cured of an acute affection of the heart with persistent lesions resembles somewhat the convalescent from facial paralysis. If he does some light work or walks on a level surface, all is well; but as soon as the path begins to ascend, the effort which the heart must perform produces fatigue and anxiety begins, leading to distressing efforts at respiration—in a word, to dyspnœa. It seems as if the organism endeavors to compensate by frequent respirations for an oxygenation which has become insufficient because the heart does not propel a sufficient wave of blood.

The more the path rises and the more the mechanical labor increases, the more marked the dyspnœa becomes. The dyspnœa will appear whatever the character of the exercise may be, whether it requires force or agility.

This anxiety and dyspnœa are common to all thoracic affections, whether they affect the respiratory or circulatory organs. But this difference between repose and action is especially marked in diseases of the heart and great vessels.

Dyspnœa on effort exists even during the most complete tolerance, and in such cases we hear the patients say: "I do not suffer while I remain quiet, but the least exertion or excitement reminds me that I have heart disease."

This form of dyspnœa must be distinguished from other varieties.

The dyspnœa of nervous asthma differs by its sudden beginning during repose, and especially at night. The onset of the affection, instead of being gradual and progressive, passes at once to its greatest intensity. The respiratory efforts made by the patient have been well described by Professor G. Sée.[1] Inspiration is short and expiration is much more prolonged [2]

---

[1] G. Sée: Art. "Asthme," du Dictionnaire des sciences médicales, t. III., 1868.

[2] Normal expiration is always longer than inspiration, but during auscultation the sonorous period of expiration is shorter than inspiration. The end of expiration is silent in the normal condition.

than in the physiological condition, with râles which indicate that contraction of the bronchi accompanies the contraction of the diaphragm. Furthermore, the asphyxia and expectoration show that we have to deal with respiratory, not with cardiac dyspnœa.

There is another form of asthma, known as catarrhal asthma, in which the attacks of dyspnœa do not develop except on the occurrence of bronchitis. Its onset is gradual and its intensity is proportionate to the extent of the bronchial lesion.

The dyspnœa of emphysema is much more difficult to distinguish from dyspnœa on effort than any other form. I do not now refer to the variety which accompanies bronchitis and constitutes catarrhal asthma, but to that form observed in emphysematous individuals during repose.

The emphysematous patient who has lost his pulmonary contractility is really in a condition of exaggerated inspiration, even during repose. At each inspiration the diaphragm and ribs are brought into play and cause the greatest possible amount of air to enter the bronchi; it would seem as if the quantity of air which enters is insufficient, and the patient is obliged to sigh from time to time in order to introduce a certain amount of supplementary air into the chest.

This necessity will be still greater if the patient is in confined air or upon a mountain, as the air then contains a smaller amount of oxygen. The slightest catarrh, which otherwise would be insufficient to produce any disturbance, reduces his respiratory field and gives rise to dyspnœa. In confirmed emphysema, indeed, nasal or bronchial catarrh recurs continually.

The dyspnœa of emphysema is distinguished from that on effort by the fact that the former exists during repose, and that if it is more marked on effort it is not accompanied, like the latter, by palpitation and changes in the rhythm of the pulse.

A dyspnœa very similar to that just described is the variety observed in fat people, in whom the diminution in the amplitude of the thorax and of the respiratory movements produces the same result as emphysema.

The dyspnœa of chloro-anæmia resembles still more closely the dyspnœa on effort of true cardiac disease, and so much the more as the former is really a cardiac dyspnœa with palpitation. The diagnosis will then be made particularly by the examination of the heart, which enables us to recognize the signs of a valvular affection of the aorta, the mitral, or tricuspid. The differential diagnosis between anæmia and stenosis of the pulmonary artery will be made by a careful examination of the symptoms, as shown in the chapter on anæmia.

Hysterical dyspnœa will be distinguished by the absence of cardiac lesions and its usually irregular course; furthermore, by the other habitual or extraordinary phenomena of hysteria.

The dyspnœa accompanied by flatulent dyspepsia with intercostal neuralgia is extremely common, and usually regarded by the patients as a disease of the heart. As they have a more vivid consciousness of the cardiac disturbances than of the dyspeptic disorders to which the former are due, they consult the physician on account of a cardiac affection, and assert positively that they are not dyspeptic. But it will be found that these cardiac disturbances occur only after eating, and that digestion is the usual cause; that they are accompanied by yawning, eructations, and a tendency to sleep; that the digestive disturbances are more marked in the evening; that they often begin in the middle of the night with a pain in the stomach, often with nausea, sometimes even with vomiting. Instead of feeling refreshed by their night's sleep, the patients feel more fatigued in the morning than upon lying down. Finally, they present the ordinary causes of dyspepsia, viz., imperfect mastication and salivation, etc. Under appropriate treatment the cardiac disturbance is soon relieved.

The Symptomatic Palpitations of Heart Disease.

Associated with the dyspnœa on effort which occurs during the period of tolerance are the palpitations, the frequent occurrence of which indicates incomplete compensation.

The cardiac rhythm may be disturbed in several ways. It may be increased in frequency, either temporarily or permanently.

It may present from time to time an absence of a pulsation both at the pulse and at the heart; this constitutes a true intermission. Or the intermissions may exist at the pulse and not at the heart; these constitute false intermissions, which indicate a lesion of the muscular fibres; then follows arhythmia, which is merely a series of false intermissions with disturbance of rhythm. Finally, arrest of the heart's action or syncope constitutes the last term of the series.

We will first consider palpitations, the characteristics of which have been well described by G. Sée,[1] and may be recapitulated as follows:

1. The pulsations are more frequent; this is the most characteristic feature of palpitation.

2. The cardiac pulsations are often more rapid, *i.e.*, the duration of each pulsation is less than in the normal condition.

3. The cardiac contractions seem to be more intense, and they are often perceived more readily; but it has not been proven that they are, in reality, more vigorous.

4. The pulsations are accompanied usually by a modification in the timbre of the bruits which is perceptible on auscultation.

5. Palpitation generally gives rise to peculiar sensations felt by the patient, but at other times the subjective sensations remain absent.

---

[1] G. Sée: Diagnostic et traitement des maladies du cœur, p. 129.

6. Finally, the palpitations may coincide with disturbances in the rhythm, or even with intermissions of the heart; but the latter event is more rare, as the pulse is usually slow in such cases. However this may be, these phenomena are independent of the palpitations, and simply complicate the latter.

Sée adds that in such cases there is an increase in the activity of the heart, but not an increase of action. This is not entirely true, as there are violent palpitations which sometimes accompany valvular lesions, and particularly mitral insufficiency in young subjects, and which certainly possess greater energy and force than in the normal condition. If there were but one factor, the frequency of the pulsations, this would be true, since their increased frequency gives rise to a diminution in the amplitude of the oscillations of the arterial tension.[1] On the other hand, when there is an obstruction to the arterial circulation, the tension increases.[2]

According to Marey, the causes which increase the frequency of the pulsations of the heart are, first, the absence of resistance. The heart beats so much more quickly the less difficulty it encounters in emptying itself.[3] Muscular exercise is a cause of acceleration of the pulse in the normal condition,[4] and this increase of rapidity is much more marked when the heart is diseased.

True Intermittence.

In true intermissions, not alone the arterial pulsation but also the cardiac impulse are absent. These have long been recognized as compatible with health. I know of cases which have lasted twenty, thirty, and even forty years, although there was no cardiac or other disease.

Lasègue, who has made a special study of these intermissions, is less reassured with regard to them,[5] though he does not confound cardiac or true intermittences with intermittence of the pulse or false intermittence.

"When we say that cardiac intermissions are present, it is implicitly admitted that the organ performs its function in a regular manner, apart from the suspensions which occur from time to time and interrupt the series more or less periodically." The heart beats normally for a more or less considerable number of pulsations, then it is arrested for a period which is measured with difficulty, and again resumes its course. It may happen, by chance, that the suspensions appear to follow a sort of rhythm, and are repeated every four, five, six, or eight pulsations, but this periodicity is never constant. It ceases at the end of a few minutes, and even during its short duration it is not as rigorous as we might believe from

---

[1] Marey : La circulation du sang à l'état physiol. et dans les maladies, p. 187. 1881.
[2] Ibid., loc. cit., p. 190.   [3] Ibid., loc. cit., p. 334.
[4] Ibid., loc. cit., p. 342.
[5] Lasègue : Des intermittences cardiaques, Arch. Gén. de Méd., t. II., p. 641. 1872.

the account given by the patient. Immediately after the intermission the first cardiac contraction habitually appears more vigorous; sometimes it is separated from the next contraction by an interval which is shorter than that intervening between the following pulsations. As a rule, this increased impulse is imaginary.

Lasègue insists upon the fact that true intermissions are not produced by the factors which ordinarily cause palpitation, such as exercise or hysteria.

"The patients are conscious of these cardiac intermissions—they are notified by a peculiarly annoying sensation, and do not discover them accidentally upon feeling the pulse. The sensations experienced by them vary in character. The less observant, or perhaps the less sensitive, merely feel the lively impulse which follows the prolonged silence; they complain of palpitation returning at irregular intervals. In a somewhat severer degree, the cardiac repose is accompanied by a sort of anxiety which is necessarily very fleeting, but the return of pulsations characterized by an exaggerated impulse terminates the malaise, perhaps restores the patient's feeling of security." In a third category the patients describe their impressions in detail. They experience the two sensations mentioned above, viz., anxiety during the suspension and the shock upon the return of the pulsations; but in addition they experience a very marked præcordial or rather epigastric sensation. In this they differ from patients suffering from intermittence of the pulse, or false intermittence; the latter, suffering from a more or less grave organic affection, have no consciousness of their intermissions, and only perceive them upon feeling the pulse. Unlike the latter form, true intermittence is not a sign of heart disease.

In certain individuals it indicates slight atheroma, and is then present almost constantly. Temporary true intermittence is very common in the course of temporary acute affections—for example, slight anginas.

Finally, it may appear in certain diseases, which Lasègue has described as follows: These are the forms of premonitory cachectic conditions which indicate that the patient is in the stage of incubation of some more or less grave disease of the organs of nutrition—for example, an apparently spontaneous phlegmon, a general bronchitis, etc. These intermissions disappear at the close of the disease during convalescence.

Intermittence of the Pulse, or False Intermittence.

True intermittence affects the rhythm only in one respect, viz., periodicity. False intermittence is more complex. The pulse is irregular, unequal; of variable frequency, sometimes rapid, sometimes slow; and when very much changed, it becomes confused by its frequency and imperceptible by its smallness.

Intermittence of the radial or carotid pulse, when the heart has not ceased to beat, is a sign of organic disease of the heart. It indicates that the cardiac systole has been unable to propel the blood to the end of the arteries.

The insufficiency of the systole in such cases may be due to two different causes: either the mitral valve is insufficient and the effort of the ventricle is exhausted in propelling the blood in a direction contrary to its normal course, or the valves are intact and the contraction of the myocardium has been unable to furnish the arterial pulsation.

Finally, other causes may give rise to false intermittence. If the pulse is very rapid it may happen that the systole has begun before the previous diastole has had time to fill the ventricle; or perhaps a mitral stenosis has not allowed a sufficient quantity of blood to enter the ventricle.

It may be asked whether, in such cases, the want of energy of the myocardium is not due to the fact that only one ventricle contracts while the other remains at rest; but this is not so. François Franck[1] has shown that in false intermissions the force exerted by the left ventricle may be less than that of the right ventricle, but one does not contract without the other.

In fine, the causes of intermittence of the pulse are variable:
1. Mitral insufficiency (systole aborted by mitral reflux).
2. Mitral stenosis (systole aborted by incomplete filling of the ventricle).
3. Alterations in the myocardium (systole aborted by want of energy).
4. Agitation of the heart (systole aborted by premature contraction and incomplete filling).
5. The action of digitalis and chloral (systole aborted by toxic action).

To these may be added the other cardiac poisons. These are the results furnished by physiological analysis; clinically, matters are more simple. Mitral insufficiency in the young, if the myocardium is still healthy and the other organs of circulation are vigorous, does not give rise to intermittence of the pulse. The energy of the contraction of the myocardium supplements the deficiency. It may be said in practice that if true intermissions are not produced by the drugs mentioned (digitalis, chloral), a condition of degeneration of the myocardium is indicated.

So long as the circulatory organs are sufficient the intermissions are rare; but if the myocardium is already changed, if the diseased heart must overcome fresh obstacles due to secondary affections, if the patient is suffering from œdema of the limbs, i.e., from paralysis of the veins, interstitial nephritis, or other cause of increase in the aortic tension, the aborted systoles become more frequent and arhythmia results. As we have previously shown, the rhythm of the heart does not depend upon the nervous system, but is a property of the myocardium, and consequently arhythmia is evidence of a lesion of the myocardium.

---

[1] François Franck, Acad. des Sciences, 16 Avril, 1877.

Intermittence of the pulse, then, indicates an alteration of the myocardium. It shows that the powers of resistance of the patient are failing, that the disease has really become organic.

PAINFUL AFFECTIONS OF THE HEART.

NEURALGIAS AND ANGINA PECTORIS.

Bouley thought that all painful affections which are symptomatic of heart disease should be classed in one group, and the term angina pectoris reserved for those attacks the severity of which threatened life itself.

We will first discuss the latter form. The patient is seized suddenly with the most intense pain situated beneath the sternum, accompanied by considerable anxiety, a sort of vertigo which compels the patient to seek support, and a feeling of cardiac syncope which makes him believe that death is imminent.

The pain is not confined to the cardiac region, but radiates in different directions, into the left upper limb from the shoulder to the elbow, accompanied by a sensation of constriction, cold, and formication. At other times the pain radiates into the neck and the temporo-maxillary articulation. In rarer cases the pain may radiate into more distant parts—for example, into the right side of the chest, the intercostal spaces, the right breast. Exceptionally, the pain has been felt in the hypogastric region, testicle, diaphragm.

It should be noted that during the seizure the patient retains consciousness perfectly ; during this period the face is pale, the features contracted as in peritonitis, the body cold.

It appears to me to be questionable whether the pains are propagated along the spinal nerves of the brachial, dorsal, lumbar plexuses, etc. I think that we have to deal, in these cases, with propagation along those branches of the sympathetic which follow the vessels. There has seemed to me to be a remarkable analogy between migraine and angina pectoris.

In some cases angina pectoris appears to exist independently of diseases of the heart, and I have observed two remarkable cases of this kind.

One of these patients, a talented painter, has had during the past twenty-five years more than a dozen attacks of angina pectoris, and it is absolutely impossible to detect in him any sign of heart disease. This patient also suffers from herpetic lichen, and his brother and sister are asthmatic. The other patient is one of our most prominent sculptors. I treated him during an attack of angina pectoris twenty years ago, and since that time he has had four or five attacks. He presents nothing but slight emphysema, and perhaps a slight dilatation of the right side of the heart. From time to time he suffers from attacks of malarial fever contracted in Rome.

But it occurs more frequently in cardiac affections and atheroma, particularly in atheroma of the origin of the aorta and the coronary arteries. It must not be forgotten, however, that at a certain age atheroma of the arteries, especially of the coronaries, is of constant occurrence, although angina pectoris is not observed.

I do not deny the frequency of atheroma in patients who have died of angina pectoris, but if I consider the large number of patients at Bicêtre and la Salpêtrière suffering from atheroma and free from angina, I cannot convince myself that the two diseases bear an undeniable causal relation to one another.

In my opinion, we may conclude that the occurrence of angina pectoris in patients suffering from organic heart disease may act as the cause of sudden death.

There are also other sides to the question. Angina pectoris has predisposing causes, and in a patient suffering from organic heart disease, indigestion, perhaps, may be the determining cause of the attack, and consequently of death.

It is a curious fact that angina pectoris is not the habitual method of sudden death, even in affections of the aorta—for example, in Corrigan's disease.

Another point remains to be considered. It is known that tobacco-poisoning is a frequent cause of angina pectoris. In these cases does the tobacco first produce atheroma of the vessels, particularly of the coronary arteries?

If spasm of the vessels, produced by the action of tobacco upon the vasomotors, can produce angina pectoris, a previous cardiac lesion is not necessary to the development of angina. But if the individual who contracts angina pectoris is suffering from a cardiac affection, and more particularly from one which exposes him to the danger of sudden death, he is subject to the greatest possible danger from the occurrence of a paroxysm of the former disease.

The onset of angina pectoris is always sudden, but the attack continues only a little while; at times it may last several hours. It does not always kill during the first attack, if the heart is not already affected. I know an inveterate smoker, a subject to angina pectoris, who was often taken with attacks at night. His vessels and myocardium were healthy, and he resisted these attacks for more than ten years. He died finally of cerebral apoplexy.

FALSE CARDIODYNIA.

Professor Sée, struck by the frequency of false cardiodynia concerning which he was consulted, has devoted a chapter to this subject. The elements of diagnosis will be found in the chapter on nervous and symptomatic palpitation.

Cardiodynia is merely an intercostal or diaphragmatic neuralgia or myodynia, symptomatic of various nervous affections, and very often of flatulent dyspepsia.

SYNCOPE.

Fatal syncope in diseases of the heart will be discussed at a later period. We now refer to lipothymia, weakness, and syncope occurring during the course of heart disease.

Syncope is by no means an undoubted indication of a cardiac affection; it may be merely the result of an emotion, of pain, especially that which is looked forward to, as in surgical operations; anæmia, particularly rapid anæmia from hemorrhage, is a cause of syncope.

There is a syncopal form of dyspepsia, and another variety due to heat (mal de théâtre). Syncope is very frequent in convalescents who attempt to stand up; it is often produced by hysteria and nervousness.

But apart from these causes of nervous origin which may give rise to stimulation of the pneumogastric by reflex action, and, more often still, to a lack of excitation through the nerves of the sympathetic, I will repeat what I said concerning angina pectoris: To be subject to syncope when the organs of circulation are healthy need not cause fear, but it may prove fatal if the individual is suffering from an affection of the heart or aorta.

Of all lesions of the heart, those which predispose most to syncope are the lesions accompanied by fatty degeneration of the myocardium, more particularly those which induce atrophy, like the sclerotic myocarditis accompanying mitral stenosis. Individuals suffering from Corrigan's disease often die from syncope if the myocardium is degenerated, but we do not find them subject to lipothymia. The premonitory symptom of syncope is arhythmia, and arhythmia always presupposes a lesion of the myocardium.

CHAPTER XXX.

HYPERTROPHY OF THE HEART.

WEIGHT OF THE HEART.

THE following figures show the weight of the heart as furnished by different authors:

Kerkruig	218 grammes.
Meckel	312 "
Lobstein	281 to 312 grammes.
Cruveilhier	250 to 300 "
Bouillaud	250 to 281 "

These figures give merely a gross approximation, as they do not take sex or stature into consideration.

Clendenning gives the average measurements according to age, sex, and stature:

Age.	Males.	Females.
15 to 30 years	264 grammes.	260 grammes.
30 to 50 years	272 "	272 "
50 to 70 years	278 "	276 "
After 70 years	312 "	287 "

Relation of the Volume of the Heart to the Length of the Body (BENEKE).

Age.	Absolute volume of the heart in cubic centimetres.		Length of the body in centimetres.		Relation of the volume of heart in c.c. to 100 ctm. of length of body.	
	Males.	Females.	Males.	Females.	Males.	Females.
18 years	202.4	174.2	161.3	159.0	122.0	110.0
19 years	.....	202.5	.....	164.6	.....	119.7
21 years	258.3	221.0	172.0	156.7	150.2	144.2
22 to 25 years	234.0	213.1	170.9	159.2	137.3	135.1
26 to 30 years	254.7	220.9	170.9	156.5	157.7	139.1
30 to 40 years	275.2	212.1	169.9	156.4	164.4	135.4
40 to 50 years	288.8	239.8	168.5	157.6	172.3	152.2
50 to 60 years	277.6	229.9	170.5	159.0	167.6	144.9

HYPERTROPHY OF THE HEART.

CAPACITY OF THE HEART.—Beneke has made numerous investigations, in which he has taken into consideration the stature, sex, and manner of death.

AGE.	MALES.			FEMALES.		
	Volume of the heart in cubic centimetres.	Length of the body in centimetres.	Number of observations.	Volume of the heart in cubic centimetres.	Length of the body in centimetres.	Number of observations.
Still-born	22.3	50.0	6	....	....	1
11 days	....	50.4	4	21.0	50.2	4
11 days to 3 months	25.8	53.7	17	24.7	55.4	14
4 months to 1 year	33.6	65.5	10	32.2	62.3	13
2 years	44.3	72.7	11	43.4	75.4	9
3 years	50.2	81.2	12	51.8	83.5	9
4 years	60.0	93.5	4	....	....	2
5 years	....	....	1	68.1	96.2	4
6 years	75.1	103.7	5	....	109.6	3
7 years	99.0	116.1	6	77.0	104.7	2
8 to 9 years	....	121.4	6	....	....	3
9 to 11 years	111.5	122.4	8	....	....	2
11 to 13 years	....	137.4	5	....	139.5	4
13 to 15 years	....	143.5	4	....	136.9	5
15 years	130.0	144.3	7	17.7	147.3	3
16 years	177.3	157.0	9	....	153.6	3
17 years	....	156.6	3	165.0	152.5	5
18 years	202.4	161.3	7	174.2	159.0	5
19 years	....	166.8	5	202.5	164.6	3
20 years	259.7	168.6	11	....	156.7	4
21 years	258.3	172.0	3	221.0	156.7	5
20 to 25 years	234.0	170.9	39	213.1	159.2	16
26 to 30 years	254.7	170.9	28	220.9	156.5	18
30 to 40 years	275.2	169.9	43	212.1	156.4	33
40 to 50 years	288.8	168.5	53	239.8	157.6	14
50 to 60 years	277.6	170.5	39	229.9	159.0	25
60 to 70 years	257.9	172.0	29	262.6	158.7	11
70 to 80 years	292.0	167.8	16	....	....	4

According to Bizot, the development of the heart is not complete until the twenty-ninth year. After this period it stops growing in length but not in volume, and the latter increases gradually until death.

This is confirmed by the figures furnished by Clendenning, and which show that the weight of the heart arrives at the average at the age of thirty years, remains at the same figure until the age of sixty, and then increases until the end of life.

COMPARISON OF THE RIGHT AND LEFT HEART.—Bizot has shown that the average length and capacity of the right ventricle are much greater than those of the left ventricle. He has also shown that the increase of capacity in old age is real, and that it affects both cavities. The right ventricle, accordingly, is much larger than the left at all ages.

These results have been confirmed by the recent investigations of Ducastel.

Comparison of the Weight and Capacity of the Ventricles in Men not suffering from Heart Disease.

Age.	Total weight.	Weight of left ventricle.	Weight of right ventricle.	Capacity of left ventricle.	Capacity of right ventricle.
16 years	270	172	68	86	53
20 years	185	115	50	38	45
21 years	210	110	48	25	34
21 years	245	150	55	28	56
23 years	275	185	70	140	190
30 years	280	160	75	57	105
35 years	260	150	60	80	90
35 years	310	160	75	83	106
40 years	310	155	82	33	71
43 years	380	200	100	140	147
45 years	205	120	45	25	31
47 years	345	225	70	95	107
48 years	350	165	100	38	72
49 years	265	145	70	29	95
50 years	580	375	130	150	210
52 years	340	190	68	12	18
58 years	380	190	85	55	105
65 years	310	180	75	30	57
68 years	255	135	60	19	24
68 years	300	180	75	153	198
68 years	390	200	80	23	26
73 years	360	180	75	121	155
74 years	290	155	60	68	52

Ducastel's tables of the weight of the heart in individuals who have succumbed to diseases other than those of the heart, give the following results.

The weight of the left ventricle is always greater than that of the right ventricle, and gives the following average:

	Total weight.	Left ventricle.	Right ventricle.
Males	308 grammes.	173 grammes.	72 grammes.
Females	276 "	150 "	62 "

Among 23 males, in 21 the capacity of the right ventricle was greater than that of the left; among 20 females, in 20 the capacity of the right ventricle was greater than that of the left; in 1 the capacity of both ventricles was the same.

The following table shows the average proportions:

	Left ventricle.	Right ventricle.
Males	65 c.c.	83 c.c.
Females	59 "	74 "

Robin and Hiffelsheim have also found that the capacity of the right auricle is greater than that of the left auricle.

COMPARISON OF THE CAPACITY OF THE AURICLE AND VENTRICLE.—Legallois thinks that the auricle is smaller than the ventricle, and Bouillaud regards them as equal. Robin and Hiffelsheim agree with Legallois, but they think that the difference increases with age, and is especially marked on the left side.

THICKNESS OF THE WALLS OF THE HEART.

Average Thickness of the Walls of the Heart (in millimetres).

REGIONS.	Bouillaud.	Bizot.		Peacock.
		Males.	Females.	
Left ventricle { base	.....	10.12	9.43	12.0
middle	16	11.73	10.35	13.0
apex	.....	8.05	7.36	5.75
Septum { base	25–30	.....	.....	.....
middle	.....	10–12	10–12	13.0
apex	.....	.....	.....	.....
Right ventricle { base	.....	4.14	3.91	.....
middle	.....	3.0	2.76	.....
apex	.....	2.30	2.30	.....
Left auricle	2–3	.....	.....	.....
Right auricle	2–3	.....	.....	.....

All these investigations furnish nothing absolute, but they enable us to recognize in a certain measure any increase in the heart.

The average weight of the heart in the adult being 300 grms., I can be positive that hypertrophy existed if the weight exceeds this figure to any considerable extent. The volume furnishes perhaps a less certain means of judgment, because the heart may be arrested either in systole or diastole. Cruveilhier states that the heart is arrested in systole and has a much smaller volume in all those who die a violent death or rapidly bleed to death. He adds that if the fingers are introduced into the hearts of such individuals, they may be dilated with surprising facility. In fact, the manner of death and the cadaveric rigidity notably affect the results.

The thickness of the walls may also furnish sources of error. The heart of a child is found much more retracted than that of an adult; this is also true of those who have suffered a violent death or died of hemorrhage. If this condition, accordingly, is not accompanied by an increase in the total weight, it does not signify hypertrophy.

Hypertrophy may exist either with or without dilatation of the cavities. May dilatation exist without hypertrophy? For my own part, I believe that hypertrophy is always present in such cases, and this also follows from the figures furnished by Ducastel, according to which emphysema is accompanied by an increase in the weight of the right ventricle. However, further investigations are necessary on this point.

Hypertrophy may exist with apparent preservation of the normal size of the cavities, or with an increase (excentric hypertrophy) or apparent diminution of the cavities (concentric hypertrophy).

As a rule, when the hypertrophy is marked the heart is increased in volume and changed in shape. It becomes less pyramidal and more conical. The increase in volume entails displacement of all the adjacent organs. The diaphragm gradually yields under the weight of the heart, the apex descends to the sixth, exceptionally to the seventh intercostal space. Parrot has observed it in the eighth intercostal space. The right angle of the heart descends to the level of the insertion of the right sixth or seventh costal cartilage. Laterally it displaces the lungs. The apex is removed farther and farther from the median line, as the insertion of the vena cava does not allow the heart to move toward the right. The apex may touch the ribs finally at the axillary line.

After the apex has been carried a certain distance to the left, it must also be carried backward. If, for example, the apex has corresponded to the axillary line, the heart must undergo forcible torsion around a pivot formed by the two vena cavae, the apex being situated then in a plane posterior to that to which it corresponds normally. Under these conditions the right auricle tends more and more to approach the thoracic walls. If the hypertrophy affects the left ventricle alone, the right heart appears like a mere appendix of this ventricle. If, on the contrary, the hypertrophy is confined to the right ventricle, the apex is less prominent, the heart assumes somewhat the shape of an alms-bag, and upon cutting both ventricles transversely, the heart is found to be more or less symmetrical.

There is no doubt of the existence of hypertrophy if the heart weighs 500 or even 400 grms. It is exceptional to find the weight greater than 900 grms., but 1,000, 1,500, and even 1,700 grms. have been reported.

The walls may be regarded as hypertrophic if the thickness of the left ventricle exceeds 15 mm., that of the right ventricle 7 mm. The left ventricle and the septum have been observed to attain a thickness of 30 and even 40 mm. Hypertrophy of the auricles begins at 4 mm.

The columnae carneae, especially those of the first order, also increase in size in hypertrophy of the heart.

According to Hyrtl, Koelliker, Rokibansky, and others, the diameter of the muscular fibres of the heart is increased in hypertrophy. Others believe that new muscular fibres develop, but this view is entertained by few. According to Weismann and Zielonko, the fibres increase not alone in breadth but also in length. Rindfleisch, in proof of the formation of new fibres, makes the statement that certain cells present two or more nuclei.

Finally, according to a broader hypothesis, hypertrophy of the heart is produced :
1. By the growth of the old fibres.
2. By the formation of new fibres.
3. By the development of interstitial connective tissue.
4. By an abundance of adipose tissue.
5. By the development of vessels.
6. Finally, by increase of the nerves and ganglia.

Letulle [1] has shown that cardiac hypertrophy, especially that which accompanies valvular lesions, begins at an early period. It affects the fibres in an irregular manner, i.e., the hyperplasia is disseminated, but limited to the region subjected to increased work—the region being determined by the site of the obstacle to circulation. In the midst of changed fibres are found healthy ones, although Letulle has been unable to discover the law of their localization. According to Letulle it is also characteristic of the hypertrophic fibres that, instead of being polygonal like the normal fibres, they have a tendency to become cylindrical or even ovoid.

Little by little the muscular fibres compress the interstitial connective tissue and the vessels. It follows that the blood-supply to the muscles diminishes in proportion as the hypertrophy increases; hence the occurrence of granulo-fatty degeneration.

Obesity of the heart must not be mistaken for fatty degeneration. The former is characterized by the presence of a few shining granules around the nuclei of the muscles; in granulo-fatty degeneration the fibres lose their longitudinal and transverse striation, and they become pale and brittle. The nuclei are more resisting, and in advanced fatty degeneration they may be stained by picrocarmine and then appear isolated in the fatty tissue.

Renaut and Landouzy [2] have observed another lesion, viz., a sort of rupture of the fibres at the level of the striæ of Eberth. In addition, the lesions of sclerotic myocarditis are also observed, but they are presented, according to Letulle, under two forms. Sometimes the hyperplasia of the interstitial connective tissue separates the muscular fibres individually, or collects them in groups of three or four. In the second form the fibres, not being very distinct from one another, present the appearance at first sight of a hypertrophic fibre, although in reality the fibres are atrophied.

---

[1] Letulle : Recherches sur les hypertrophies secondaires. Thèse de Paris, 1879.
[2] Renaut and Landouzy, Soc. de biologie, 1877.

We have seen, in describing the myocardium, that the muscular fibres are destitute of sarcolemma. They are united by loose connective tissue, which contains the vessels and nerves.

Under the pericardium this connective tissue is loose; under the endocardium it is so close that it presents no meshes. The stroma of the heart is connected almost directly with the primitive fibres, and if we remember that the connective tissue forms the origin of the lymphatics, it follows that the muscular fibres are directly surrounded by the lymphatic meshwork. In hypertrophy this tissue undergoes the changes mentioned in the chapter on sclerotic myocarditis.

At a later period the interstitial tissue itself undergoes fatty degeneration, especially at the surface beneath the pericardium.

Letulle describes two forms of this cirrhosis, the perivascular and perifascicular. The former is observed particularly around the arteries of medium and small calibre. Hence the patch of cirrhosis radiates in an irregular manner, separating the secondary bundles in the vicinity and pressing in between a certain number of primary bundles which are isolated and made to atrophy.

Fascicular cirrhosis is characterized, in the beginning, by a narrow band of perifascicular tissue, showing embryonic nuclei. In proportion as the cirrhosis advances the embryonic cells diminish in number and size; as the fibrous bands increase in thickness, the muscular fibres which they surround soon undergo atrophy.

Hypertrophy of certain muscular fibres is associated, accordingly, with cellular hyperplasia and muscular atrophy in other parts.

As the hypertrophy progresses chronic arteritis, atheroma, and calcareous degeneration are found in the vessels of the heart.

CHAPTER XXXI.

HYPERTROPHY OF THE HEART.—(*Continued.*)

The first symptom which attracts attention in a patient suffering from hypertrophy of the heart is the elevation of the ribs by the præcordial impulse. This elevation is particularly well marked in the region of the apex, but extends often to the entire cardiac region; the sternum itself is sometimes elevated. This symptom is better marked when the patient stops breathing; it may be absent either because the lung covers a large part of the heart, or because the fatty degeneration of the myocardium considerably diminishes the heart's energy. On the other hand, the phenomenon of impulsion may be produced by nervous palpitations.

In addition, a notable prominence of the ribs and sternum is observed in the region corresponding to the hypertrophic heart, particularly in very marked and chronic cases. This is produced by the projection of the ribs and even of the sternum, and also by the enlargement of the intercostal spaces. But before this projection is interpreted as the result of cardiac hypertrophy, we must be satisfied that it is not a congenital deformity or the effect of curvature of the spine.

The hand, when applied to the thorax, detects a vigorous impulse, which is soon increased by the emotion produced in the patient by the examination of his heart. The impulse is felt either at the apex alone or more frequently over the entire surface corresponding to the ventricle.

Examination shows that the apex is lowered to the sixth, seventh, or even the eighth intercostal space; it is found most frequently in the sixth space. On the other hand, on account of the fixed and immovable attachment of the inferior vena cava, the heart is carried altogether to the left, the apex being situated 11, 12, even 16 ctm. from the median line. Next the transverse axis of the heart begins to turn around the vena cava as a pivot, and is carried backward more and more. The apex is then covered more and more by the lung, and when the organ is very much enlarged, the apex beats less forcibly against the thorax than at the beginning of hypertrophy.

If the hypertrophy is not confined to the left heart, the right or hepatic angle of the cardiac triangle is also lowered, and corresponds to the insertion of the right sixth or seventh cartilage. If dilatation of the right auricle occurs at the same time, the right border of the heart is removed to a distance of 3 or even 4 ctm. from the median line.

In order to obtain the measurement of the heart in the manner which I have previously described, prolonged percussion is necessary, and this is often impossible on account of the anxiety of the patient.

Auscultation often shows that the intensity of the sounds has increased, and at times they may be heard at a distance. In certain cases in which Laennec and others have heard the pulsations at a distance, the stomach was filled with gas and acted as a sounding-board. In other cases the impulse of the heart against the fluids and gases contained in the stomach has produced a peculiar bruit, called auriculo-metallic. In a case of pyopneumothorax under my observation, each beat of the heart produced a succussion sound.

The area in which the heart sounds are heard is very much increased upon the anterior surface of the chest; in such cases, also, the heart sounds are readily heard posteriorly.

In certain cases the transmission of the heart sounds occurs not alone through the left lung, but they may be heard behind the right lung, although no lesion of the lungs is present to render them a better conductor of sound.

Auscultation usually discloses changes in the timbre of the bruits; they are deeper or more sonorous, but, as a rule, no blowing murmur is present if the arteries are healthy.

What are the conditions which give rise to a blowing murmur? When a blowing murmur is heard it is situated at the apex, is systolic and soft; this is a blowing murmur of mitral insufficiency, produced by dilatation of the cavities and induration of the walls. When this insufficiency is produced it is impossible to approximate the columnæ carneæ of the first order, which, by their apposition, close the mitral valves. If the three principal pillars of the left ventricle cannot approach in systole, this results in a patulous condition of the mitral canal and consequently insufficiency from dilatation of its valves.

Now, as the muscular tissue does not pass in a single moment from physiological conditions to the pathological state, it follows that when this lesion occurs the systolic blowing murmur of insufficiency is produced especially when the heart is fatigued, and does not result when the heart is in repose. It is, therefore, right to suppose that this special blowing murmur, produced by insufficiency from dilatation of the cavities, will be an intermittent blowing murmur for some time before becoming continuous, while the insufficiency due to sclerotic endocarditis gives rise to a constant blowing murmur which does not disappear.

This peculiar quality of intermittence, added to an increased size of the heart, enables us to make the diagnosis. At the same time it is natural that this murmur should be a murmur of mitral insufficiency, and that at the beginning it should be intermittent, only appearing when the heart is fatigued. This is illustrated by the following observations:

OBSERVATION LXI.—P. L——, aged sixty-two years, entered the hospital January 18, 1879. Upon admission the patient presented the signs of mitral insufficiency. Apex beat in the sixth intercostal space, 16 ctm. from the median line. The upper border of liver corresponds to insertion of fifth cartilage; vertical border of heart is 2½ ctm. from the median line. Blowing murmur heard at apex and propagated toward the axilla; bruit is systolic, begins with the ventricular systole, lasts throughout the entire systole, covers the second sound, and is prolonged into the greater period of silence (paradox bruit). No appreciable click of the semilunars. Pulse small and regular.

Diagnosis.—Hypertrophy of left heart and mitral insufficiency.

January 30th.—A rough systolic murmur is heard over the aortic orifice and is prolonged into the right second intercostal space. The second sound is not distinct; the beginning of a diastolic murmur is heard. The diagnosis was then made: dilatation with roughness of the origin of the aorta, with slight insufficiency.

Two years ago the patient had been taken sick suddenly, while walking, with an attack of suffocation. He remained at home a week, then returned to work, but was compelled to give up permanently at the end of three weeks. Upon admission he was suffering from considerable dyspnœa and œdema of the lower limbs. The heart was in the condition described above.

Under the administration of digitalis the dropsy and dyspnœa diminished; it was then found that the mitral insufficiency observed in the beginning was only temporary, and that the organic lesion was situated at the aorta. April 2d the patient left the hospital much improved, but returned on May 3d. The dyspnœa has returned accompanied by pulmonary congestion. The symptoms gradually grew worse and the case terminated fatally.

Autopsy.—The heart weighs 845 grms. (including the aorta up to the subclavian artery); the hypertrophy affects mainly the left heart. The pulmonary artery is healthy. The aorta is atheromatous and contains projecting calcareous patches at its origin. The valves are flexible but roughened and thickened; one of the valves is attached to the wall by small filaments which prevent its complete closure. The main arteries are dilated.

The walls of the left ventricle are 35 mm. thick (normally 13 mm.), and its cavity is very much enlarged. The mitral valve is healthy, but the pillars cannot be brought into complete apposition; there is insufficiency from dilatation. The muscular tissue is yellowish and firm.

OBSERVATION LXII.—Duke de la R—— suffered for a long time from gout, then dyspepsia, and finally dyspnœa. He long presented the signs of considerable hypertrophy of the heart without a blowing murmur. Toward the close of life a blowing murmur appeared at the apex but could not be heard at every auscultation. He died finally of uræmia.

At the autopsy the heart was found markedly hypertrophied (excentric hypertrophy) without valvular lesion. There is mitral insufficiency from dilatation.

The following are the measurements of the heart:

Weight.................................... 920 grammes.
Circumference of pulmonary artery............. 73 mm.
"   " aorta........................ 78 "
Thickness of walls of left ventricle............. 36 "
"   "   "   right "   .... 13 "

The tissues were firm but had the appearance of fatty degeneration. The liver was markedly cirrhotic; kidneys very small, contracted.

OBSERVATION LXIII. *Mitral Insufficiency from Dilatation.*—J. C——, aged fifty years, entered the hospital February 17, 1882. For twenty years the patient has had pains occasionally in the feet and knees; creaking felt in the right big toe and both knees.

At the present time he is suffering from marked feeling of oppression; œdema of the lower limbs. Pulse small and irregular. Apex beat in sixth intercostal space, 16 ctm. from the median line. Upper border of hepatic dulness at fifth rib. A systolic murmur at the apex, propagated toward the axilla. The liver projects beyond the false ribs and is painful on pressure. Urine scanty and albuminous. Digitalis administered.

February 21st.—Heart more regular; murmur not heard.
March 2d.—Murmur has reappeared.
March 18th.—The two heart sounds present the galloping rhythm; persistence of the murmur, especially toward the axilla.
March 23d.—Slight retraction of the apex observed at the beginning of systole. On account of this phenomenon on the one hand, and the intermittence of the murmur on the other hand, I diagnosed valvular insufficiency from dilatation, and inclined to the idea of pericardial adhesions.
March 30th.—Œdema increased; respiration very difficult.
April 4th.—Murmur has disappeared.
April 7th.—Patient died.
Autopsy.—Mitral insufficiency without valvular lesion. Heart weighed 600 grms. Left ventricle walls 27 mm. in thickness, pulmonary orifice 95 mm. in circumference, aortic orifice 85 mm. in circumference. Heart muscle very firm. Passive congestion of liver and kidneys.

To recapitulate: When a patient is suffering from great hypertrophy of the heart, if a soft systolic blowing murmur is heard at the apex, and if this murmur is intermittent and is not accompanied by the small pulse of mitral disease, we may make a diagnosis, with tolerable certainty, of mitral insufficiency from dilatation. The same remarks hold good with regard to tricuspid insufficiency, but the latter results much less frequently from hypertrophy than from dilatation. It will be discussed under the head of dilatation of the heart without hypertrophy.

Such are the signs of cardiac hypertrophy before degeneration of the muscular tissue has occurred. Changes in the myocardium finally occur and give rise to diminution in the vigor of the contractions of the heart and consequently to intermittence of the pulse, arhythmia, and finally asystole.

On the other hand, hypertrophy predisposes to pulmonary and cerebral congestion and hemorrhage.

The differential diagnosis must first be made from nervous palpitation, which presents symptoms similar to those of hypertrophy, viz., dyspnœa, increased pulsations of the heart, flushes of heat, turgescence of the face; the impulse of the heart is felt not only at the apex but over the entire ventricular surface. The diagnosis will be established by measurement of the heart and the determination of the cause of the disease. The causes of palpitations include excitement, nervous irritability, nostalgia, hysteria, hypochondria, diseases of the brain and spinal cord or their membranes, dyspepsia (particularly the flatulent form), poisoning by alcohol, coffee, tea, and tobacco.

In cases of hypertrophy, on the other hand, we will find changes in the heart, vessels, or other viscera.

The differential diagnosis from pericardial effusion is easy on account of the smallness and frequency of the pulse in pericarditis. Nevertheless, a mistake is possible, and in such cases puncture usually kills the patient by entering the heart.

CHAPTER XXXII.

THE CAUSES OF HYPERTROPHY OF THE HEART.

HYPERTROPHY of the heart is never primary; when the cause is not found in the heart itself, it is present in other organs.

Corvisart believes that frequent excitement, especially the depressing passions, will give rise to hypertrophy. Masturbation and excessive coitus are also regarded as causes of palpitations and hypertrophy. In these cases the hypertrophy is symptomatic of a nervous affection.

Prolonged and arduous labor will also produce hypertrophy in porters, miners, smiths, etc. Forced marches also act in the same manner, as does exercise on horseback. These facts are in accord with the theory of Corvisart, who thinks that hypertrophy is always the result of the effort of the organ to overcome abnormal resistances.

Others think that it is the result of inflammation by propagation or of a nutritive irritation which ends in the proliferation of the tissue. However this may be, whenever the circulatory equilibrium is lost and a new obstacle is presented, the heart is excited by a reflex act to make an effort to overcome this obstacle.

THE VARIOUS FORMS OF HYPERTROPHY CAUSED BY CARDIAC LESIONS.

Hypertrophy is partial at first, and this justifies Corvisart's theory that the part which hypertrophies first is that which makes the effort to overcome the obstacle created by a lesion. Thus if the obstacle is situated in the course of the arterial blood, the left ventricle first hypertrophies; if the obstacle is found in the course of the venous blood, the right ventricle first enlarges.

The determination of hypertrophy of the left or right heart follows in the simplest manner by the method for measuring the heart which I have indicated. When the left side alone is heavier and is depressed, left hypertrophy is present; when the right angle alone is depressed, there is right hypertrophy; when both are lowered, general hypertrophy.

The causes of hypertrophy of the left heart are:
1. Stenosis of the orifice of the aorta.
2. The prearterial aortic stenosis described by Vulpian.

3. Insufficiency of the aortic valves.
4. Stenosis of the aorta in a part remote from its origin.
5. Chronic endaortitis, which destroys the elasticity and consequently the function of the vessel at its origin.
6. Aneurisms of the aorta.
7. Mitral stenosis. The hypertrophy then affects not only the ventricle but also the left auricle.

The causes of hypertrophy of the right heart are :
1. Congenital or acquired stenosis of the pulmonary artery.
2. Congenital stenosis of the aorta and pulmonary artery.

The most active cause of general hypertrophy of the heart, more active even than valvular lesions, is furnished by pericardial adhesions, particularly by obliteration of the pericardial cavity.

Before passing to the consideration of the extracardiac causes of hypertrophy, we will describe the characteristics of cardiac dilatation.

CARDIAC DILATATION.

We have stated that the determination of hypertrophy requires the coincidence of at least two of its principal characteristics, viz. : 1, increase in volume with increase in weight; 2, increase in volume with thickening of the walls ; 3, increase of weight with thickening of the walls. If we find increased volume without increased weight, or even with diminished weight, we say that dilatation is present in the same manner as if there were increased volume without increased thickness of the walls, better still with diminished thickness.

The dilated heart presents a volume greater than that of the normal heart; it encloses a greater quantity of blood, so that when the latter is removed the organ does not weigh more than normally, despite its volume. If, after having emptied the heart of blood, water is injected into it, we will be astonished at the quantity which may be introduced into the organ. When it is still in situ the heart pushes back the lungs, but particularly the diaphragm until it projects at the epigastrium.

The walls of the heart are thinned ; the thickness of the left ventricle may diminish from 12 mm. to 5, 4, even 1 mm. The walls of the right ventricle diminish still more in thickness than those of the left.

Examination of the interior of the heart gives the idea of a heart reduced to its two enveloping membranes. The columnæ carneæ of the first and second order no longer project; those of the third order are effaced. It seems as if the heart were unfolded. The orifices are enlarged and the valves become insufficient.

The fossa ovale is enlarged and the valve of Vieussens no longer closes the foramen Botalli. A communication between the two auricles is again effected.

Dilatation of the heart is usually general; when unequal, it affects the ventricles more than the auricles. It involves the right heart more frequently than the left. Among 45 cases of cardiac dilatation, Lancereaux found 39 confined to the right heart. In two cases there was obliteration of the pulmonary artery and in two others communication between the two sides of the heart; in some cases the auricles were more affected than the ventricles.

Dilatation, then, is a sort of distention of the heart without the ability to return upon itself. Its causes are of two kinds: on the one hand, obstruction to circulation; on the other hand, changes in the myocardium, either from age or fatty degeneration.

Does temporary dilatation occur without asystoly? Beau admits it in chlorosis, hydræmia, typhoid fever, the eruptive fevers, yellow fever, intermittent fever during the attack. Panot also admits it and says that it is accompanied by a blowing murmur with the first sound at the base, and due to tricuspid insufficiency. This view is also held by Professor Fabre, of Marseilles.

But despite the authority of these writers I cannot admit the occurrence of this temporary dilatation. In some cases the bruit at the base is found in the left second intercostal space; it is situated in the pulmonary artery, is due to anæmia, and measurement of the heart shows no increase in size. In endocarditis a temporary insufficiency occurs from atony of the papillary muscles, but this is a mitral insufficiency. In typhoid fever we have to deal with changes in the myocardium—with a true myocarditis—without notable increase in volume, etc. We are compelled, therefore, for the present, to admit the existence of chronic dilatation alone.

Its symptoms are of two classes. On the one hand, the physical signs of increased volume, which are furnished by measurement. These enable us to observe that both angles of the cardiac triangle are lowered, that the lower border of the heart has become more or less completely horizontal, that the apex is farther removed from the base.

The functional symptoms are furnished by the pulse, but they are confounded with those of degeneration of the myocardium. The pulse is feeble and irregular.

But the course and progress of cardiac dilatation are shown better by the venous stasis which appears in the neck, face, liver (the volume of which is increased thereby), in the intestinal vessels and the veins of the lower limbs (where it produces œdema). At the same time the venous circulation grows weaker, little blood is carried to the lung, and the pulmonary veins become congested.

At a later period the dilated tricuspid valve yields and then appears the real sign of its insufficiency, viz.: the reflux in the jugulars. However, auscultation does not disclose a blowing murmur at the xiphoid cartilage; the blood which distends the cardiac cavities is almost motionless and does not give rise ordinarily to a pathological bruit. It is only in exceptional cases, when, despite the distention, the heart still possesses a certain amount of energy, that the contraction of the right heart produces sufficient vibration of the blood to give rise to a blowing murmur. In such an event the murmur is situated near the xiphoid appendix and along the lower border of the heart. Its situation might lead us to believe that it is produced as well in the hepatic vein as in the auricle, because blood which is forced back into the right auricle never gives rise to a blowing murmur to the right of the sternum. The murmur in question is temporary, being heard one day and then not for a number of days.

Parrot [1] has described, under the term asystolic bruit, a murmur which appears to be due to tricuspid insufficiency rather than to asystoly. In the seven cases reported by this writer he heard a blowing murmur in the left fourth intercostal space in all the patients except the fourth, in whom the murmur was heard in the fifth interspace. All these patients presented more or less distinct insufficiency and degeneration of the myocardium. Parrot attributes the murmur to dilatation and insufficiency due to asystoly. I confess that my investigations on this subject have not led to the same result. I have found that when asystoly develops, pre-existing murmurs diminish in intensity, and, on the other hand, where there was no pre-existing murmur, none is produced at the time of asystoly. If Parrot's theory were true, all patients with cardiac disease who die of asystoly should present this peculiar murmur before death.

THE EXTRACARDIAC CAUSES OF HYPERTROPHY AND DILATATION OF THE HEART.

These causes are very numerous and are furnished by the organs of circulation and the viscera.

INFLUENCE OF AORTIC LESIONS UPON THE HEART.

1. *Stenosis and Obliteration of the Aorta.*—Congenital stenosis of the aorta appears to be more frequent in the female sex than in the masculine. During life it gives rise to the signs of chlorosis (Rokitansky, Virchow). It is associated very frequently with insufficient development of the genital organs (testicles, penis, uterus, ovaries), but this coincidence is not constant (Virchow).

Two conditions may be presented by the heart. Either it is in a condition of arrested development and is then small in proportion to the size of the aorta;[2] or the heart is normal and reacts to the small size of the

---

[1] Parrot: Sur un bruit de souffle cardiaque symptomatique de l'asystolie, Arch. Gen. de Méd., Avril, 1865, t. 1er, p. 385.

[2] Dumontpallier: Gaz. Méd., 1857.

artery by becoming dilated and hypertrophied (Legrand,[1] Barth,[2] Andral[3]). In such cases, despite the cardiac hypertrophy, the circulation remains very slow and results in brown induration of the lungs with dilatation of the right side of the heart. When the stenosis of the aorta occurs at a later period hypertrophy always takes place.

2. *Atheroma of the Arteries.*—Atheroma of the arteries is a frequent cause of cardiac hypertrophy,[4] not always on account of the narrowing of the vessels (which occurs in certain cases) but especially on account of the loss of elasticity of the arteries. The elasticity of the vessels adds nothing to the sum of impelling forces, but it diminishes the resistance to the entrance of blood into them.[5]

Ch. Roy[6] has found that the extensibility of the arteries is greatest when the pressure within them is equal to that to which they are subjected during life.

The degree of elasticity is approximately the same in the aorta and its branches. It is readily modified by diseases which affect nutrition. The pulmonary artery presents an extreme degree of elasticity.

In cases of atheroma of the aorta and the arteries, dilatation of the left heart with hypertrophy usually occurs, generally in proportion to the vascular lesion; sometimes even dilatation with atrophy occurs when the patient is in a condition of marasmus.

It is not only atheroma of the aorta which produces hypertrophy. In Polotebnow's four cases, in some atheroma was present in the arteries of the upper limbs, in others in the arteries of the lower limbs. In 23 cases of chronic endarteritis with sclerosis (without nephritis), A. L. Galabris[7] found a marked hypertrophy of the heart 13 times. Chronic endarteritis and sclerosis of the coronary arteries do not produce hypertrophy of the heart, but atrophy and fatty degeneration of the myocardium.

3. *Aneurism of the Aorta and Large Arterial Trunks.*—The heart is often hypertrophied in aneurism of the aorta, but not constantly. Sénac[8] had stated that when the aneurism causes no obstruction to the circulation of blood, it does not produce hypertrophy. Pitres[9] accepts this view but

---

[1] Legrand: Du rétrécissement de l'aorte. Paris, 1834.

[2] Barth: Des rétrécissements congénitaux de l'aorte, Thèse de doctorat, 1837.

[3] Andral: Clin. Méd., 3ᵉ édit., t. iii., p. 62. 1834.

[4] Polotebnow: Berl. klin. Wochenschr., 35, 1867.

[5] Marey: La circulation du sang à l'état physiol. et dans les maladies, p. 161. 1881.

[6] Ch. Roy: The Elastic Properties of the Arterial Walls, Journ. of Phys., Cambridge, 1880.

[7] A. L. Galabris: On the Connections of Bright's Disease with Change in Vascular System. Thesis, London, 1873.

[8] Sénac: Traité de la structure du cœur, etc., t. ii., p. 407.

[9] Pitres: Des hypertrophies et des dilatations cardiaques indépendantes des lésions valvulaires, Thèse d'agrégation, p. 14. 1878.

states that although hypertrophy is not constant it is nevertheless very frequent. Among 58 cases of aneurism which he collected in the "Bulletins de la Société Anatomique," the heart was dilated or hypertrophied 53 times, atrophied once, and normal only 4 times.

The following is a *résumé* of his observations:

Aneurism of the arch of the aorta, 38 cases:

Hypertrophy, considerable	10	cases.
" slight	15	"
" moderate	6	"
Dilatation, simple	4	"
Atrophy	1	"
Normal	2	"

Aneurism of the thoracic aorta, 13 cases:

Hypertrophy, considerable	1	case.
" slight	9	cases.
" moderate	1	case.
Dilatation, simple	1	"
Normal	1	"

Aneurism of the abdominal aorta, 7 cases:

Hypertrophy, considerable	1	case.
" slight	2	cases.
" moderate	1	case.
Dilatation, simple	2	cases.
Normal	1	case.

While Wunderlich and Bamberger regard hypertrophy as constant, Stokes shares Sénac's opinion, and believes that when valvular changes are not present the aneurism does not entail an excess of labor upon the heart.

Axel Key has found that in eighteen cases of aneurism of the ascending aorta and the arch, hypertrophy was not observed; slight dilatation of the left ventricle was noted once. Quincke and Frederici have found that the heart is usually not hypertrophied.

The following table gives the results of Pitres' investigations concerning the influence of aneurisms of the arteries of the limbs:

Aneurism of the brachio-cephalic trunk alone, 3 cases:

Hypertrophy, slight	1	case.
" moderate	1	"
Normal	1	"

Aneurism of the internal carotid, 1 case:
Hypertrophy, moderate........................ 1 case.

Aneurism of the carotid, subclavian, and brachio-cephalic trunk, 1 case:
Hypertrophy, moderate........................ 1 case.

Aneurism of the primary iliac arteries, 1 case:
Hypertrophy, considerable .... 1 case.

Aneurism of the renal artery, 1 case:
Hypertrophy, considerable...................... 1 case.

It follows from all these investigations and from my own observations that the aneurism itself very often produces a slight but rarely a considerable hypertrophy.

4. *Pregnancy.*—Woman appears to be endowed with a greater hæmatopoietic power than man. Every month she loses more or less blood and usually is not very much affected thereby, even if the amount lost reaches 550 or 600 grammes. It would be interesting to know whether this loss is repaired only after the menses, or if the menstrual flux is not preceded by a sort of plethora. The prodromata which are manifested for a few days in some women, the swelling and tension of the breasts, the heat or congestive flushes of the head, and on the other hand the apparent congestive condition which develops when the menses are delayed, the relief which follows the menstrual discharge, etc., lead to the idea that the menses are preceded by a sort of plethora.

When pregnancy occurs menstruation ceases and a part of the elements of nutrition are supplied to the fœtus. What effect is produced upon the blood and circulation?

Heidenhain's experiments [1] show that the weight of the blood in rabbits compared to the weight of the body increases during pregnancy from 5.55 per cent. to 6.84 per cent. This increase does not occur until the second half of pregnancy and reaches its maximum at the moment of parturition.

Does this occur in women? This is difficult to prove, but the best argument in its favor is found in the fact that pregnant women often complain of a sort of plethora and that they are relieved by bleeding. The proof of increased tension has been sought in the characteristics of the pulse, but the traces furnished by the sphygmograph indicate rather the variations of tension than the tension itself. The actual condition of our knowledge, then, allows us to assume that during pregnancy there is a certain increase in the quantity of blood and a probable increase of arterial tension.

---

[1] Heidenhain: Arch. f. Heilk., Neue Folge, Bd. I, p. 536.

The researches of Andral and Gavarret, Becquerel and Rodier, and others show that the blood contains during pregnancy more water and less albumen, and in the latter months a little more fibrin. Quinquaud's [1] investigations have furnished the following results:

	Healthy women.	During pregnancy.
Hæmoglobin	83.50 per 100	62.50 per 100
Absorbent power of the hæmoglobin for oxygen	160 c. ctm.	100 c. ctm.
Solid substances	80	74

The blood of the pregnant female is therefore relatively impoverished. But with regard to the production or aggravation of cardiac hypertrophy, the most important conditions are the mechanical ones, and it seems to follow that the heart, having a larger amount of blood to move, and the arterial tension being increased, must be subjected to increased effort on account of pregnancy. This increased effort first falls upon the left heart. At a later period, toward the close of pregnancy, many women experience a dyspnœa which indicates diminution of the respiratory area and increased labor on the part of the right heart.

The question now arises, Is the heart hypertrophied or merely dilated during pregnancy? In 1828 Larcher [2] stated that almost all females who died during pregnancy or after delivery presented increased thickness of the walls of the left ventricle. In 1859 this writer stated that, judging from the results of 130 autopsies made upon women who had died of puerperal fevers, he could affirm that "the heart, in the human species, is normally hypertrophied during the course of gestation."

But, as we stated above, in establishing the anatomical characteristics of hypertrophy, the thickness of the walls does not suffice for the determination of this condition; there must also be an increase in weight proportionate to the increase in the thickness of the walls.

Blot concludes that during pregnancy there is an increase of more than a fifth of the total weight of the heart. This hypertrophy is confined almost exclusively to the left ventricle, and is remarkable from the fact that it is temporary like the uterine hypertrophy. I must confess that Blot's investigations have not convinced me, as they were made upon only twenty females, and the weight of the hypertrophied heart did not exceed 291 grms., a very slight increase above the average weight.

Duroziez also concludes, as the result of percussion according to Piorry's method, that the heart is hypertrophied during pregnancy. I am still less convinced by the results of this investigation. In the first

---

[1] Quinquaud: Chimie pathol. Delahaye, 1880.
[2] P. Ménière: Arch. Gén. de Med., 1828, t. xvi., p. 389.

place the method of measurement is very inaccurate, and in the second place it indicates merely a displacement of the heart by the distended abdomen.

The measurements which I have made upon pregnant women have been rendered very difficult on account of the weight and size of the breasts. The displacement of the heart upward makes the lower border more horizontal, and thus causes a greater separation of the apex from the median line. Letulle, by employing my method for the measurement of the heart, has obtained the following figures:

	Number of observations.	Distance of apex from median line.
Non-pregnant women, 18 to 30 years old...	12	6 to 9 ctm., average 7½
" " 30 to 50 " " ...	13	6 to 9 " " 8¼
Pregnant women, 18 to 30 years old...	8	8 to 11 " " 9½
Women, less than 24 hours after delivery...	18	8 to 10 " " 9¼

I have come to the conclusion that in pregnancy, especially toward the close of the period, the heart is pushed upward and its apex removed farther from the sternum; in addition, there is a greater tension, at first in the left cavities, then in the right, and also dilatation of the heart. This dilatation is accompanied perhaps by a very slight hypertrophy, which is probably acquired only after a number of pregnancies.

Apart from dilatation and hypertrophy, does pregnancy entail a greater liability to the production of diseases of the endocardium? The numerous cases of heart disease which have come under my observation include hardly any which have been contracted during pregnancy. I am convinced that not more cases of acute endocarditis or pericarditis develop during pregnancy than at other times.

But if pregnancy and parturition do not give rise to acute affections of the heart, the same cannot be said concerning the puerperal condition. Articular and cardiac affections form part of the class of puerperal diseases.

J. Bouley admits in puerperal affections a sort of rheumatism, entirely analogous to non-puerperal acute rheumatism, and purulent infection giving rise to articular and cardiac lesions which resemble rheumatism.

Casanova and Bucquoy have shown the frequency of valvular endocarditis during the puerperal state.

Pregnancy, and particularly the puerperal condition, may give rise, accordingly, to endocarditis, pericarditis, and myocarditis with a tendency to suppuration, even if the heart was previously healthy. But this action is much more marked if the patient has suffered previously from a cardiac affection. We will attempt to give the measure of the danger.

If the patient is suffering from a cardiac affection—usually a chronic valvular endocarditis—if the general condition is good and the other organs of circulation are healthy, she will do very well during the first months of pregnancy. Bad symptoms must be looked for after the fifth month. These symptoms may be confined to some nervous disturbances, an exaggeration of the dyspnœa on effort, and palpitation. These phenomena will be so much more marked the more pronounced the cardiac hypertrophy. Acute pulmonary congestion, passive congestion, and œdema may also be produced under such circumstances.

The patients will be more or less exposed to these dangers according to the character of the affection. According to Spiegelberg,[1] the patients who are most exposed during this period are those suffering from aortic insufficiency, while those suffering from mitral disease have more to fear during the puerperal period. Loehlein,[2] on the contrary, thinks there is more danger from mitral lesions during pregnancy. It is difficult to arrive at any positive conclusion on this subject on account of the small number of observations in which the distinction between the character of the cardiac lesions has been made. But if hypertrophy is considerably advanced and the auxiliary organs are involved, if granulo-fatty degeneration has begun, there is danger that pregnancy will be followed by grave consequences. Epistaxis, hæmoptysis, and hæmatemesis may supervene; cerebral hemorrhage has been observed several times by Ollivier. Finally, the granulo-fatty degeneration may make rapid progress and lead to dropsy, asystoly, and death.

MacDonald[3] thinks that the patient is liable to great danger during the second stage of labor, *i.e.*, from the efforts at expulsion. I have already indicated the dangers of the puerperal period; these have been insisted upon particularly by Spiegelberg and MacDonald.

It is evident then that pregnancy constitutes a real danger to patients suffering from cardiac disease; afterward the cardiac affection of the mother becomes a source of danger to the infant. But I do not think that we should go to such an extreme as M. Peter, who says: If a patient suffering from cardiac disease belongs to the female sex, she should not marry; if married, she should not be allowed to become pregnant; if she has been delivered, she should not nurse the child.

I append the table compiled by Porak from 84 observations:

The stationary condition has been noted 21 times . . . . 25.00 per cent.
Temporary aggravation, 4 times. . . . . 4.76   "
Persistent aggravation, 51 times. . . . . 60.71   "
Improvement during child-bed, 22 times . . . . 26.19   "
Labor aggravated the symptoms, 11 times. . . . . 13.09   "

[1] Spiegelberg: Mittheilungen aus d. Gynækol. Klinik. (Monatschr. f. Geburtsk., 28, p. 439). Handb. d. Geburtshuelfe, 1879.

[2] Loehlein: Ueber d. Verhalten d. Hneuerzens b. Schwangeren u. Woechneriunen (Zeits. f. Geburtsk., 1 B., 3 H., 454, 1876).

[3] MacDonald: The Bearings of Chronic Disease of the Heart upon Pregnancy, Parturition, and Childbed. London, Churchill. 1878.

Death occurred :

Before delivery, 5 times	5.95	per cent.
During delivery, 2 times	2.38	"
During child-bed, 25 times	29.76	"
After temporary improvement, 8 times	9.52	"
By progressive aggravation, 17 times	20.29	"

It must not be forgotten that this table refers to pronounced heart disease which had already been recognized by the physician, and that the slighter ones may often pass unnoticed.

CHAPTER XXXIII.

HYPERTROPHY AND DILATATION SECONDARY TO DISEASES OF OTHER ORGANS.

LESIONS OF THE KIDNEYS AND URINARY PASSAGES.

BRIGHT was the first to notice the frequent coincidence of cardiac hypertrophy and diseases of the kidney. Among 101 bodies which presented the lesions of albuminous nephritis he observed hypertrophy of the left ventricle 52 times, and in half of these cases there was no valvular lesion which would account for the hypertrophy. Bright believed that the hypertrophy of the left heart was the result of the albuminous nephritis, and he offered two explanations :

" The two most natural solutions of this problem," he says, "appear to be the following : either the altered blood is an irregular and abnormal excitant of the heart itself, or it modifies the circulation in the capillaries and small arteries in such a manner as to demand of the heart an excess of energy, in order to propel the blood through the most remote subdivisions of the vascular system."

Three points must be considered in this statement of Bright : 1. The coincidence of renal affections and hypertrophy of the heart, affecting chiefly the left ventricle. 2. A theory which makes the cardiac lesion symptomatic of the renal lesion. 3. An explanation of the pathogenetic action of the renal lesions upon the heart.

Now, of these three points the first has been demonstrated again and again by the authors who have written upon the subject ; the two others are disputed.

Rayer stated that the lesion of the heart was primary, and that the renal lesion was the result of the cardiac hypertrophy. This, however, is a mistake : the renal lesion in question is interstitial nephritis (contracted kidney), very different from the kidney changed by hypertrophy of the heart.

Kelsch[1] has shown that the lesions of the " heart kidney " involve the epithelium, particularly of the convoluted tubes. The cells are swollen

---

[1] Kelsch : Revue critique et recherches anatomo-pathologiques sur la maladie de Bright, Arch. de Physiol., 1874, p. 722.

and granular; there is not the least trace of connective tissue proliferation. The heart kidney can only simulate the large white kidney—a form of disease which is associated only exceptionally with cardiac hypertrophy.

We must return, therefore, to Bright's theory, which has been accepted by the majority of writers. But all are not agreed with regard to the mode of action of interstitial nephritis as a cause of cardiac hypertrophy. Gilewski[1] assumes a reflex action of the kidneys upon the heart, the condition of the kidneys causing palpitations which lead to cardiac hypertrophy. This explanation has not been accepted, because generally this hypertrophy only gives rise to moderate palpitation. O. Weilling has enunciated a still more problematical theory which need not be considered.

Under certain aspects Bright's theory was a chemical one, since the first result of the renal lesions was an alteration of the blood. But the view that a fluid charged with urea traverses the capillaries with greater difficulty than another fluid has been disproven by the experiments of Potain, who found that fluids loaded with urea traversed the capillary tubes as readily as those which were not so loaded.

The latter part of Bright's hypothesis remains, viz., that the hypertrophy of the heart is the result of the reaction of this organ against the interference with circulation in the small vessels.

In 1856 Traube[2] adopted this theory and showed that the heart hypertrophies in proportion as the kidney atrophies, thus rendering the theory more explicit. Pitres makes a résumé of the theory in the following terms:

1. In the final period of Bright's disease a large number of arterioles and capillaries in the kidneys are destroyed; the capillary field is thus diminished and the tension increased in the aortic system, which contains the same quantity of blood in less space.

2. At this time the kidney, being destroyed in part, becomes unable to fulfil its eliminatory functions, the blood remains charged with a certain amount of water, which the glands, other than the kidneys, are unable to expel, and the mass of blood, already cramped in the narrowed vascular system, is increased still further, thus again increasing the intra-arterial tension.

3. The hypertrophy of the heart results from this excessive tension and from the excessive energy which the organ is obliged to employ in order to struggle against it.

Although accepting the view of an action of the kidney upon the heart, many authors oppose the theory of exaggerated tension in the aortic sys-

---
[1] Gilewski: Ueber d. muthmassliche Ursache d. Hypertrophie in Morbus Brighti, Wien. Med. Wochenschr., 60, 1869.

[2] Traube: Ueber den Zusammenhang von Herz und Nierenkrankheiten. Berlin, 1856.

tem. Lancereaux[1] has observed more than once that the kidney has been destroyed and reduced to a fibrous sac full of purulent fluid or cheesy matter, although the heart was not hypertrophied.

On the other hand, arguments are not lacking to prove that an obstacle to the renal circulation, apart from interstitial nephritis, may result in cardiac hypertrophy. Virchow has seen a case of congenital dropsy of the kidney in which the glomeruli were reduced to a third of their normal volume, and in which the left ventricle was hypertrophied without any other lesion to explain it. Rokitansky has noted the coincidence of absence of one kidney with ventricular dilatation,[2] and Storch[3] has observed hypertrophy of the left ventricle under such circumstances. Quite a number of similar cases have been observed by various writers.

This theory has been attacked recently by Debove and Letulle. These writers have recalled the fact that Gull and Sutton believe that the obstruction exists not alone in the kidney, but in all the arterioles,[4] which present a peculiar change that diminishes their elasticity (arterio-capillary fibrosis), a change which Debove and Letulle regard as real. According to these writers, the change in the heart is not produced by the lesion of the kidneys and capillaries, but it is produced at the same time as the latter by a similar cause and by the same diathesis.

Debove and Letulle rely upon the result of their histological investigations. They have observed cardiac sclerosis with atrophic changes in the muscular fibres of the heart. The new formation, according to them, begins in the vessels; it is at first a periarteritis, then as this lesion progresses it becomes difficult to recognize exactly its point of departure. The sclerosis of the kidney and heart, accordingly, consists of a capillary periarteritis, identical with that found in the small vessels.

On the other hand, these writers recognize the fact that, clinically, the renal phenomena are predominant in the beginning, and that it is only toward the close of the disease that the cardiac sclerosis produces the ordinary symptoms of cardiac cachexia.

However, future investigations must decide the pathogenetic relations of these two affections.

Whatever may be the correct theory, it is a remarkable fact that this hypertrophy presents a peculiar clinical history, which I shall now endeavor to describe.

A patient presents himself complaining of dyspnœa on exertion, and examination of the heart shows hypertrophy. The apex has descended into the sixth intercostal space, and is 10 to 12 ctm. from the median

---

[1] Lancereaux: Nouv. Dict. des Sciences Méd., Art. "Rein."
[2] Rokitansky: Wien. Zeitschr., 1859.
[3] Storch: Hospitals Tidende Reports, 1863.
[4] Debove et Letulle: Recherches anat. et clin. sur l'Hypertrophie cardiaque de la Néphrite interstitielle (Arch. Gén. de Méd., Mars, 1880).

line ; the right angle of the cardiac triangle is not lowered, the right auricle is not dilated. There is no blowing murmur present, but an alteration in the rhythm. Three sounds are heard with the rhythm of the bruit de galop or the bruit de caille (quail). The urine is found to contain a small amount of albumen. The lungs and liver are not congested, the lower limbs are slightly œdematous. The diagnosis is then made of interstitial nephritis with hypertrophy of the left heart.

No blowing murmur is heard in the beginning. It is not until a later period that mitral insufficiency from dilatation occurs and is characterized by the phenomenon upon which I have previously insisted, viz., that before becoming constant the murmur long remains intermittent, appearing on certain days, disappearing on others. Then follow the secondary phenomena of hypertrophy with mitral insufficiency ; pulmonary and hepatic congestion are produced, and are soon followed by dropsy and cardiac cachexia, leading to asystole.

We must again refer to the bruit de galop, and will first establish the conditions under which it is presented.

Usually the rhythm is composed of three bruits, the first two being approximated, the latter more remote, and this is the rhythm of galop. In rarer cases the first sound is isolated, the two others are approximated, and this is the rhythm of the *cri de la caille* (cry of the quail). Both rhythms are mere varieties of the same sound ; they are found in the same patient if we listen to a series of cardiac revolutions. Furthermore, these sounds always present their maximum in the second intercostal space near the semilunar valves, and are always heard near the lower border of the heart. They are due to a reduplication of the first sound, the second portion or added bruit being more or less delayed in systole, in an analogous manner to what I have described concerning the reduplication of the sound in mitral stenosis.

I said above that the two sounds are always heard near the lower border of the heart. If careful attention is paid, it will be found that the first of these sounds has its maximum of intensity at the left border of the sternum, near the xiphoid appendix, and that the following one is situated more to the left. If the rhythm is examined, it is found that the first precedes the impulse of the apex, and the second coincides with the impulse. It will also be found that the less marked the reduplication is the more the second sound approaches the first ; finally, that when the reduplication is replaced by a murmur, this is always systolic.

It follows that the first sound is that of the tricuspid valve, the second sound that of the mitral valve more or less delayed by the obstacles to the arterial circulation, and that this sound, when delayed very much, approaches so closely to the second sound as to create the appearance of reduplication of the semilunar click.

When, as sometimes happens, reduplication of the second sound does occur, its maximum is heard at the level of the second intercostal space, either at the right border or behind the sternum. The rhythm then is not that of the galop but of the "cry of the quail." In such cases the second sound is louder than the third, and is produced by the closure of the aortic valves, which precedes that of the pulmonic valves on account of the tension of the aorta.

Finally, in certain cases there are four sounds, two at the venous and two at the arterial valves; it may then be said that they succeed one another in the following order: tricuspid-mitral, then aortic-pulmonic. This is not the theory of Potain,[1] who thinks that the first bruit in the rhythm du galop is a præsystolic murmur; this theory has been adopted by his pupils, Exchaquet, Cuffer, Rendu, etc. Potain, relying upon the cardiographic traces, thinks that the first bruit occurs at the moment of the præsystole, i.e., at the moment of the contraction of the auricles. But upon this very trace it is seen that the ascent formed by the ventricular systole does not precede the pulse as it does in the normal condition, and that there is a delay in the ventricular systole. I am inclined to believe that, in this case, the systole is made in two periods, for it is often found delayed. In Exchaquet's trace No. 3, for example, the impulse of the apex, corresponding to the height of the ventricular systole, does not occur until after the pulse. Moreover, how can hypertrophy of the left heart give rise to such a powerful contraction of the right ventricle that the sound of the passage of the blood, which meets with no obstacle, gives rise to a dry bruit, not blowing, and entirely analogous to a valvular click? Potain himself recognizes the fact that this theory is not entirely satisfactory.

Hypertrophy and Dilatation due to Diseases of the Respiratory Apparatus.

Pulmonary diseases, by reducing the field of hæmatosis and particularly by presenting an obstacle to the pulmonary circulation, react upon the heart, especially on the right side. As shown above, when the obstruction is situated near the heart at the orifice of the pulmonary artery, the right heart hypertrophies and acts like the left heart in the presence of obstructions situated in the course of the arterial blood.

This action of pulmonary lesions upon the right heart will be so much more marked the greater their duration. Among these lesions the first rank is occupied by emphysema, particularly when accompanied by recurrent catarrh.

Next to emphysema comes chronic bronchitis. Laennec said that a neglected catarrh may be the cause of the most serious disease of the heart.

---

[1] Potain: Du rhythme cardiaque appelé bruit de galop (Soc. Méd. des Hôpit., 23 Juillet, 1875).

According to Kreysig, diseases of the lungs have perhaps a greater influence on the heart and pulse than diseases of the heart itself.

Dilatation of the right heart is constant in interstitial pneumonia—pulmonary sclerosis with bronchial dilatation.

Among the chronic lesions of the lung, however, phthisis does not give rise either to cardiac hypertrophy or dilatation. The observations of Bizot, Peacock, etc., have shown, indeed, that the heart is more frequently atrophied in phthisis. A certain number of cases of chronic pleurisy with adhesions have been reported which were complicated with dilatation of the right heart. This is especially well marked when the pleurisy is accompanied by interstitial pneumonia and dilatation not alone of the large bronchi but also of the smaller branches.

May the cardiac lesions in question be produced by acute diseases of the lungs—for example, by pneumonia? I do not think that any case has been reported which proves that an attack of pneumonia may give rise to chronic dilatation of the heart, unless endocarditis or some other form of inflammation of the heart is present.

A word in regard to the temporary dilatation of the right heart. This is observed in croup, stridulous angina, broncho-pneumonia, pneumonia, capillary bronchitis—in a word, in asphyxia. It occurs even in the temporary asphyxia produced by paroxysms of whooping-cough and in toxic asphyxias.

The attempt has been made to determine experimentally the effect of obstruction to circulation in the pulmonary artery. François Franck reports that the experiments of Poiseuille and others show that the most unfavorable condition to the passage of blood through the pulmonary vessels is the condition of insufflation of the lung. This condition of insufflation represents very well the mechanical obstruction caused by emphysema. Franck has shown experimentally that compression of the pulmonary artery does not allow the right ventricle to be emptied completely.

The Heart of Hunchbacks.

It has long been known that spinal curvatures produce profound disturbances of the respiratory and circulatory functions.

Under the influence of flexion of the vertebral column not alone in lordosis, but also in lateral curvatures, the thorax is depressed anteriorly and forms a sort of fold beneath the false ribs.

The supraumbilical region is always retracted; the liver, being pushed into the thorax, presses upward the diaphragm and the heart. The capacity of the thorax is always diminished. The lung situated on the side of the convexity of the curvature when this is lateral is pushed from without inward, and its posterior border is converted into a narrow rim.

This lung is usually the seat of pathological changes—emphysema, condensation and atelectasis, carnification, and even interstitial pneumonia.

The heart is usually increased in size; the right heart is generally dilated and gives to the organ the shape of a wallet. It is usually distended by blood, either fluid or clotted, and this dilatation extends to the pulmonary artery. The aorta and innominate artery are short. They are no longer symmetrical, and the obstruction to circulation either on one side or the other causes inequality of the radial pulses. This is also true of the pulmonary artery, the two branches of which become unequal on account of the unequal development of the lungs.

Hunchbacks are usually short of breath, subject to pains in the chest, to neuralgias of the cerebro-spinal or sympathetic nerves. From time to time they suffer from angina pectoris. Exceptionally, however, their thoracic cavity is not diminished in size, and they are very robust.

According to Schneevogt, the quantity of air inspired or expired is smaller than in the normal condition. This is due, on the one hand, to the fact that the thoracic capacity is less than normal, and on the other hand to the fact that the thoracic movements are restricted by the diminished mobility of the bones composing the thorax.

According to Jules Guerin, "at times there is no dilatation of the thorax on the two sides, at times it is incomplete on the right or left side. The respiration is exclusively diaphragmatic or abdominal in a large number of cases."

The patients suffer from dyspnœa on the least effort, and this is aggravated by the slightest affection of the respiratory passages. The mildest bronchitis may prove fatal to them.

Stoll found that in the Trinité Hospital the hunchbacks died of pneumonia, asthma, pleurisy, and phthisis.

Examination of the heart is characteristic. The apex usually beats in the fifth intercostal space, sometimes in the sixth, a little more infrequently in the fourth; in twenty cases I found it in the third space. The distance of the apex from the median line is generally increased. The hepatic angle is often lowered and the vertical border is 4 ctm. from the median line.

The apex of the heart is elevated, the right angle is depressed, the inferior border is horizontal, the right auricle is dilated.

These details of the measurement of the heart are in accordance with dilatation of the right heart.

The pulse is generally rapid, and at times it presents true intermissions. In considerable curvatures it is unequal on the two sides. The interference with circulation is observed, especially in the lesser circulation. It results in stasis in the pulmonary artery and the right heart, and increased tension in the jugular, hepatic, and mesenteric veins.

When the dilatation has been carried to the point of tricuspid insufficiency, reflux occurs in the jugulars.

Auscultation usually shows the maximum of intensity of the bruits at the level of the ensiform cartilage. There is sometimes a change of rhythm characterized by slowness of the systole, producing equality of the two periods of silence and the bruit of a pendulum, which I have discussed *apropos* of dilatation of the right cavities.

Measurements taken upon twenty Hunchbacks.

Sex.	Character of the curvature.	Situation of the prominence.	Heart. Situation of apex.	Heart. Distance from the median line.	Liver.	Vertical border of heart.	Depression.
			Space.	Ctm.	Rib.	Ctm.	Ctm.
M.	Lordosis of Pott's disease.	3d dorsal vertebra.	5th	10	5th	4½	1
F.	" " "	4th " "	6th	10½	4th	2	0
M.	" " "	6th " "	3d	8⅓	4th	4	0
M.	" " "	5th " "	5th	11	6th	4	1
F.	" " "	7th " "	5th	10	4th	..	3
F.	" " "	1st lumbar "	5th	9	7th	2	0
F.	" " "	1st " "	5th	7	6th	6	0
M.	" " "	1st " "	5th	10½	...	4½	0
F.	" " "	2d " "	5th	11½	5th	4½	1
F.	Rachitic scoliosis........	Right dorsal.......	4th	7	4th	6½	0
M.	" "	" "	5th	10	5th	3	0
M.	" "	" "	5th	11	5th	5	0
M.	" "	" "	5th	10	4th	4½	1
M.	" "	Left "	5th	...	5th	2	0
M.	" "	" "	5th	10	5th	5	0
M.	" "	" "	5th	11	6th	4	0
F.	" "	" "	5th	11½	5th	2	0
M.	Rachitic kyphosis	Dorsal.............	5th	10	5th	3½	0
M.	" "	"	5th	11	6th	2½	1
M.	" "	7th dorsal	6th	9½	6th	5½	0

At the origin of the great vessels it is sometimes found that the bruits of the pulmonary artery have as great intensity as those of the aorta. In addition, anæmic murmurs are sometimes found in the pulmonary artery.

Very rarely the tricuspid insufficiency gives rise to a murmur.

Nevertheless, these patients sometimes remain for a long time without presenting the signs of cardiac dilatation. When they do occur, cyanosis appears in the face, hands, and knees, and then becomes general. Then the venous plexuses are congested, particularly in the liver. However, there is usually little ascites and little albumen in the urine. Œdema of the limbs does not occur until long after the cyanosis. The final termination is asphyxia and asystole aggravated by hydropericardium.

Hypertrophy and Dilatation of Gastro-hepatic Origin.

I have already stated that palpitation is sometimes symptomatic of a dyspepsia which is so slight that the patients are astonished when informed that the digestive functions are really at fault.

In other cases clearly marked dyspepsia leads to cardiac disturbances of two kinds, viz., palpitations and syncope. These palpitations have been accurately described by Abercrombie;[1] they present the following characteristics:

1. The pulse remains regular and the action of the heart normal in the intervals between the attacks.
2. The cardiac disturbances are connected evidently with the gastric disorder, and are improved by treatment of the latter.
3. The symptoms are most apt to be produced after meals and while the patient is resting.
4. They are not aggravated but rather diminished by exercise.
5. They are not excited by bodily movements.

Stokes also states that in certain cases the cardiac disorders result from an affection of the liver, and that they may be relieved by the administration of an emetic.

According to Garrod, gouty palpitations are usually the result of the dyspepsia, but they may be caused sometimes by a vice of the blood.

Syncope is often the result of gastric disturbances, even though it is not produced by pain. If unaware of the occurrence of this form of dyspepsia, we are often alarmed at the frequently repeated attacks of syncope. Fortunately this form of dyspepsia is not difficult to cure by the use of bitters, quinine, and proper diet.

Temporary affections of the heart may be produced by diseases of the liver.

Gangolphe[2] states that a mitral blowing murmur is frequently present in jaundice, and that it must be attributed to temporary insufficiency produced by a condition of atony of the myocardium from contact with the biliary acids. I have never found this murmur, though I have often looked for it.

Murchison says that in cases of prolonged hepatic disorder it is not rare to observe symptoms of enfeebled circulation. Potain has indicated more clearly the secondary cardiopathies of hepatic origin. According to him, in a certain number of patients in whom the heart is large and the hepatic functions disordered, the course of the disease appears to show that the hepatic disturbances have preceded those of the heart.

---

[1] Abercrombie: Pathological and Practical Researches on Diseases of the Stomach, etc., p. 81. Edinburgh, 1828.

[2] Gangolphe: Du souffle mitral dans l'ictère. Thèse de Paris, 1875.

In this event the right heart is particularly affected and appears to suffer from dilatation. Potain has found the apex separated further from the median line without being lowered, and the cardiac dulness enlarged.

Auscultation reveals a sharp semilunar click at the level of the left second intercostal space, i.e., over the pulmonary artery. In addition, Potain has noticed an abnormal bruit, giving rise to the rhythm of the bruit de galop, the maximum of which is found near the xiphoid appendix, and not over the apex like the bruit de galop in Bright's disease. The sensation of this bruit is often noticed better by the hand than by the ear. The pulse is weak and compressible, while in the hypertrophy of renal origin it is hard and tense.

According to Rendu,[1] the cardiac disorders due to hepatic affections undergo three successive phases.

In the first stage functional disturbances are observed, the pulsations of the heart increase in frequency and intensity, and their rhythm changes. These phenomena are temporary and the cardiac functions again may become normal.

In the second stage there is a certain dilatation of the organ, indicating that the cardiac muscle is enfeebled and the organ distended. Auscultation reveals a bruit de galop with its maximum at the xiphoid appendix and a second sharp bruit at the level of the pulmonary valve. Finally, a systolic tricuspid bruit is heard from time to time.

In the third stage—that of confirmed tricuspid insufficiency—the disease may be improved, but the tricuspid lesion may reappear and become definitive. It then results in all the secondary phenomena of heart disease, and terminates in asystole.

Last winter I observed a patient whom I may place in this category, though the case is slightly different from that described above.

OBSERVATION LXIV.—M. A. G—— has suffered for a long time from pulmonary catarrh, which causes shortness of breath from time to time. In September, 1880, while walking fast, he was suddenly seized with an attack of cardiac suffocation. When he returned to Paris I found the catarrh in its ordinary condition. His heart, slightly dilated, presented the following characters:

Apex beat in the fifth intercostal space, 10 ctm. from the median line; the upper border of liver corresponds to insertion of fifth cartilage, and lower border of heart is almost horizontal.

Auscultation reveals neither retardation of the second sound in the pulmonary artery, nor a xiphoid bruit de galop, nor a tricuspid bruit, but a considerable arhythmia, indicating an alteration in the myocardium; at times, a certain tendency to cyanosis. The pulmonary catarrh is insufficient to explain the cardiac disturbance; there is nothing abnormal in the urine. The tongue is always thickly coated.

---

[1] Rendu: De l'influence des maladies du cœur sur les maladies du foie et reciproquement. Acad. de Méd., 1881.

There is inability to mount the stairs on account of cardiac dyspnœa, œdema of the base of the lungs, and slight œdema of the legs.

The enormous congestion of the liver at first led me to employ drastics, which caused a rapid disappearance of the dropsy and hepatic congestion, and enabled the patient to walk up five flights of stairs without difficulty.

Since that time there has been considerable improvement, and whenever the symptoms have begun to reappear similar treatment has caused a restoration of health. Slight arhythmia, however, still persists.

In this case of cardiac disturbance without valvular lesion the hepatic origin seems evident, and this view is confirmed by the results of treatment.

Rendu has remarked that the greatest effect upon the heart is not produced by the most serious diseases of the liver. For example, cancer of the liver, whether primary or secondary, has appeared to him to be destitute of action in this respect, and this is confirmed by my own experience. The same may be said of other organic lesions, even of suppurative hepatitis or grave icterus from rapid cellular degeneration. Rendu has observed a more frequent effect upon the heart from hypertrophic cirrhosis with jaundice than from cirrhosis with atrophy.

Finally, the observations of Stokes, Potain, and Rendu lead to the belief that the action of the liver upon the right ventricle is especially marked in cases of biliary lithiasis.

By what mechanism does the liver act in this manner upon the heart? Gangolphe, Fabre, and others believe that the biliary acids have a direct action upon the cardiac muscle. In opposition to this view, Rendu adduces the fact that the cardiac lesion is not proportionate to the intensity of the jaundice. I will add that I have ausculted many jaundiced patients without finding the murmur described by Gangolphe. On the other hand, Rendu has observed the cardiac phenomena precede the icterus by several days.

Murchison looks for the cause in the poisoning of the blood by substances which the liver no longer excretes, and thinks that this poisoning is manifested by changes in the myocardium. According to Potain the blood, not being purified by the liver, irritates the capillaries in the lungs and creates an obstacle to the pulmonary circulation, which leads to increased work on the part of the right ventricle. This action of the blood is effected through the medium of the hepatic nerves reacting upon the capillaries of the lung. Although physiological experiments offer some support to this theory, it needs further confirmation.

CHAPTER XXXIV.

THE SEQUELÆ OF HYPERTROPHY AND DILATATION.

EFFECT ON MENSTRUATION, GESTATION, AND DELIVERY.

DISORDERS of the genital functions may be produced even during the period of tolerance of cardiac lesions.

It has been found that children suffering from congenital affections of the heart often are imperfectly developed. They are relatively small and flabby.

It has been found that in congenital stenosis of the aorta the infants have a peculiar predisposition to hemorrhages. At least, Virchow has found that the aorta is narrow and small in all cases of hæmophilia. This predisposition to hemorrhages is shown often by profuse and prolonged menstruation. At other times the menses are scanty or absent, but there is almost always arrest of development in such cases.

Menstruation is also affected if the cardiac disease is acquired after puberty. The menses are usually more frequent and abundant; this continues until the period of cachexia, when amenorrhœa supervenes. This does not hold good of lesions of the arterial orifices which are accompanied by anæmia.

The tendency to menorrhagia is shown still further by the continuance of menstruation during pregnancy. Porak reports five cases of hemorrhage during pregnancy before the expulsion of the fœtus and not due to an abnormal insertion of the placenta. But uterine hemorrhages in women suffering from cardiac disease must be especially dreaded during delivery after the expulsion of the child. These patients must be watched more carefully than others, as hemorrhage sometimes occurs two, three, even ten hours after delivery, sometimes during the following night. At other times, on the contrary, the loss is not excessive but it is prolonged indefinitely.

Another result of this tendency to menorrhagia is abortion or premature labor. Porak found that among 214 pregnant women suffering from cardiac disease 126 reached full term, and in 88 pregnancy terminated before full term, thus showing an enormous proportion of premature deliveries. This is probably due to changes in the placenta.

Another frequent accident is the death of the fœtus. Porak observed this 119 times, *i.e.*, in a tenth of all patients suffering from heart disease.

Finally, if the child, not being affected to such a great extent, has escaped abortion and premature delivery and is born alive, it is usually puny. The disease is not necessarily fatal, and Devilliers and Regnault have observed the birth of healthy children to patients suffering from anasarca due to cardiac disease.

The dangers are not equal in every form of heart disease. Spiegelberger thought that the period of greatest danger to patients suffering from aortic lesions was the end of the pregnancy, and to patients suffering from mitral disease, during the puerperal state. Porak remarks that no case of sudden death during delivery has been observed in women suffering from Corrigan's disease.

Mitral lesions are tolerated remarkably well by pregnant women (G. Sée). MacDonald states that there is a great difference between sclerotic endocarditis with insufficiency and that accompanied by stenosis. But in such cases the more or less advanced stage of the disease must be taken into consideration. I have had occasion recently to see a young woman suffering from mitral stenosis, in whom a double pregnancy has caused a remarkable progress of the cardiac cachexia.

CHAPTER XXXV.

SECONDARY AFFECTIONS OF THE RESPIRATORY PASSAGES.

The influence of diseases of the heart upon the pulmonary circulation naturally varies according as the lesion affects the right or left heart. But we must go still farther. On account of the solidarity of the two ventricles, which does not permit them to contract separately, if there is a lesion of the left heart, for example, this side of the heart, reacting to surmount the obstacle, will contract with greater energy. But although this energy may suffice to maintain an equilibrium with the obstacle, it induces greater energy of contraction on the part of the right heart, and consequently an excessive supply of blood to the lungs. Accordingly, it is not rare to find active congestion of the lungs during the period of tolerance and compensation. This hyperæmia leads to frequent hæmoptysis and pulmonary hemorrhages. Gendrin[1] makes the following statements:

"Pulmonary congestion and hemorrhage are manifested during the first period of diseases of the heart, *i.e.*, in that period of the disease during which the organism has not suffered any great deterioration from the prolonged functional disorder, either of the heart or lungs. These congestive phenomena are found to develop especially in those young subjects who have not yet suffered from anasarca and still maintain a healthy color of the skin."

Observation LXV. *Pulmonary Congestion in the course of Heart Disease; Mitral Stenosis?*—R. P——, aged fifty-nine years, entered the hospital December 14, 1864.

Upon admission had a cardiac affection characterized by extreme irregularity of the beats and a murmur perceptible at times, and which was recognized sometimes as beginning immediately before the first sound, accompanying and prolonging it.

In January she began to complain of pains in the abdomen and right hypochondrium due to congestion of the liver, which was enlarged. Marked œdema of the lower limbs.

January 20th.—Signs of pulmonary congestion on right side; feeble respiration at the apex; moist râles with exaggerated voice sounds at the middle of the same side, with pain over the same region for the past two days. Respiration normal on left side. Absence of cough and expectoration.

---

[1] Gendrin: Leçons sur les maladies du cœur, p. 217. 1841–42.

January 22d.—Respiration vesicular throughout the lungs.

March 26th.—Expectoration of bloody sputa; very weak; pulse very feeble and frequent; heart impulses 144 per minute. Tympanitic sound over entire chest except at base of left lung, where it appears normal. The respiratory sound is sniffling, and expiration prolonged. A few subcrepitant râles at base of left lung.

March 31st patient died.

Autopsy.—Right lung congested; left lung heavy and compact at the base, which does not crepitate and is the site of a blackish-red infiltration. This portion of the lung sinks in water.

Heart very large; walls hypertrophied and cavities dilated. No apparent insufficiency or stenosis of the orifices. Thickening and opacity of the mitral valve, in which are felt a few small indurated nodules. The valve is supple despite the thickening.

The liver is congested. Other abdominal organs healthy.

Insufficiency of the aortic valves rarely gives rise to hæmoptysis in young subjects. Such individuals, on the contrary, are pale and anæmic in appearance.

We cannot deny the occurrence of active congestion of the lungs as well as of passive congestion in these conditions.

The latter occurs frequently; it results from the tension produced in the pulmonary veins by the obstacles to the flow of blood through the left cavities.

Hemorrhages are caused by stenosis of the aortic orifice, and by mitral lesions, by stenosis more frequently than by insufficiency. They may occur at the beginning of these affections, even when the cardiac lesion is not yet recognizable. J. Bouley has divided the hemorrhages which occur in heart disease into three varieties: 1, prodromal or premonitory hemorrhages; 2, symptomatic hemorrhages; 3, final hemorrhages. This distinction is entirely true, and I have had frequent occasion to verify its correctness.

The hemorrhage may enable us to recognize affections of the heart which had hitherto remained latent. Peter reports two observations in his clinic. At other times the hæmoptysis, as well as pulmonary apoplexy and infarctions, are due to pulmonary thrombosis and embolism. Ranvier has shown that in such cases the rupture of the vessel occurs above the embolus. These hemorrhages may be produced even in cases of fatty emboli which have developed in the course of fractures and osteomyelitis. In other cases the hemorrhage may be due to compression of the pulmonary vessels by an aortic aneurism.

Finally, the hæmoptysis may be connected with the development of tubercles either in cases of congenital or acquired stenosis of the pulmonary artery, in compression of the pulmonary artery by an aneurism of the aorta, or even in certain cases of mitral stenosis. Frommhold has observed tuberculization in one-tenth the cases of mitral stenosis.

But of all the pulmonary phenomena the most frequent one is passive congestion leading to pulmonary œdema, even before the condition of the kidney has induced albuminuria and anasarca. Gendrin makes the following remarks in this connection:

"It is generally held that œdema of the limbs is the first sign of anasarca in diseases of the heart; this is not true, however, at least in the largest number of cases. We have always found œdema of the lungs in those whose limbs were infiltrated as the result of heart disease, and we have often noted the former condition as a prodrome of the latter. In the majority of cases the œdema of the lungs has appeared and disappeared several times before the infiltration has become apparent in other parts."

"We regard pulmonary œdema, accordingly, as the most frequent complication of heart disease, and the one which appears before all others. It is recognized by a dry cough, evident diminution of resonance at the base of the thorax, and moist crepitant râles in this region."

The passive œdema of pulmonary congestion is an hypostatic œdema which appears only at the base of the thorax and almost at the same level in both lungs, while the œdema of renal disease is scattered over both lungs (Lasègue[1]).

"Upon examination with the stethoscope during the attack, full respiration is found almost all over the chest. The resonance appears normal, the frequency of respiration is not increased. In one or more places which must be looked for carefully, crepitant râles are heard. The voice retains the normal resonance, and auscultation is not modified by coughing. If one of these patches is studied separately, the fine crepitant râles are found to diminish toward the periphery, either that they diminish in number or appear to be more distant. These spots have no fixed position; they never occupy an entire lobe."

Hypostatic œdema generally occupies only the lower third of the lungs; if it rises higher, it is almost always accompanied by hydrothorax.

The œdema is produced especially by mitral lesions (stenosis and insufficiency); aortic lesions do not produce it until the hypertrophy has arrived at the period of fatigue, and particularly of degeneration. This is also true of other cardiac and pericardial lesions; when the valves are healthy pulmonary œdema is caused by cardiac asthenia.

In chronic endocarditis there are other lesions which aid in the production of œdema. Dittrich has shown that in such cases the capillaries of the lung undergo fatty degeneration, and on the other hand Buhl has found that the pulmonary capillaries become varicose and fill a part of the pulmonary alveolus, thus furnishing another cause for dyspnœa. This condition of the capillaries leads to small hemorrhages in the alveoli and the interstitial connective tissue.

Hyperplasia of the interlobular tissue soon follows, and this condition

---

[1] Lasègue: Des bronchites albuminuriques (Arch. gén. de Méd., Janvier, 1879).

is known as brownish-red induration of the lungs. The transformation of the red coloring matter of the blood into pigment invades the entire lung and gives it a reddish-yellow color. The lungs then become firm, heavy, and destitute of air. This condition may be recognized during life if the patients expectorate sputa containing pigment.

Finally, phthisis is the common, or at least frequent, termination in congenital stenosis of the pulmonary artery, and even in acquired stenosis. Frommhold has also observed it in some cases of mitral stenosis; finally, this termination has been observed by Habershon and Hanot in compression of the pulmonary artery by an aneurism of the arch of the aorta.

It remains for me to speak of a phenomenon which develops at the close of life and is known as the Cheyne-Stokes respiration ; this occurs in the following manner :

For about a quarter of a minute (10 to 15 seconds) respiration is completely arrested ; then the movements are gradually resumed, at first feebly, but soon exceeding the normal amplitude. Then the amplitude of the respiratory movements diminishes gradually until they cease and, respiration is again arrested. The period of respiration occupies on the average 40 seconds, during which about 28 respirations are effected. The arrest occurs in expiration. After another pause the respiration begins in the same manner.

This phenomenon has been described by Stokes in the following striking manner :

"There are few phenomena which are more remarkable and characteristic, whether we consider the prolonged suspension of respiration, which occurs without pain to the patient, or study the inspirations at the moment of their greatest violence, when the patient holds the head back, raises the shoulders and contracts each of the respiratory muscles by a supreme effort, without the slightest *râle* or other sign of mechanical obstruction to the entrance of air into the chest. Upon auscultation it is found that the vesicular murmur becomes stronger and stronger. When the paroxysm has reached the height of its violence intense puerile respiration is heard.

"After this series of, so to speak, desperate inspirations, the respiration grows weaker from a sort of exhaustion, then ceases entirely.

"The patient may remain so long in this condition that those around him believe that he has drawn his last breath ; then the inspirations return, at first feeble, then stronger and stronger, and so it continues."

This phenomenon is never produced except during the last weeks of life, and may last several days. Biot observed it for a period of six weeks in one case.

According to Stokes, the phenomenon always indicates fatty degeneration of the myocardium. According to Von Dusch, however, it is not a symptom proper of heart disease, and may occur in the course of cerebral diseases, tumors of the brain, basilar meningitis, uræmic coma, and is also found in pericarditis.

What is the pathogeny of this symptom? Traube has formulated the following theory: There is a diminution in the supply of blood to the medulla and consequently in the excitability of the respiratory centre, which does not receive sufficient carbonic acid. The respiratory movements are therefore arrested, the asphyxia grows marked, and the carbonic acid, then becoming more abundant, again stimulates the medulla to produce the respiratory movements anew. As the stimulus produces exaggerated respiratory movements the blood becomes arterialized, and the quantity of carbonic acid becoming insufficient, respiration again ceases.

This theory is ingenious, but if it were true we would never respire in any other manner than that observed in the Cheyne-Stokes respiration.

If we wish to account for the phenomenon we must recall the physiological conditions of the influence of the nervous system upon the respiratory movements. As these are not yet definitely fixed, it will be useless to attempt an explanation of the Cheyne-Stokes phenomenon.

CHAPTER XXXVI.

SECONDARY AFFECTIONS OF THE LIVER (HEART LIVER) AND THE DIGESTIVE TRACT.

In the normal condition the tension in the vessels containing arterial blood is much greater than in those containing venous blood. The pressure in the left ventricle is about three times greater than that in the right ventricle. In the aorta the pressure remains almost constantly as great as the maximum pressure (during systole) in the left ventricle.

Sharpey has found that in the pulmonary artery a certain rapidity of circulation is effected by one-half the force requisite in the aortic system.

When the heart is diseased a certain period arrives in which the passage of the blood through the cavities becomes more difficult, and the tension in the right cavities increases. This takes place particularly in auriculo-ventricular lesions, still more in arterial lesions, and even more in lesions of the pulmonary artery than in those of the aorta. On account of the increased tension dilatation of the right heart occurs and, later, insufficiency of the tricuspid.

In the superior vena cava the force of gravity assists the venous walls, but in the inferior vena cava and hepatic veins the elasticity and contractility of the vessels is soon overcome, so that the circulation in the hepatic and portal veins is soon interfered with. Another factor then aggravates this condition, viz., the weakness of the heart, which has a tendency to equalize the tension in the two systems.

In addition, the hepatic and abdominal vessels are more sluggish than the vessels of other organs, and do not need to be interfered with in order to be overcome by the hydraulic pressure. These organs become diseased in their turn, either on account of insufficient innervation or because local lesions are produced in them.

We may thus understand the fact that the digestive or other disturbances associated with diseases of the heart sometimes are much more marked and serious than the pulmonary disorders which almost always precede them. It may be added that the determining cause of the cardiac lesion—rheumatism, gout, alcoholism, syphilis, etc.—may also have a direct action upon these organs and precipitate the period at which they would become affected, if they were subjected merely to the mechanical disturbances of the circulation. We may also add that in the same manner as hepatic affections may give rise to diseases of the heart, so they aggravate the latter although they are merely their sequelæ.

The lesion first presented by the liver is the hyperæmia so well described by Frerichs.

The hepatic vessels are first gorged with blood; they gradually dilate and their walls become hypertrophic. Thence the stasis is propagated to the portal vein and to the organs from which it derives its source. It then develops a series of functional and nutritive anomalies.

The liver increases in size in all directions, the capsule is stretched, and the consistency of the parenchyma increases. In the centre of each lobule the point corresponding to the dilated interlobular vein has a brownish-red color, and often gives off sinuous radiating ramifications which are formed by the dilatation and congestion of the branches passing to the central vein. The circumference of the lobule, on the contrary, is deprived of blood, and the rootlets of the portal vein are pale and have a brownish color. These appearances are known under the name of nutmeg liver.

When the hyperæmia is very marked the hepatic veins are dilated even as far as their anastomoses with the capillaries of the portal vein, and the reddish-brown color is perceptible throughout almost the entire tissue.

The biliary secretion is little modified in quantity and quality as the result of this stasis. The mucous secretion of the canals increases and sometimes impedes the passage of the bile, giving rise to slight jaundice. At length the nutrition of the organ is changed and the cells near the hepatic veins become infiltrated with fat, pigment granules, and crystals of hæmatoidin. This change is manifested to the naked eye by grayish patches. At a later period the parenchyma atrophies and the liver approaches atrophic cirrhosis in appearance. Cornil and Ranvier[1] have found that the central cells disappear and are replaced by new-formed connective tissue (central sclerosis) with periphlebitis and thickening of the external wall of the central vein. At other times true atrophic cirrhosis is produced, not by the cardiac lesion but by alcoholism, the cause of the general atheroma. The cells of the periphery of the lobule remain intact.

At a certain period the stasis in the hepatic vein extends to the portal vein, and consequently to the vessels of the intestines, the spleen, pancreas, uterus, kidneys, etc.

The congestion of the gastro-intestinal mucous membrane is shown by its reddish color; its tissue is infiltrated with a serous fluid and is softened; its secretion is usually arrested. Hemorrhagic extravasations occur in its tissues, sometimes even ulcerations.

The mesentery is congested, the lymphatic glands are large and vio-

---

[1] Cornil and Ranvier: Manuel d'Histologie, p. 885.

let, and hemorrhages occur around them. The pancreas is infiltrated, the spleen grows harder, and an ascitic effusion occurs into the abdomen.

The same lesions are produced in the gastric mucous membrane, which reddens, swells, and presents ecchymoses.

Digestion and absorption become difficult under these conditions.

Palpation and percussion show an increase in the size of the liver. The patient complains of gastralgia and eructations, sometimes of nausea. The hemorrhoidal veins are swollen from time to time. Constipation is usually present, rarely diarrhœa.

The stasis in the vena cava causes swelling of the veins of the lower limbs, then œdema, which begins around the malleoli and gradually spreads upward. It is sometimes increased by another cause, viz., albuminuria.

The congested liver may contain an enormous amount of blood, and may be dilated almost like erectile tissue.

From time to time, in patients suffering from cardiac disease, attacks of dyspnœa occur which are accompanied by enormous and temporary enlargement of the liver—an enlargement which recurs at each attack of dyspnœa. At other times the liver, distended by the congestion which accompanies tricuspid insufficiency, is transformed into a sort of erectile tissue, through which the cardiac pulsations are transmitted. In these cases there is, at the same time, a cervical and hepatic venous pulse.

These changes in the liver and digestive tract impede digestion and absorption, and aid in hastening the course of the cachexia if they cannot be improved by treatment.

CHAPTER XXXVII.

SECONDARY AFFECTIONS OF THE KIDNEYS (HEART KIDNEY).

SECONDARY lesions in the kidneys may follow either acute or chronic affections of the heart.

During the course of acute endocarditis fibrinous deposits may be detached from the valves and transported into the vessels. In such cases the emboli are arrested most frequently in the kidneys. This is shown by a sudden pain in the region of the kidney, temporary hæmaturia, and, a little later, albuminuria. At the autopsy infarctions are found with their usual appearances: pyramidal shape, with the base turned toward the periphery, paleness of the tissue, red circle around it, etc.

If the endocarditis has occurred in the course of a septic disease, small miliary abscesses may be found as the result of the capillary embolism.

In the course of chronic affections affairs take a different turn.

During the period of compensation disorders of the kidney may occur— for example, in aortic insufficiency. These will consist of inflammations of the kidneys, but without albuminuria. At a later period, when the hypertrophy no longer compensates the lesions and the cardiac impulse grows weaker, the venous tension increases and the entire venous system suffers a notable stasis, which leads to nutritive changes in the viscera.

The venous stasis in the kidneys at first causes a scanty excretion of urine with an abundance of uric acid and deposits of urates, with albuminuria at the end of a certain period. At the same time we may find hyaline casts and a few blood globules, but no true hæmaturia.

Upon autopsy the "heart kidney" presents the following appearances: it is enlarged, hyperæmic, and its consistence is increased. The capsule is thickened; the separation of the cortex from the pyramids, which is at first difficult, becomes more marked in time, the pyramids becoming more deeply colored at their base and paler at their apex.

If the congestion is intense the microscope shows hemorrhages, either from rupture or diapedesis, into the interior of the glomeruli. The blood extravasated between the vessels and the capsule distends it, then is infiltrated into the convoluted tubes of the cortex, then into the loops of Henle, then into the straight tubes and collecting tubes.

After a while the blood in the interior of the tubes undergoes various modifications; hæmatin granules are formed and infiltrate the desquamated cells. The blood is excreted with the coagulated fibrin and appears in the urine in the shape of hyaline casts containing red globules or covered with pigmented cells. The casts sometimes have a yellowish color from the presence of the coloring matter of the blood (Cornil and Ranvier). At a still later period the kidney presents the lesions of diffuse interstitial nephritis.

Finally, in acute myocarditis it is not rare to find a similar affection develop in the kidneys (parenchymatous nephritis). Cases have been observed by Buhl,[1] Stein,[2] and Demme.[3]

In the same way that interstitial nephritis produces cardiac hypertrophy although the heart was primarily healthy, so when the kidney lesion has developed secondarily it reacts upon the heart disease and hastens the course of cardiac cachexia.

---

[1] Buhl: Zeitschr. f. rationelle Medicin., xiii., 32.

[2] Stein: Untersuchungen u. Myocarditis. Munich, 1861.

[3] Demme: Beitraege f. Anat. u. Diag. d. Myocarditis. Schweizer Zeitschr. f. Heilkunde, I., 79 and 461.

CHAPTER XXXVIII.

SECONDARY AFFECTIONS OF THE NERVOUS SYSTEM.

INDEPENDENTLY of palpitations, syncope, and painful affections as symptomatic of heart disease, we have to consider the nervous disorders secondary to hypertrophy during the course of the disease. We will first consider cephalalgia, which may be of several kinds.

1. It may be fixed, neuralgic in character, associated with the anæmic disturbances which are produced by aortic stenosis. Sometimes the cephalalgia is dull and predisposes to drowsiness, indicating venous stasis; this is often associated with a condition of venosity of the blood produced by the secondary pulmonary lesions.

2. The attacks of vertigo present two varieties. One variety is due to an insufficient supply of blood to the brain, caused by aortic stenosis or by degeneration of the myocardium with or without hypertrophy. At a later period during cardiac cachexia the attacks often depend on cerebral œdema.

3. Insomnia and nightmare are very common in patients suffering from heart disease during the period of compensation, and still more during the period of cardiac cachexia. The nightmare almost always consists of dreams in which the patient falls down precipices, is attacked without being able to defend himself, or takes part in murders, etc.

4. Hallucinations are not uncommon in patients suffering from cardiac cachexia with anæmia, passive congestion, or cerebral œdema.

5. Maniacal excitement and hallucinations during the day have long been observed among these patients. In 1818 Nasse stated that the insanity of cardiac patients is an impulsive variety, leading to violence, crime, or suicide. Morel believed that insufficiency of the cardiac valves produced feelings of terror with a tendency to suicide. Burmann has noted hypochondriacal anxiety associated with cardiac affections.

The following statistics have been collected by Limbo:

Esquirol, 11 cases of heart disease among 68 cases of melancholia	16 per cent.
Webster	12 "
Bayle	16 "
Calmeil, 31 cases of heart disease among 100 insane	31 "
Lawrence	12 "
Vienna Asylum, 75 cases of heart disease among 602 insane	12 "
Voppel, 12 cases of heart disease among 75 insane	16 "
Thyermann	14 "
Dufour, 44 cases of heart disease among 61 insane	72 "
	22 "

This average is undoubtedly somewhat exaggerated, but Morel, Griesinger, Marie, and Burmann, who do not furnish figures, have recognized the frequent coincidence of these affections. It must be remembered, also, that in a certain number—for example, in alcoholic cases—the lesion of the brain is not the result of the heart disease, but that both are due to a common cause.

6. Temporary hemiplegia from cerebral anæmia without change of tissue appears from time to time in the course of cerebral affections, when the myocardium is altered either by myocarditis or by hypertrophy which has reached the stage of fatty degeneration. The following is a good illustration:

OBSERVATION LXVI.—A woman, sixty-four years of age, was seized suddenly, in the beginning of 1877, with palpitations and dyspnœa caused by myocarditis of a chronic form. The heart was not very large; no murmur. Great disturbance of cardiac rhythm; pulse feeble; dyspnœa and vertigo on the slightest exertion.

Immediately after having gotten off a chair she was seized with momentary vertigo, which I recognized by her fixed stare. I saw the mouth deviate to the left. I asked her to give me her right hand, but she was unable, as it was completely paralyzed; leg in same condition; sensibility very obtuse. Pulse very feeble and irregular. At the end of three-quarters of an hour sensation was restored and power of motion began to return. When I returned on the following morning power was entirely restored.

This accident occurred on April 10th. The arhythmia continued to increase, and on July 24th she died in a few hours during an attack of apoplexy.

Some authors have mentioned the occurrence of convulsions after attacks of vertigo, but it is probable that they were produced by uræmia rather than by simple anæmia of the brain.

7. Cerebral apoplexy occurs frequently in diseases of the heart, partic-

ularly in those forms which lead to great hypertrophy. According to the most recently accepted theory the increased tension in the vessels is due less to the cardiac hypertrophy than to the changes in the vessels from endarteritis deformans.

8. Cerebral softening may result from embolism or thrombosis.

In a general way it may be said that embolism occurs most frequently in relatively young subjects, while thrombosis, which presupposes vascular lesions in the brain, is rather a disease of old age.

Cerebral embolism occurs during the course of endocarditis, from thrombosis in the ventricles or after thromboses of the large arteries.

While thrombosis has no site of election and may occur in any part of the brain, emboli pass more frequently through the left carotid than the right.

Embolism usually gives rise in the brain to white softening, and the infarction is surrounded less frequently by a hemorrhagic zone than in other organs. If the embolism is the result of an infectious disease, an abscess will be produced. In other cases it may happen that the thrombus becomes organized and the vessel again becomes permeable (Joffroy, Murchison, Ogle).

Disorders of the Special Senses.

Disturbances of Vision.—The most frequent disturbances of vision which occur in the course of diseases of the heart are the circulatory derangements which are recognizable by the ophthalmoscope.

Retinal hyperæmia sometimes appears during the course of valvular lesions. The arterial circulation is diminished; the veins appear gorged with blood, and dropsy sometimes is made evident by a grayish reflex along the vessels. These lesions are much more marked the greater the intensity of the cyanosis.

Retinal apoplexies may also occur. According to Sichel, they appear generally at the periphery of the retina near the equator of the eye, but give rise to scarcely any interference with vision. They are found usually in one eye alone.

When these hemorrhages are small they may disappear or be transformed into grayish patches and completely absorbed.

In one case Limbo observed a hemorrhage into the vitreous body.

The retinal apoplexies often are followed in a little while by hemorrhages into the brain.

I will refer also to the pulsations in the retinal arteries, which are so remarkable in cases of aortic insufficiency.

Spinal Phenomena.

In 1879 Charcot and Vulpian called attention to the frequent coincidence of locomotor ataxia and sclerotic and atheromatous changes in the aortic valves. This association has since been noted by Rosenbach and Beyer, Grasset, Letulle, Hanot and Jaubert, Buch, and Dreyfus-Brissac.

Among 38 cases of this kind, in 18 the lesion affected the aortic valves exclusively, in 7 the mitral valve exclusively, and in 7 both the aortic and mitral valves.

Dreyfus-Brissac has remarked that these subjects were not well advanced in years, and that generally they had not suffered from rheumatism.

I also observed in 1868 and 1869 a patient who belongs to this category. At the age of seventeen years, during the course of an acute articular rheumatism, this individual contracted an insufficiency of the semilunar valves. He tolerated it well and experienced no functional disturbance. In 1868 he was seized with ocular disorders, and shortly after by ataxic symptoms of a very rapid course, so that at the end of two years he had passed through all the stages of tabes, then had paraplegia, and finally convulsions. He had never had any symptoms of syphilis. Despite the rapid and progressive increase of the nervous disorders, the cardiac rhythm was not disturbed for a single moment until death. In this case the aortic insufficiency had preceded by six years the ocular disturbances of ataxia. Can it be held that the ataxia was a sequel of Corrigan's disease? It remains for further observation to decide whether there is any real connection between the two lesions.

CHAPTER XXXIX.

BASEDOW'S DISEASE (ANEURISMAL BRONCHOCELE).

I will discuss Basedow's disease merely in order to conform to general usage, for it is neither a disease of the heart nor is it symptomatic of a cardiac affection. Stokes has furnished the following detailed description of this malady:

I. Under certain circumstances the action of the heart may be permanently overexcited, the contractions becoming rapid, irregular, and more vigorous than usual. This condition is accompanied by enlargement of the thyroid gland, exaggerated impulses of the arteries of the neck, and increased size of the eyeballs.

II. This affection is not accompanied by fever or any signs of inflammation of the heart.

III. The disease is more common in the female, associated with hysteria, neuralgia, and uterine disturbances.

IV. It develops at any period after puberty.

V. It presents periods of exacerbation and remission.

VI. The thyroid enlargement is entirely independent of the causes which ordinarily produce endemic goitre.

VII. This hypertrophy is accompanied ordinarily by a diastolic pulsation.

VIII. This is associated with the ordinary signs of aneurismal varix.

IX. The murmur and purring thrill occupy the entire gland or only certain parts. They vary in intensity in different parts or according to the stage of the disease.

X. These signs disappear during the last stages, when the gland becomes firmer.

XI. The jugulars and large veins traversing the tumor present various murmurs, either while the disease is progressing or even after it has lasted for a long time.

XII. It is possible that the sensation known as globus hystericus is due to the temporary manifestation of the disease at its onset.

XIII. The exaggerated pulsation of the arteries in the neck cannot be explained either by regurgitation into the heart or by an afflux of blood toward the brain. It can be looked upon merely as the proof of a general irritation of the arterial system.

XIV. Blowing murmurs and a double impulse are often found in the carotids.

XV. The enlargement of the eyeballs is not accompanied necessarily by a change of vision; it does not appear to produce any tendency to inflammation of the deeper tissues of the eye or of the conjunctiva.

XVI. This enlargement varies according to the progress of the disease. It may disappear in great part if not completely.

XVII. When the disease terminates fatally, the anatomical lesions are found to consist of dilatation and hypertrophy of the heart, increase in the calibre of the inferior thyroid arteries, and dilatation of the jugulars.

XVIII. There are cases in which an ordinary bronchocele, which has existed for a long time, undergoes the influence of the more recent appearance of an organic or nervous affection of the heart.

XIX. The disease appears to consist essentially of a functional affection of the heart, which may be followed by organic changes.

Very little can be added to this exact description by Stokes. The onset of the disease is gradual; it begins sometimes with palpitation, sometimes with exophthalmia, at other times by a general agitation.

The palpitations are often violent and accompanied by true intermissions of the pulse; sometimes false intermissions are produced, if organic heart disease is present.

To these symptoms must be added the nervous disturbances; the patients are agitated, their character changes, they become more irritable.

All the vessels of the thyroid gland are dilated, but Stokes believes that the inferior thyroid arteries are chiefly affected. In one case under my observation, on the contrary, the superior thyroid arteries were mainly involved. The principal phenomenon consists in a vascular dilatation which occurs in paroxysms and produces the pulsations and diastolic bruit in the dilated arteries. However, even during the intervals the thyroid body remains larger and firmer than normal. Inflammation of the interstitial connective tissue occurs at a later period.

Hitherto no lesion has been detected to which we can attribute the functional disturbances. Numerous lesions have been found, such as softening at the base of the brain, obliteration of the central canal of the spinal cord, hyperplasia of the connective tissue around the sympathetic, induration of the middle and inferior cervical ganglia, etc. But these lesions are merely complications of the disease, not its real cause.

One of the three symptoms which constitute the disease, viz., aneurismal bronchocele, exophthalmia, cardiac agitation, may be absent. Exophthalmia is the symptom absent most frequently.

The course of the disease is not regular; sometimes it leads to gradual and progressive exhaustion of the patient, sometimes it is transformed into another disease. In one case I observed its metamorphosis into diabetes; in another, into parenchymatous nephritis. According to Friedreich and Charcot, about three-fourths of the cases end in recovery.

OBSERVATION LXVII.—C. S——, aged thirty years, entered the hospital February 10, 1881. Two years ago the patient suffered from acute articular rheumatism, and the palpitations which began at that time have increased progressively in severity. Ten months later the thyroid region was suddenly enlarged and has remained so. Six weeks ago the eyes began to be prominent. Upon admission, examination of the heart gave the following results: the entire præcordial region is elevated at each systole, the pulsations being visible in the second, third, and fourth intercostal spaces; a systolic thrill is felt on palpation. The dimensions of the heart appear to be normal. There are no cardiac murmurs, and the rhythm is unchanged.

In the anterior region of the neck is observed a uniform projection, very soft to the touch and presenting systolic pulsations. This enlargement is bilobed and is adherent to the thyroid cartilage. Upon palpation this tumor presents a very marked systolic thrill, and a systolic blowing murmur is heard over its entire extent. This murmur has its greatest intensity at the right superior angle of the tumor, where the superior thyroid artery is felt to be very much enlarged.

The eyes protrude from the orbits and are incompletely covered by the lids; the pupils react normally; vision is intact. Appetite and digestion good. The patient is emaciated, and the skin has a waxen pallor. He fell into a condition of collapse and died at the end of a week.

Autopsy.—The thyroid tumor is formed by hypertrophy of the lateral lobes; it is firm and resisting. No alteration in the larynx or trachea; the œsophagus is not compressed. The superior thyroid arteries are as large as the external carotid; the veins present nothing abnormal. The internal jugulars are dilated; the other vessels of the neck are normal. Both recurrent laryngeal nerves are surrounded by the tumor, particularly the right superior tracheal branches. The heart is normal in every respect.

CHAPTER XL.

NEW GROWTHS AND PARASITES.

FIBROMATA.

LUSCHKA[1] found a fibroma as large as a hen's egg in the tissue of the left ventricle of a boy six years old. Albers[2] found one as large as a grapeseed in the anterior wall of the left ventricle. Kollmeier observed[3] one 6 ctm. in length in the auricular septum, and projecting through the mitral orifice into the ventricle. The *Gazette des Hôpitaux* of 1872 (No. 101) contains the history of a fibrous polypus of the septum which projected into the aorta and separated the semilunar valves in such a manner as to produce insufficiency. Wagstaffe[4] found, in a girl of three months, a fibrous tumor as large as a hen's egg, situated within the septum and filling up in great part the cavities of the heart.

CALCULI.

Calcified and sometimes ossified patches are found occasionally in certain parts of the heart near the valves. At other times concretions are found in abscesses of the myocardium which have recovered by absorption.

Heschl[5] found partial calcification of the myocardium in a case of Bright's disease. Lueken[6] observed calcification of the anterior papillary muscle of the tricuspid valve. This concretion, 3 ctm. long and 2 ctm. broad, was buried deeply in the myocardium and prevented the closure of the auriculo-ventricular orifice by the valve.

---

[1] Luschka: Virch. Arch., Bd. viii.

[2] Albers: Atlas d. Pathol. Anat., iii., t. x., Fig. 1.

[3] Kollmeier: Fibroese Neubildung im Herzen, Virch. Arch., xxiii., H. 3 and 4, 1862.

[4] W. Wagstaffe: Fibrous Tumor of the Heart, Trans. of the Path. Soc., xii., p. 121. 1871.

[5] Heschl: Theilweise Verkreidung d. Herzmusculatur (Oest. Zschr. f. prakt. Heilkunde, xiv., 1861).

[6] Lueken: Die pathol. Neubildungen d. Myocardium (Zschr. f. rat. Med., xxiii., H. 3, 1865).

Lipomata.

Apart from fatty degeneration of the myocardium and fatty deposit upon the heart, only a single case of fatty tumor of the myocardium, quoted by Albers,[1] is known.

Cysts.

These cavities usually are nothing more than abscesses or the sites of parasites.

Myomata.

These tumors are almost always congenital. Virchow[2] and Recklinghausen[3] have observed some of various sizes.

Cancers.

Cancer is the most frequently observed neoplasm in the heart. Koehler[4] found 6 cases among 9,118 autopsies, Tauchon[5] 6 cases among 8,289 autopsies, Willigk[6] 9 cases among 4,547 autopsies.

Cancer of the heart is usually secondary, propagated from adjacent organs, the mediastinum, œsophagus, etc. At other times it results from metastasis. Lober[7] has seen a cancer of the testicles, bladder, and kidneys give rise to a tumor in the septum of the right ventricle, and an embolism in a branch of the pulmonary artery.

Bodenheimer[8] has collected 45 cases of cancer of the heart which were distributed as follows: left ventricle, 7 cases ; right ventricle, 3 cases ; right auricle, 2 cases ; the others were situated in the myocardium and projected either into the pericardial or cardiac cavities, involving more or less the orifices and the valves.

Epithelial cancer has been observed once by Paget,[9] after extirpation of a tumor of similar character. It was situated at the apex of the right ventricle in the interventricular septum.

---

[1] Albers: Virch. Arch., Bd. x.
[2] Virchow: Virch. Arch., Bd. xv., xxx.
[3] Recklinghausen: Myoma Cordis (Monatschr. f. Geburtsk, Bd. xx., 1862).
[4] Koehler: Ueber Krebs, Stuttgart, 1853.
[5] Quoted by Schroetter, Krankheiten d. Herzens, p. 300.
[6] Willigk: Prager Vierteljahrschrift, 1856.
[7] Lober: Contribution à l'étude des maladies du cœur (Thèse de Paris, 1877).
[8] Bodenheimer: Beitr. z. Pathol. d. Krebsartigen Neubildungen am Herzen (Diss. inaug. Berne, 1865).
[9] Paget: Surg. Path., vol. ii., p. 449.

Tubercles.

In acute miliary tuberculosis, tubercular granulations are found commonly in all the tissues. They occur in the heart in the form of gray granulations, but may also appear as yellow tubercles. On rare occasions we find cheesy masses, the result of chronic tuberculosis of adjacent organs. Recklinghausen,[1] Townsend,[2] and Waldeyer[3] have reported examples. The latter writer observed a tubercular mass situated in the right auricle and compressing one of the pulmonary veins. In three other cases the tubercle was attached to the wall of the left auricle. In all the others they had developed in the infundibulum of the pulmonary artery.

Parasites.

The parasites found in the heart are the trichinæ and hydatids. Among 160 cases in which Davaine noted echinococci in organs other than the liver, in 10 they were found in the heart, more frequently in the right ventricle than in the left. The size of the hydatid varies from the head of a pin to an orange. The cyst develops usually between the fibres of the myocardium, but in certain cases it develops upon the internal wall and projects into the cardiac cavity. They sometimes are set free and are carried into the circulation. Coote[4] describes a case in which the cyst of the echinococcus developed upon the anterior wall of the left ventricle, and imparted such a volume to the heart that it extended from the right third rib to the left eighth rib and pushed the lungs backward.

Oesterlen found that the sac ruptured 6 times among 21 cases.

Trichinæ are found in the myocardium as in all the other muscles.

Clots and Polyps of the Heart.

We will first consider the differential qualities of ante-mortem and post-mortem clots in the heart. Those formed during the agony, or after death, present two varieties, one soft and gelatinous, another with a fibrous elastic texture.

---

[1] Recklinghausen : Virch. Arch., Bd. xvi.

[2] Townsend : Dub. Journ. Med. Science, I., 1832.

[3] Waldeyer: Tuberculose d. Myocardium, Virch. Arch., xxxv., 1866.

[4] Coote: Med. Times and Gaz., February, 1854. *Vide* also Williamson: Echinococcus in Herz u. Lungen, Allg. Med. central Ztg., 74, 1864. Ch. Kelly : Hydatid Cyst in the Heart, Trans. of the Path. Soc., xx., p. 145 ; 1870. W. Moxon : Hydatid of the Heart Obliterating by its Pressure the Coronary Sinus, Trans. of Path. Soc., xxi., p. 99 ; 1871. Peacock : Hydatid Cyst Imbedded in the Walls of the Heart, Trans. of Path. Soc., xxiv., p. 37 ; 1873. Goodhart: Cured Hydatid Cyst in the Walls of the Heart, Trans. of Path. Soc., xxvii., p. 72; 1876. Brodowski : Ein Fall einer Apfelgrossen Echinococcusblase in d. rechten Herzkammerwand, Medicina, Bd. iv., H. 24 ; 1876. Mettenheimer : Ein Fall von Echinococcus des Herzens, Memorabilien, viii.

1. Sometimes the clots form a soft, blackish jelly of a homogeneous consistence, which cannot give rise to any difficulty in diagnosis.

2. At other times the clots have a fibrous elastic texture, a firm consistence, and a white or rosy color. They are formed of stratified layers, arranged in a manner which depends upon the position of the cadaver; in the lower parts are found the blood globules, above a more or less thick layer of fibrin. The clots contain a large quantity of serum, and are very elastic; they fill the cavities of the heart incompletely and are prolonged into the large vessels.

The clots formed during life have a much firmer consistence, almost like that of muscle. Their color varies according to age and to the proportion of blood globules which they contain. They adhere to the walls of the heart.

The recent clot is grayish, semi-transparent, elastic; the old clot is opaque, yellow, and friable.

In the heart the clots formed during life bear the name of polypoid concretions, blood polyps, fibrinous clots. They are of two varieties, one autochthonous, the other embolic.

According to Bouillaud, autochthonous clots are more frequent in the right cavities than in the left, and also more frequent in the auricles than in the ventricles. This is due to the fact that venous blood presents a greater tendency to coagulation than arterial blood. Legroux, on the contrary, states that the clots are formed particularly in the left ventricle and auricle.

They may be either free or adherent. They adhere either to the valves or to the walls of the heart, either in masses or in layers. The centre of the mass sometimes breaks down and presents a purulent appearance. When present in the shape of layers they may cause adhesion of the valves to one another.

These clots are usually single, unless they are very small and scattered over the edge of a valve. In such an event they are not larger than a millet-seed. Under other circumstances their volume is variable, and they may fill a large part of the cavity in which they are contained.

They are either free, especially when embolic in their origin, or they may have been adherent and then become free, or they may be interlaced among the tendons of the columnæ carneæ.

The formation of these clots is not explained sufficiently by the stasis of blood; they are the result of the exudations of endocarditis.

The diagnosis of heart clot is almost impossible. When an embolus is detached and occludes a vessel, the symptoms of sudden obstruction of this vessel may sometimes lead us to suspect its origin in the heart.

When grave symptoms suddenly indicate profound disturbance of the cardiac functions, with respiratory distress, extreme cyanosis or pallor, bloody expectoration, coldness of the extremities, and tendency to syncope;

when the movements of the heart are disordered, and these symptoms are rapidly followed by dropsy, we may suspect the formation of clots in the heart. The diagnosis is more certain if we notice at the same time the disappearance of a clot in any of the veins, or the symptoms of arterial embolism.

Richardson[1] and Gerhardt[2] have claimed recently to be able to recognize by the nature of the dyspnœa whether the thrombus is situated in the right or left side of the heart. But the signs which they regard as characteristic are far from possessing such importance. Cyanosis, disappearance of the clot, and pulmonary infarctions they consider characteristic of thrombosis of the right heart; pallor of the patient and embolism of an artery of the brain or limbs, of thrombosis of the left heart.

---

[1] Richardson: Lecture on Fibrinous Deposition in the Heart (Brit. Med. Journ., 1860 and 1873).

[2] Gerhardt: Ueber Blutgerinnung im linken Herzohre (Wurzburg Med. Ztg., Bd. iv., p. 150; 1864.—Ibid., Thrombosis cordis dextri, p. 221; 1864).

CHAPTER XLI.

ANEURISMAL TUMORS OF THE THORACIC AORTA (FALSE ANEURISMS.)

THE term aneurism refers to a dilatation of the arteries which occupies either a part of the contour of the vessel or its entirety. The latter is known as fusiform dilatation, or true aneurism. This is the variety which I have described as Hodgson's disease. At other times there is a rupture of the vessel with a larger or smaller tumor, and a usually contracted opening, known as the neck. This is the sacculated aneurism, the old false aneurism. The old classification has been abandoned, and to-day only three varieties are recognized: fusiform aneurism with atheroma (Hodgson's disease), sacculated aneurism, and dissecting aneurism, a variety of the former. The size of aneurisms varies greatly. Their walls are formed by the changed walls of the vessel. These changes are almost always the result of a chronic inflammation of the internal tunic.

The inflammatory process consists of thickening of the walls by the proliferation of connective tissue, atrophy, fatty degeneration, ulceration or calcification of the walls.

When the aneurismal sac is of large size the three layers of the walls are only found in a few places; sometimes traces of two can alone be discovered; finally, in the most advanced parts of the aneurism, no traces of the membranes are visible. The walls are then formed of the adjacent organs, connective tissue, muscles, etc., or even of the integument alone.

In a case in which the aneurism was merely covered by the integument, Burkhardt found the following histological condition. The subcutaneous cellular and adipose tissue had disappeared almost entirely; there were no blood-vessels in the subcutaneous layer, while the superficial parts of the integument were traversed by vessels which were full of blood. In the most prominent part the epidermis was converted into an eschar stained with blood; in other places it was replaced by a meshed tissue similar to connective tissue, with fibrinous deposits in the lacunæ. The sudoriparous glands were atrophied but the sebaceous glands were intact.

The interior of the sac contains stratified clots, the density of which increases the more we approach the periphery.

The sac may become inflamed and thus inflammation may be produced which will give the walls greater power of resistance. As a rule the inflammation is adhesive and gives rise to the formation of fibrinous exudation (Broca), but the inflammatory process sometimes terminates in suppuration. Lebert has twice found pus within the walls.

What are the causes of the coagulation of blood in the aneurism?

As Bruecke has shown, one of the first causes of this coagulation is the fact that the blood is not in contact with the normal walls of the vessels. Every solid foreign body in contact with the blood causes its coagulation, and atheromatous and especially calcareous patches upon the walls act in the same manner.

The aneurismal sac, by its growth, compresses the thoracic organs with which it comes in contact and causes their destruction and absorption; if it comes in contact with hollow organs it ruptures into their cavity. This feature will be considered later under the head of prognosis.

The secondary changes induced by aneurisms are hypertrophy of the heart with dilatation and, consequently, valvular insufficiency, both of the aortic and mitral valves. Either obliteration or dilatation of the arteries may develop as sequelæ.

DIAGNOSIS OF ANEURISMS OF THE THORACIC AORTA—EXAMINATION OF THE SYMPTOMS.

Tumefaction.—The growth of aortic aneurism pushes before it the tissues and elevates the ribs, and the first phenomenon which becomes apparent is the projection of the ribs. This prominence corresponds to the most projecting part of the tumor toward the exterior. The aneurism, destroying the tissues before it, forms a veritable tumor which projects between the ribs, or later, through the ribs, when these have been destroyed. At first the tumor appears about at the level of the origin of the aneurism, but as it enlarges its weight causes it to descend, and it is only arrested by the diaphragm. Thus an aneurism of the ascending aorta may rest finally upon the liver and project at the level of the third and fourth ribs. An aneurism of the lower part of the arch may occupy solely the left side of the chest and give rise to a tumor, not in the second intercostal space but at the level of the seventh, eighth, and ninth ribs, as we shall show by an illustration later on.

Pulsation.—The characteristic feature of the aneurism is its pulsation. It is evident that the tumor pulsates in all directions, *i.e.*, at each systole there is a movement of expansion which resembles more or less the pulsation of the main artery. These expansile movements are much more marked in the thorax when the ribs have been eroded and the muscles absorbed.

Pressure upon the veins usually leads to increased size of the superficial venules, which become varicose. At a later period the skin is inflamed and becomes gangrenous and rupture occurs either externally or into an internal cavity or even into the subcutaneous tissue.

In the intervals between the pulsations fluctuation may sometimes be detected during diastole.

The pulsations are so much more marked the more prominent the tumor is and the thinner its walls. If a graphic trace of the pulsation is taken, it will be found that it is delayed so much longer after the impulse of the apex the farther removed the orifice of the aneurism is from the heart. The pulsation sometimes appears to be double when the orifice of the aneurism is situated near the valves; a pulsation then occurs at the moment of the impulse of the apex, and another one at the closure of the valves. At other times the double pulsation is simply the result of oscillation due to the elasticity of the pouch.

OBSERVATION LXVIII. *Aneurism of the Ascending Aorta; Aortic Insufficiency.*—G——, aged thirty-six years, entered the hospital March 17, 1877. The patient had enjoyed good health hitherto; no rheumatism, syphilis, or alcoholic excess. At the age of fourteen he suffered from parenchymatous nephritis.

Four years ago he had a violent fall upon the right side of the chest in the mammary region; the pain disappeared at the end of a few days. Two years ago a dull pain began in the right side of the chest, without appreciable cause, and radiated into the arms, particularly the right. These pains were soon associated with very violent palpitations, which increased under the influence of excitement or effort. It soon became impossible to do any hard work, and extreme dyspnœa developed forthwith when the patient lay upon the left side. For almost a year he noticed a more marked protrusion of the right thorax at the level of the fourth and fifth ribs. The voice became muffled, there was difficulty in swallowing, and the neuralgic pains increased in severity.

Condition on Admission.—The patient is pale, like individuals suffering from aortic disease; respiration is rapid, and the dyspnœa is increased on the slightest effort. The thorax is elevated by pulsations and a prominence is visible to the right of the sternum at the level of the third, fourth, and fifth ribs, extending to the mammary region.

If the hand is applied to the tumor it is raised by the vigorous and rhythmical pulsations. Percussion reveals dulness over the entire tumor, passing inferiorly into hepatic dulness. Auscultation discloses a double blowing murmur, much more marked with the second sound than with the first; its maximum of intensity is situated at the level of the third intercostal space, one centimetre from the right border of the sternum.

The apex-beat is situated in the sixth intercostal space, outside of the mammary line. The size of the heart is increased; a blowing murmur with the second sound is heard at the base of the heart and along the aorta. The pulse is bounding and equal on the two sides.

Apart from these local symptoms the patient presents no thoracic disturbances. In addition to the dyspnœa and the painful pulsations, the aneurismal tumor gives rise to symptoms in the vicinity. The sternal pain is prolonged into the arms, particularly the right; the hoarseness of the voice and dysphagia are intermittent, and appear at irregular intervals.

All these symptoms render the patient unable to make the slightest effort; he is scarcely able to walk a few steps in the garden.

A sphygmometer enables us to note that the pulsations of the tumor and those of the heart alternate with one another.

Auscultation enables us to observe a separation between the limits of the cardiac murmurs and those of the tumor; the latter present their maximum of intensity in the third intercostal space, one centimetre from the right border of the sternum. Percussion allows us to affirm that there is a tumor of the ascending aorta, which probably presents its point of departure at the level of its extrapericardial portion.

According to our examination, the aneurismal sac, having its apex in the second intercostal space and its base at the level of the convex surface of the liver, occupied the right side of the aorta, extending over the third, fourth, and fifth intercostal spaces, and communicated with the aorta by an orifice very close to the aortic valves.

This diagnosis was completed by the examination of the heart. The apex-beat was situated in the sixth intercostal space, outside of the mammary line; furthermore, the blowing murmur with the second sound and the trace of the pulse showed that the increase in the size of the heart was due to aortic insufficiency.

Auscultation.—As a rule, a sacculated aneurism of the aorta gives rise to two murmurs which are usually blowing in character. Of the two murmurs, the first is the most constant; the second is sometimes absent, and it may happen that there is none at all. When the murmurs are present they are generally very rough and more or less prolonged, but they terminate very abruptly.

Their location depends upon the site of the aneurism. It is characteristic of them that they are situated outside of the ordinary positions of pathological heart murmurs. They are rarely heard over one spot alone. As a rule, they cover an ellipsoid surface which indicates very well the direction of the blood entering the sac at each systole. In such cases the maximum of the murmur is heard at the end of the ellipse which is nearest to the orifice of communication with the vessel. It is evident that the mode of propagation and the diffusion of the murmur depend on the condition of the clots in the sac, its walls, and the greater or less conducting powers of the intervening organs.

With regard to the period at which the murmurs are produced, there is no doubt concerning the first sound, which corresponds to the cardiac systole, consequently to arterial diastole. With regard to the second sound, it is very difficult to determine whether it corresponds to the second period of the impulse during systole, or to the cardiac diastole. I am strongly inclined to believe that it is produced at the latter period.

The timbre of the murmur depends undoubtedly on the narrowness and rigidity of the orifice, and upon the rugosities which may be present. There is much more chance of finding a second murmur if the first one is very rough and vibrating, and if a distinct purring thrill is perceptible.

It may be said, therefore, that the first sound is produced by the entrance of blood into the sac, by a rigid and rough orifice of medium size, and that it will be absent if the orifice is very large and destitute of resistance. It may also be said that the second sound is produced by the exit of blood from the sac, and that it is necessary to its production that the orifice be vibrating and the sac possess a certain amount of elasticity, either in itself or on account of the organs which it compresses.

The transmission of these abnormal sounds outside of the sites of their production depends upon the conducting power of the adjacent organs. In such cases the bones are good conductors, and for this reason the vertebral column very often transmits the sounds.

We must now consider the effects of the aneurism upon the pulse.

The Pulse in Aneurism of the Thoracic Aorta.—Aneurism of the aorta modifies the pulse in various ways. If it is situated near the origin of the innominate artery, a difference will be observed between the two pulses. The pulse on the side of the aneurism is commonly smaller than that on the opposite side, but sometimes the reverse holds true. F. Franck[1] observed the latter condition in two patients, and he explains the increased amplitude of the radial pulse on the affected side by the vasomotor paralysis of the vessels of this limb. Marey,[2] on the other hand, thinks that it is due to the wave furnished by the aneurism at the moment of reflux —a dicrotic wave which may be larger than that observed in the normal condition.

Franck has pointed out another characteristic which is much more important, because more constant; this is the retardation of the pulse on the side of the aneurism. This delay is indicated very clearly by Marey's polygraph. If care is taken to register the traces of the apex of the heart and the two pulses at the same time, it will be found that the retardation of the pulse on the affected side may amount to from five to seven hundredths of a second. Franck's discovery is very useful in locating the exact situation of the aneurism.

Marey has found also that, in the sphygmographic traces of the pulse of patients suffering from aortic aneurism, the undulations produced by the respiratory movements are much more marked than in the normal condition.

---

[1] F. Franck: Recherches cliniques et expérimentales sur la valeur comparée des signes fournis par l'examen du pouls radial dans les anévrysmes du tronc brachio-céphaliques de l'aorte et de l'artère sous-clavière (Journ. de l'Anat. et de la Physiol., xiv., Mars et Avril, 1878).

[2] Marey: La circulation du sang à l'état physiologique et dans les maladies, p. 641. 1881.

Effects Produced by Aortic Aneurisms upon Adjacent Organs.

In proportion as the aneurism enlarges, it pushes before it the adjacent organs; these undergo a sort of absorption, while the sac itself may be subjected to an inflammatory or gangrenous process, and discharge its contents into the tissues or cavity of the adjacent organ, or even rupture externally.

The situation of the origin of the aneurism determines which organs will be compressed. In like manner, the character of the disorders produced by this compression or rupture will serve as a point of departure for the diagnosis of the situation of the orifice of communication.

At the level of the origin of the aorta, the aneurism may press upon the heart itself and rupture into its cavities. Thus rupture has been observed into each of the cavities of the heart.

Aneurisms which are situated a little higher, between the origin of the aorta and the aortic sinus, may compress the pulmonary artery and open into this vessel. Crisp has observed four cases, Laveran has observed one.

When the aneurism compresses the pulmonary artery, secondary phthisis follows as in cases of primary stenosis of this vessel.

The aneurisms which develop between the origin and sinus of the aorta may project into the pericardium, and as they are not supported by other organs, they often rupture before they attain large dimensions. Death is usually rapid from the distention of the pericardium, the blood in this cavity compressing the heart and arresting its action. The patients generally die in less than an hour. Rupture into the pericardial cavity may be produced also by aneurisms of the descending aorta, particularly by dissecting aneurisms. Cases of this kind have been reported recently by Renault,[1] Cornil and Martin,[2] MacBride,[3] etc.

A little higher, at the level of the sinus of the aorta, the aneurism first displaces and compresses the superior vena cava; it may even be obliterated, as I have observed in two instances. The aneurism then attacks the pleura and presses outward, then downward, until it rests upon the diaphragm, as in the case reported above, and in which the autopsy proved the entire accuracy of the diagnosis.

Stokes relates that Dr. Mayne observed and diagnosed a case of rupture of an aortic aneurism into the superior vena cava. Apart from the interference with the venous circulation, a remarkable sign was noted, viz., a humming bruit analogous to that heard in arterio-venous aneurisms. This bruit was situated at the level of the insertion of the right second cartilage. The autopsy revealed a button-hole opening, with sharp and irregular edges, into the vena cava.

---

[1] Renault: Union méd., December 2, 1872, p. 859.
[2] Cornil et Martin: Soc. de Biol., May 16, 1874.
[3] MacBride: Glasgow Med. Journ., February, 1873.

An analogous case was observed by Chabond;[1] in this case an aneurism of the aorta communicated with the innominate vein.

It often happens that these aneurisms, or others situated farther from the heart, rupture into the pleural cavity. Cases have been reported recently by Coyne,[2] Besnier,[3] McCall Anderson,[4] Coupland.[5]

Aneurisms situated a little farther from the heart come in contact with and compress the arterial vessels. This affects the pulse, either on account of compression of the vessel or vasomotor paralysis. The aneurism may also compress the innominate vein or the other veins of the mediastinum. This gives rise to more or less localized œdema, and varicose veins appear on the surface of the chest.

When situated still more remotely, the aneurism comes in contact with the trachea and may then give rise to various symptoms.

At first, if the aneurism is a good conductor of sound, it transmits the tracheal bruits to the spinal column, and then we hear, at the upper part of the back over the spine, a prolonged, blowing expiratory murmur, which Trousseau had noticed in his clinics, and which I have had the opportunity of observing on two or three occasions. I have made mention of this phenomenon in discussing Hodgson's disease.

If the aneurism compresses the trachea, it first produces stenosis of this organ with dyspnœa, bronchial respiration, etc. ; then it may open by small fissures, whereupon repeated small hæmoptyses are observed. But, on the contrary, if the opening is large, a foudroyant hemorrhage occurs which may either appear externally or fill the air-passages.

At other times the laryngeal nerves are compressed or destroyed. In these cases the compression may affect either one or both recurrent nerves, but it is not surprising that the left recurrent should be affected most frequently, since this nerve passes under the arch of the aorta while the right nerve passes under the right subclavian artery.

In a case under my observation I observed with the laryngoscope the paralysis of the left vocal cord.

OBSERVATION LXIX. *Aneurism of the End of the Arch of the Aorta at the Origin of the Descending Aorta.*—M. B——, aged forty-three years, entered the hospital January 26, 1878, suffering from aneurismal dilatation of the heart and descending aorta. The disease began three years ago, January 1, 1875 ; the patient, on waking, found that he could not move his left arm which was bent under the head ; the forearm was cold. He entered the hospital at that time, when anæsthesia of the skin was found, and contracture of the fingers when he attempted to make a muscular effort. These symp-

---

[1] Chabond : Lyon médical, p. 478. 1873.
[2] Coyne : Mouvement médical, 15, October 12, 1872, p. 141.
[3] J. Besnier : Bull. de la Soc. anat., p. 233. 1873.
[4] McCall Anderson : Glasgow Med. Journ., November, 1872.
[5] Coupland : Path. Soc., Lancet, February, 1873.

toms improved very rapidly, and the patient left the hospital at the end of ten days. However, he could not resume work until the end of eighteen months, and even then the power of the limb was not restored entirely.

At the end of two years a fresh paralysis supervened; he suddenly became aphonic and remained so for three months. At the end of this time the voice returned, but its character had changed; it had become more shrill and often became hoarse. About the same time the patient was seized with a slight dry, frequent cough, but without dyspnœa.

Some time afterward new symptoms appeared, viz., difficulty in swallowing solids, which was often followed by nausea and vomiting. About ten months before admission to the hospital he began to complain of dyspnœa. During the past six weeks he has experienced acute pains in the right side, then in the left shoulder. Three weeks before admission he suffered from a phlegmonous angina, which terminated in an abscess of the left tonsil.

Present condition: The patient is pale and emaciated; the left radial pulse is entirely absent, nor could pulsations be felt in the left axillary or subclavian. Two finger's breadths above the clavicle pulsations were noticed, which were evidently produced by a transverse and sinuous arterial branch, that could be followed to the edge of the trapezius.

The apex of the heart could not be found on palpation; upon auscultation it was found in the fifth intercostal space, 10 ctm. from the median line.

A systolic elevation, synchronous with the right radial pulse, was observed, upon careful examination, in the left second intercostal space, 3 or 4 ctm. from the sternum. Expansile movements were not very evident.

Auscultation revealed a very loud systolic blowing murmur, with its maximum of intensity in the left second intercostal space, very near the sternum. This murmur extends, on the left, to 8 ctm. from the left border of the sternum, and on the right, to 1 ctm. beyond the right border of the bone. Above, it is propagated into the left first intercostal space, as far as the sternal insertion of the sterno-mastoid; below, it is lost in the left third intercostal space.

At the apex is heard a soft systolic murmur, separated from the upper foyer by a space of 4 ctm., in which the two murmurs are feebly audible.

Upper border of liver corresponds to insertion of right sixth costal cartilage; the border of right auricle corresponds to edge of sternum.

The cardiac dulness continues above into the second intercostal space; on the left, to 8 ctm. from the edge of the sternum; on the right, to the edge of this bone.

A slight prolongation of the bruit is heard in the arterial vessels of the neck. Posteriorly a very rough systolic murmur is heard along the spine, from the first dorsal vertebra to the middle of the dorsal region; the maximum of its intensity corresponds to the first three dorsal vertebræ. Examination of the lungs shows very feeble respiration throughout the entire left lung.

Diagnosis.—The most important point is the presence of a pulsating tumor, the maximum of expansion being in the left second intercostal space, outside of the pulmonary artery.

The pulsations of this tumor are synchronous with the pulse. The tumor is therefore an aneurism of the arch of the aorta. Its situation, to the left of the sternum, leads us to believe that it springs from the descending portion of the arch. On the left side, the lung is pushed to 7

ctm. beyond the sternum. The orifice of communication of the sac with the aorta appears to be situated immediately below the origin of the subclavian. The carotid is normal, and we may attribute to a lesion of the origin of the left subclavian the following phenomena:
1. The complete or almost complete absence of arterial pulsations from the subclavian artery to the radial; 2, the ready muscular exhaustion; 3, the spasms. The left recurrent nerve appears to be involved, thus accounting for the aphonia. Examination with the laryngoscope shows incomplete paralysis of the left vocal cord.

October 3, 1878.—The pulsations of the tumor are more noticeable to the patient; the dysphagia has increased and the patient states that the food is arrested at the level of the first portion of the sternum. Acupuncture of the sac was resorted to three times, the patient feeling greatly improved thereby. The dyspnœa notably diminished, the deglutition of fluids and solids was effected more readily. Toward the end of January, 1879, acupuncture was again performed, the needles being introduced with great difficulty. The patient, feeling better, then left the hospital.

Upon inquiry, I found that he had resumed his work, but that a month later he sank into a very serious condition, and died in a few days.

In a case observed by G. Johnson,[1] in which the left recurrent and pneumo-gastric nerves alone were compressed, the right recurrent being intact, all the muscles of the larynx were found atrophied. The laryngoscope had shown paralysis of both vocal cords; aphonia and dyspnœa were present. There was also compression of the œsophagus with dysphagia. Tracheotomy was performed, and although this relieved the patient, he died two days later.

In other cases, Stokes has observed laryngeal spasm as the result of irritation of the pneumogastrics, and these are the only cases in which tracheotomy is proper.

Compression of the bronchi may lead to other phenomena, for example, dilatation behind the obstruction by the accumulation of the products of secretion. Desplats[2] has reported a case of this kind in which death was due to a foudroyant hæmoptysis, and the bronchi were found distended and filled with pus.

An extremely important point is the occurrence of phthisis secondary to aneurisms of the aorta, a relation which has been well established by Hanot.[3]

Stokes said:[4] "Of all the general morbid conditions which accompany aneurism of the aorta, pulmonary phthisis is the most common." And he adds: "In these cases, the phthisis often presents equivocal, irregular symptoms, and its course is slowly progressive."

---

[1] G. Johnson: The Lancet, December 7, 1872, and January 4, 1873.
[2] Desplats: Union Méd., August 26, 1879.
[3] Hanot: Du rapport entre l'anévrysme de la crosse de l'aorte et la pneumonie caséeuse.
[4] Stokes: Traité des Maladies du Cœur et de l'Aorte, p. 590. Translated by Senac.

The following table, illustrating this connection, has been collated by Fuller:

Guy's Hospital	27 aortic aneurisms,	3	phthisical.
Fuller	7 " "	1	"
Pathological Society	25 " "	0	"
Rokitansky	108 " "	5	"
Crisp	8 " "	0	"
Crisp	32 " "	2	"
Green	12 " "	4	"
Walshe	14 " "	2	"
	233	17	

Hanot's observations tend more strongly in the same direction. In certain cases he has found that the phthisis is the result of compression of the pulmonary artery, again confirming the statement made by me concerning stenosis of this vessel. In other cases, he admits Habershon's theory that the phthisis is the result of compression of the pneumogastric nerve. This theory has been accepted by Herard and Cornil, Bucquoy, Vulpian, Reynaud.

The œsophagus is often compressed by aneurisms, and the patients then suffer from dysphagia. Catheterization of the œsophagus may enable us to recognize the presence of a tumor. Rupture of the aneurism into the œsophagus has been observed on several occasions.

Leudet has reported several cases in which the œsophagus, although not compressed, presented ulcerations and gangrenous patches. Hanot believes—and I accept his theory—that these ulcerations, and even the gangrene, may be the result of trophic disturbances produced by irritation of the pneumogastric nerves by the aneurismal tumor.

Compression, or rather irritation of the pneumogastrics, like division of these nerves, will then give rise to nutritive disturbances of the larynx, lungs, and œsophagus. Hanot thinks that certain forms of pericarditis are also due to the same cause.

The bones must be numbered among those organs which may be compressed, irritated, and particularly eroded. Sometimes the ribs are absorbed over a certain area, leaving only a soft tumor; sometimes the spinal column is eroded. The vertebræ then become painful on pressure or during motion; then the irritation may extend to the spinal meninges and the cord, or compression of the cord is made manifest by paralysis. The aneurism may open into the spinal canal.

Certain aneurisms of the descending portion of the arch merely push back the left lung, involve no important organ, and cause disturbance by their volume alone. In such cases expansile pulsations are perceptible, but no blowing murmur. This occurred in a patient under my observation, in whom the autopsy revealed an enormous aneurism with a large opening, starting from the thoracic part of the descending aorta. The following is the history of this case:

OBSERVATION LXX. *Aneurism of the Descending Aorta.*—D——, aged fifty-six years, has been complaining for eighteen months.

On the left side of the heart is found a spherical or ovoid tumor, the inner portion resting on the spinal column, the upper border extending to the seventh dorsal spinous process. The lower border extends to the first lumbar vertebræ. The spinal column is curved to the right at the level of the sixth, seventh, and eighth dorsal vertebræ; above the tumor is a slight curvature in the opposite direction.

The tumor, which begins internally and posteriorly at the spine, occupies the entire posterior thoracic wall, the lateral thoracic wall, and elevates the ribs anteriorly. It presents very evident expansile pulsations. The left ninth, tenth, eleventh, and twelfth ribs are not recognizable near the spine, and they have probably been eroded in this situation.

Percussion shows dulness posteriorly over the entire extent of the tumor. The lateral portion is also dull, but its upper part is curved, and presents a resonant portion, indicating that the lung is interposed between the tumor and chest-wall. Auscultation shows complete silence posteriorly, except that a distant murmur is heard in the inferior and internal part.

Anteriorly the ribs are not very much deformed, but there is an elevation *en masse* of the lower part of the left side of the thorax. There is a violent impulse on the left border of the xiphoid cartilage. No blowing murmur is heard in any situation.

The following is a recapitulation of the signs of aneurisms of the arch of the aorta:

1. A tumor in some part of the thorax.
2. Purring thrill outside of the heart.
3. Expansile pulsation.
4. Dulness on percussion.
5. Difficulty in reducing the tumor, with pain and threatenings of angina pectoris.
6. Upon auscultation, one or two murmurs outside of the localities of cardiac murmurs.
7. Bad general condition, emaciation, and loss of energy.
8. Signs of irritation or even of compression and erosion of adjacent organs; interference with circulation in the vena cava and other venous trunks; irritation or compression of the pericardium, pleura, laryngeal and pneumogastric nerves, trachea, bronchi, œsophagus, etc.
9. Difference in the radial pulses, and particularly delay in the pulse.
10. Cardiac cachexia.

Differential Diagnosis of Aortic Aneurisms, and some Diseases which may Simulate Them.

When a patient presents all the signs mentioned above, the diagnosis is readily established, but such cases are rarely observed. There are a number of possible causes of error.

Pulsations may be due to various causes:

1. To Hodgson's disease, *i.e.*, to a true aneurism; this mistake would not possess much importance.

2. The pulsations may be communicated by the aorta to a solid tumor, particularly to a carcinoma. But this tumor is generally a poor conductor of sound, and is ordinarily silent on auscultation.

The opposite mistake may also be made, and an aneurism regarded as a cancerous or glandular tumor. This is due to the fact that sufficiently careful search has not been made for the pulsations and the murmurs on auscultation.

The course of an aneurism is generally slow; it is accompanied by an increase in the hypertrophy of the heart, which has usually preceded the aneurism, and has developed at the same time as the other vascular lesions, except in traumatic cases.

Aneurisms often terminate by a perforation, and then death usually occurs suddenly, within twenty to forty minutes. But if the perforation is very minute, small hemorrhages sometimes continue for a long time, indeed, they have been observed for several years.

Nothing is more difficult to establish than the prognosis of aortic aneurism. Stokes has well expressed himself in the following manner: "Given an aneurism, it is impossible to foretell in what direction it will grow, the duration of the life of the patient, or the character of the final symptoms. It cannot even be affirmed that death will be the result of the aneurism, as it is sometimes caused by an acute or chronic disease of an entirely different nature, especially if judicious treatment has arrested the progress of the aneurismal affection. In patients who do not present any disease of the heart, and in whom the aneurism does not compress any important organ, such as the trachea or œsophagus, death may take place in several ways (if the sac does not rupture). The patient may succumb to the effects of the compression exercised by the aneurism upon the surrounding parts; he may be worn out, so to speak, by the pain, deprivation of sleep, and an inflammatory fever. He may be carried off by some intercurrent acute disease, pneumonia, cholera, typhoid fever, phthisis, pulmonary gangrene, or a cerebral affection."

ANEURISM OF THE PULMONARY ARTERY.

Aneurism in this situation is very rare, as the pulmonary artery is scarcely ever subject to chronic endarteritis. Among 915 cases of aneurism, Crisp found only 4 of the pulmonary artery. Goldbeck[1] collated 20 cases in his inaugural thesis.

These aneurisms are situated usually upon the trunk of the vessel. We do not refer now to those which form in the interior of the lungs, and constitute one of the phenomena of pulmonary phthisis.

Aneurisms of the trunk of the pulmonary artery are usually small, from the size of the head of a pin to that of a hen's egg. When the tumor attains a certain size, it projects into the left second intercostal space, and raises the left second cartilage. In Lebert's[2] case the tumor projected even into the first intercostal space. Feeble expansile movements have been noticed in the second intercostal space, sometimes even in the first; Dowse also detected a diastolic thrill.[3] Very extensive dulness has sometimes been noted; auscultation reveals a loud systolic murmur, with thrill (Dowse). Lebert heard a double murmur in the left second intercostal space, and a diastolic murmur in the right second space.

These aneurisms usually rupture into the pericardium. They may be complicated by other cardiac lesions and lead gradually to cardiac cachexia.

OBSERVATION LXXI.—A laundress, aged thirty-seven years; disease began with acute articular rheumatism seven months previously.

She presented œdema of the lower limbs, ascites and jugular reflux. Dulness in the left second and even the first intercostal spaces, with pulsation. Auscultation shows a systolic murmur at the apex of the heart. In the left second space the murmur was systolic and diastolic; in the same space, on the right side, it was diastolic in character. The patient sank into a condition of collapse and died.

The pulmonary artery was spherical and contained clots; the valves presented vegetations. The pulmonary artery had a circumference of 110 mm., the aorta 70 mm. (Lebert).

ANEURISMS OF THE CORONARY ARTERY.

Aneurisms of the coronary artery are found in young subjects. They are usually multiple; from three to twenty have been found in one individual.

There is usually no arterial sclerosis but endocarditis of the left ventricle. These aneurisms, which are associated with aneurisms of the cerebral arteries, are probably due to embolism.

---

[1] Goldbeck: Dissert. inaug. Giessen, 1868.
[2] Lebert: Berl. Klin. Woch., 20, p. 273, 1876.
[3] Dowse: Brit. Med. Journ., December 5, 1874, p. 710.

Quincke[1] found twenty in a young girl, varying in size from the head of a pin to a bean. The patient suffered from an old mitral affection, with a relapse characterized by fresh vegetations upon the valves. The aneurisms were either fusiform or sacculated, and were situated upon the branches of the third or fourth order. All contained clots, though the vessels were still permeable. The coronary arteries were dilated at their origin, but not sclerotic. This observation gives a good idea of what has been found by others in such cases.

The aneurisms are generally superficial and rupture either into the tissue or into the cavity of the pericardium, more rarely into the myocardium. Aran has observed rupture into the ventricle.

The symptoms are merely those of hæmopericardium and terminate rapidly in death.

Observations of this variety of aneurism are rare and we may add the following to those which have been quoted above: Peacock (*Monthly Journal*, March, 1849); Lombard (*Gaz. Médicale*, iii., 644); Richard (*Bulletin de la Société Anatomique*, xxi., 40); J. V. Ogle ("St. George's Hosp. Rep.," 1867, p. 285); Feignaux (*Journ. de Bruxelles*, January, 1859, p. 30); Crisp ("Trans. of the Path. Soc.," xxii., p. 106, 1871; "St. Bartholomew's Hosp. Rep.," vii., p. 147, 1871).

Diseases of the Heart in the Fœtus—Congenital Diseases of the Heart.

THE FŒTAL HEART.

At the beginning of the evolution of the embryon, the circulatory apparatus is represented by a few cells. Some of these are soon colored and constitute the blood-globules; others arrange themselves in parallel series and are the first rudiments of the circulatory passages. Still others begin to contract and give rise to a movement which does not cease with life.

The development of the heart may be divided into three stages, viz., the vitelline, allantoic, and pulmonary.

In man, the cardiac cavity becomes curved and the first pulsations appear toward the fifteenth day. The heart, which is straight in the beginning, soon becomes curved like the letter S. The upper part furnishes the two aortic arches; the lower part receives the omphalo-mesenteric veins.

The heart soon divides into three cavities; the upper one, which furnishes the aortic arches, becomes the bulb of the aorta; the lower one, which receives the omphalo-mesenteric veins, becomes the auricle, and the middle one the ventricle.

---

[1] Quincke: Krankheiten der Gefaesse, p. 419.

During the second or allantoic stage, a septum forms which divides the ventricle into two. Then the two auricles become distinct; the auriculo-ventricular orifices are mapped out, and a septum appears in the aortic trunk, dividing it into the aorta and pulmonary artery.

At the moment when the auricular cavity is divided into two, the vena cava, which is still single, is provided, at its entrance into the auricle, with two valves. The one on the left side is the valve of the foramen ovale, that on the right side is the Eustachian valve. These two valves form an arch which passes on to meet an arch proceeding from the posterior wall. The space between the two arches forms the foramen ovale.

At a later period, when the vena cava has divided into the superior and inferior branches, the Eustachian valve is arranged in such a manner as to conduct into the foramen ovale the blood which comes from the inferior vena cava, while the blood from the superior vena cava is directed toward the same orifice by a projection known as the tubercle of Lower.

When the septum forms in the aortic bulb and divides it into the aorta and pulmonary artery, one portion remains in its original condition and allows communication between the two vessels. This opening, which persists into extra-uterine life, is known as the arterial canal or canal of Botalli.

At birth the placental circulation is replaced by the pulmonary circulation.

The Forms of Congenital Lesions of the Heart.

Isolated Persistence of the Foramen Ovale.

At the lower and inner portion of the auricular septum in the adult is a depression known as the fossa ovale, which marks the site at which, in the fœtus, is situated the foramen ovale. This depression is continuous below and posteriorly with the inferior vena cava, thus indicating the way followed by the blood in passing from the inferior vena cava almost directly into the left auricle.

In front of and above the fossa ovale is found a muscular projection with its concavity posteriorly and inferiorly. This is the valve of Vieussens, which is continuous below with the Eustachian valve and, in its development, gradually closes the foramen ovale. These two arches, crossed like a pair of scissors, often allow the passage of the handle of a scalpel, which can thus penetrate from the right into the left ventricle. In these cases, there is a communication between the right and left sides of the heart from an anatomical point of view, but, at the moment of auricular systole, the two half-rings are applied to one another, and close the opening, so that there is no admixture of the arterial and venous blood.

But this complete development of the two valves does not exist in all subjects, and it is often found (perhaps more frequently in females than in males) that a slit or orifice of communication between the auricles persists. Sometimes this is very small, but in certain cases it permits the introduction of the little finger.

As a rule, these vices of conformation produce no disturbance. It is not until a later period, if this sort of auricular insufficiency has fatigued the heart like mitral insufficiency, that the organ, being in a condition of asystole may give rise to cyanosis, as in acquired diseases of the heart affecting the venous orifices.

In certain cases the diagnosis may be suspected. Two years ago I saw a little girl suffering from palpitation and hypertrophy of the heart, in whom the beginning of asystole was indicated by œdema and venous stases, and a soft, deep blowing systolic murmur was present to the right of the sternum, in the second intercostal space. This was due probably to a persistence of the foramen ovale, but I merely present this as an hypothesis, as I have lost sight of the patient.

PERSISTENCE OF THE DUCTUS ARTERIOSUS.

It is a mistake to regard as pathological the persistence of the ductus arteriosus in the new-born. The anatomo-pathological investigations of Billard [1] show that the canal is not obliterated, as a rule, until two weeks after birth. In truth, the canal may not close until three weeks or a month after delivery.

Table showing the Period of Closure of the Canal.

Age.	Number of cases.	Author.	Open. Percentage.	Partly closed. Percentage.	Obliterated. Percentage.
Birth	19	Billard.	68	22	10
2 to 7 days..........	100	Billard.	60	25	15
8 days...............	20	Billard.	15	30	55
10 to 20 days........	21	Brière.	..	33	66

It is evident from this table that the permeability of the canal becomes rarer as we approach the twentieth day.

The persistence of the ductus arteriosus after the first month is pathological, and other vices of conformation usually are found associated with it.

---

[1] Billard : Traité des maladies des enfants nouveau-nés, p. 573. 1833.

When the persistence of the ductus is the sole lesion, life may be prolonged, as is shown by the following table:

Case reported by Sanders................	4½ months.
" " Almagro[1]................	19 years.
" " Bernutz[2]................	23 "
" " Babington[3].............	32 "
" " Duroziez[4]...............	48 "
" " Luys[5]..................	52 "

In these six cases, despite the admixture of the two kinds of blood for the lower limbs, while the upper limbs received only arterial blood, no disproportion was observed between the upper and lower extremities.

There were no very marked cardiac disturbances in these patients, but peculiar respiratory disorders were noted. It must be remarked that the blood which reached the lungs was less abundant at each pulsation and more venous.

In the first months of life the cardiac disturbances in all these patients have been slight. The infant observed by Sanders presented no cyanosis and but slight cardiac hypertrophy. In all the patients who have survived for a long time considerable hypertrophy has been found, as in insufficiency of the pulmonary valves.

The peculiar respiratory disturbance is very noticeable in these patients. It begins at the end of the second or third year, and is particularly well marked after exertion or crying. At the slightest movement the patients are seized with suffocation, with a violet discoloration of the integument. But despite the constant admixture of the arterial and venous blood, cyanosis is not constant; it occurs in paroxysms and then disappears. In the case observed by Duroziez, cyanosis was not produced until the occurrence of the final symptoms attended with asystole.

Auscultation has revealed blowing murmurs in these cases, but they have not been well defined. In two cases there was no blowing murmur; Sanders' patient presented a systolic blowing murmur; in the cases of Almagro and Bernutz there were two blowing murmurs, but it must not be forgotten that the aorta was narrowed in both.

The course of this disease is the same as that of organic affections of the heart, characterized by insufficiency of the auriculo-ventricular orifices. The lesion is probably due to relative narrowness of the ascending or descending aorta.

---

[1] Almagro: Thèse de Paris, 1862.
[2] Bernutz: Arch. de Méd., xx., p. 451. 1849.
[3] Babington: London Med. Gaz., May, 1847.
[4] Duroziez: Soc. de biol., 1862.
[5] Luys: Bull. de la Soc. anat., Juin, 1855.

ANEURISM OF THE ARTERIAL CANAL.

The process of obliteration of the arterial canal is effected by the successive organization of stratified clots; and if the two ends of the canal are retracted more than the middle, it thus assumes the appearance of an aneurism. This has been observed by Billard and Thore.[1] In a case seen by Luellich,[2] in an infant two weeks old, the aneurism measured 15 mm. in diameter, was filled with clots, and opened 3 ctm. farther down in the descending aorta.

CONGENITAL AFFECTIONS OF THE RIGHT HEART.

1. SIMPLE STENOSIS OF THE PULMONARY ARTERY WITH CLOSURE OF THE FORAMEN OVALE.

The following is the history of a case reported by d'Heilly.[3] The affection in this instance must be considered as probably congenital, in view of the irregular development of the patient:

OBSERVATION LXXII.—M. X——, aged twenty-one years, entered the hospital September 1, 1863; died on October 15th. Never had any serious illness; since infancy has suffered from shortness of breath and palpitations; caught cold very easily in the winter. Three years ago he had several attacks of spitting of blood and epistaxis. For three months past the dyspnœa and palpitations increased.

Condition on admission: aged look, emaciation of arms and chest, contrasting with increased size of lower parts; face congested but not cyanotic. Very marked dyspnœa, frequent palpitations, slight cough. Obscure sound on percussion over the upper two-thirds of chest posteriorly, complete dulness inferiorly. Auscultation of chest shows feeble respiratory murmur mingled with fine, moist râles; complete silence posteriorly and inferiorly.

Marked prominence of the præcordial region; irregular and very violent pulsations felt upon applying the hand. Percussion shows considerable hypertrophy. A very feeble double blowing murmur is heard at the base of the heart, but not very distinctly; the murmur disappears at a distance from the base. Feeble radial pulse. Percussion of the liver shows no increase in volume; evidences of fluid in the abdomen; urine contains a considerable amount of albumen.

After numerous alternations of improvement and relapse, the patient was seized with a syncopal attack and died on October 15th.

Autopsy.—The pericardium occupies a considerable portion of the thorax; it is distended by an enormous amount of fluid, pushing back the lungs, which crepitate slightly and are œdematous. Some crude tubercles at the apex of the right lung. Slight amount of fluid in pleural cavities.

---

[1] Thore: De l'anévrysme du canal artériel. Arch. Gen. de Méd., 1850.

[2] Luellich: Arch. d. Heilkunde, 1876.

[3] D'Heilly: Des oblitérations et rétrécissements congenitaux de l'artère pulmonaire. Thèse de Paris, 1864.

The auriculo-ventricular orifices are normal; aortic orifice slightly narrowed. The heart is considerably hypertrophied, the right ventricle being relatively thicker than the left and very much dilated.

The orifice of the pulmonary artery is occupied by a horizontal diaphragm having the shape of a spherical cap, the concavity being directed toward the ventricle, the convexity toward the lumen of the vessel.

The artery measures 67 mm. in circumference. Its lower surface is smooth and regular; the upper surface presents three diverging projections, the traces of the separation of the valves. In the centre of the diaphragm is an opening 4 mm. in diameter. This orifice is covered by a sort of circular rim which projects into the cavity of the vessel. The upper surface of this operculum, apart from the diverging projections, is perfectly smooth.

The pulmonary artery, above its orifice, presents the normal dimensions.

The inner surface of the heart presents no traces of inflammation and the septa are intact. The venæ cavæ are large and manifestly dilated.

This observation furnishes the following data:

1. The lesion of the pulmonary artery is isolated.
2. Closure of the foramen ovale.
3. Absence of cyanosis until the end of life.
4. A double blowing murmur at the origin of the pulmonary artery, indicating both stenosis and insufficiency.
5. Life lasted twenty-one years.
6. Death from syncope.
7. The presence of crude tubercles at the apex of the lungs.

2. STENOSIS OF THE PULMONARY ARTERY WITH PERSISTENCE OF THE FORAMEN OVALE AND OBLITERATION OR EVEN DISAPPEARANCE OF THE DUCTUS ARTERIOSUS.

These cases are more frequent than the preceding, and like them are undoubtedly developed during the last period of intra-uterine life.

In these cases the stenosis is situated generally at the origin of the pulmonary artery and is caused by adhesion of the valves. This gives rise to a convex septum projecting toward the lumen of the artery and perforated in the centre by a rounded opening, so that stenosis and insufficiency are almost always present at the same time. The calibre of the artery may be normal. The infundibulum is usually small, and the right ventricle is the seat of concentric hypertrophy with thickening of the walls and considerable diminution of its cavity. The auricle retains its normal dimensions and communicates more or less freely with the right auricle through the open foramen ovale.

This condition is compatible with life, as in the following case reported by Bertin, in which the patient lived to the age of fifty-seven years:

OBSERVATION LXXIII.—M. V—— has presented a peculiar appearance since early childhood. As soon as she engaged in any work her color changed to a violet-red; respiration was always obstructed. At the age of forty-seven years she ceased menstruating and began to complain of palpitations with acute pain in the præcordial region. She often had to rest while walking; the lips and face became bluish even while walking slowly; had severe attacks of epistaxis. July 1, 1821 (she was then fifty-seven years of age), she complained of a species of spasm in the left hand and foot. She soon noticed a great difficulty in the movements of these limbs; soon after motion and sensation were lost on this side but consciousness was retained.

On the third day after this attack she entered the hospital and presented the following symptoms: face of a violet-red color, eyes prominent and brilliant, lips bluish, respiration obstructed; pulse small, readily compressed in left arm but hard and strong in right arm; complete loss of motion and sensation in limbs of left side. During the night the paralyzed limbs were seized with convulsions, accompanied by greater difficulty of respiration. The lips assumed a rose color; the cardiac impulses were tumultuous. Consciousness not lost during the attack; symptoms subsided almost entirely in the morning. A number of milder attacks occurred from July 5th to 12th. July 12th patient suddenly lost consciousness, the pupils dilated, respiration was more and more obstructed, and the paralysis extended to all parts of the body. The symptoms gradually grew worse, and death occurred toward noon.

Autopsy.—An abscess was found in the anterior portion of the right cerebral hemisphere. The heart was extremely large. The right auricle was very thick and contained several ounces of blood; the fossa ovale was very deep; the foramen ovale was open, about four lines wide, and permitted communication between the auricles. The right auriculo-ventricular orifice was narrow; the capacity of this ventricle was little larger than that of a pigeon's egg; its walls varied from 11 to 16 lines in thickness. The valves were small, but their chordæ tendinæ large. At its origin from the right ventricle the pulmonary artery presented a horizontal septum, convex toward the artery, concave toward the ventricle; it was perforated in the centre by a circular opening 2½ lines in diameter. Three small folds were apparent upon its convexity, but no trace of its separation into three valves could be detected. Above this septum the artery offered no peculiarity. The left auricle had the usual dimensions. The left ventricle, the capacity of which was much greater than usual, had very thick walls. The ductus arteriosus was small and entirely obliterated. (Bertin, in "Recherches Anatomiques sur l'Encéphale," by Lallemand, II., p. 7.)

The question arises, with regard to this and the preceding forms of disease, whether the pulmonary stenosis was really congenital or developed shortly after birth. Kussmaul states that the stenosis may be regarded as congenital under the following circumstances:

1. When death occurs soon after birth.
2. When cyanosis appears early, and the symptoms of pulmonary stenosis are observed soon after birth.
3. When the foramen ovale remains open, and especially when the ductus arteriosus remains permeable.
4. When the patent foramen ovale is very wide.
5. When the stenosis of the pulmonary artery is produced by an evident vice of conformation.
6. When the pulmonary artery is narrow and its walls very thin.

7. When the right ventricle is very small and contracted on itself.

The symptoms and course of this affection are the same as those of acquired pulmonary stenosis, which have been described above. It terminates frequently in pulmonary phthisis, though this is not necessarily fatal.

3. STENOSIS OF THE INFUNDIBULUM OF THE PULMONARY ARTERY WITH INCOMPLETE DEVELOPMENT OF THE VENTRICULAR SEPTUM AND COMMUNICATION BETWEEN THE TWO VENTRICLES.

In these cases the stenosis may affect the base of the infundibulum, and the latter, being thus separated from the right ventricle, appears to form a small supplementary ventricle. The pulmonary artery, which receives little blood, is not well developed, and its valves often present vices of conformation, so that, for example, only two valves are present. The walls of the artery are thin, often looking more like a vein.

Below the obstruction, on the contrary, the right ventricle is hypertrophied. The foramen ovale is not always open.

Stenosis of the infundibulum is usually the result of intra-uterine myocarditis, and closely resembles what I have previously described as præarterial pulmonary stenosis. Cases of this kind are reported by Peacock [1] and Oldham.[2]

Kussmaul [3] believes that this form of stenosis may be due to a sort of hypertrophy of one or more muscular bundles at the base of the infundibulum.

4. CONGENITAL STENOSIS OF THE PULMONARY ARTERY, OR EVEN COMPLETE OCCLUSION OF THIS ORIFICE WITH PERSISTENCE OF COMMUNICATION BETWEEN THE VENTRICLES.

This is by far the most frequent of congenital affections of the heart. It would not be difficult to collate more than a hundred observations.

This affection, which is produced during the first three months of intra-uterine life, often remains stationary after its development, and if the inflammatory process ceases it may become stationary and be compatible with very long life.

Among 64 cases of congenital stenosis with patency of the ventricular septum, Kussmaul has noted the following duration of life :

From 0 to 1 year .... 8 cases.
" 1 to 5 years .... 14 "
" 5 to 10 " .... 19 "
" 10 to 20 " .... 14 "
" 20 to 30 " .... 9 "

---

[1] Peacock: Report of the Proceedings of the Pathological Society of London, 1847–48.
[2] Lebert-Schroetter: Angeborene Herzkrankheiten, p. 746. 1879.
[3] Kussmaul: Ueber angeborene Enge u. Verschluss d. Lungenarterienbahn. Zeitschr. f. rat. Med., Leipzig, 1866.

In complete or almost complete obliteration of the vessel the duration of life is diminished, as is shown by the following table of 24 cases, also collated by Kussmaul:

<pre>
From 0 to 6 months .... 10 cases.
  " 6 months to 1 year 4   "
  " 1 to 5 years....................... 5   "
  " 5 to 10 " 3   "
     To 21   " .... 1 case.
     To 37   " 1   "
</pre>

ANATOMICAL LESIONS.

The pulmonary artery is narrowed; its calibre may be reduced to that of a goose-quill or crow-quill, and indeed the stenosis may progress to complete obliteration. The walls of the vessel are usually thinned and resemble the walls of veins.

When the stenosis is not marked the valves may be found normal in number and development. However, an arrest of development is often observed, two valves being present instead of three.

As a general thing the valves are adherent at one part of their edge and free in the inner portion. They thus form a sort of cone, or rather cupola, with the convexity directed toward the artery.

If the valves are free in their centre they form a narrowed orifice which diminishes the amount of blood passing through it in a given time; but they may then close without insufficiency. But it often happens that the valves have been inflamed and their edges thickened and indurated, thus forming a sort of diaphragm pierced by a more or less narrow opening, and giving rise to stenosis and insufficiency. Finally the valves may be very rough and even calcified.

If the obliteration of the orifice is almost complete the artery may be obliterated almost entirely and converted into a simple permeable strand, sometimes even into a strand which is impermeable as far as the bifurcation.

The condition of the heart varies. Sometimes it is atrophied, sometimes enlarged; at other times one side of the heart is dilated, the other thickened and contracted. When the heart is opened it is often found that the stenosis is not confined to the orifice but affects the infundibulum, and the muscular fibres of the latter have been found increased in size in such cases. The tricuspid valve is often opaque and thickened, a proof that endocarditis is frequently the primary lesion; at other times the valve is imperfectly formed or even rudimentary. The right auricle is almost always enlarged and its walls thickened.

The left ventricle is usually smaller and thinner than the right, but the opposite condition is observed occasionally, particularly in cases of atresia or almost complete obliteration. The aorta is generally dilated and issues from one or the other ventricle, very often from both. It always retains the normal position with reference to the pulmonary artery.

The perforation of the septum is situated constantly at the base, *i.e.*, near the aortic orifice. It is often large enough to allow the passage of the little finger, but sometimes will merely admit a stylet. Its edges are more or less roughened. The remainder of the septum is often muscular, but it is sometimes reduced to the endocardium.

The foramen ovale remains patent in three-fourths of the cases. The size of the opening varies extremely, and at times it is so large that there is, so to speak, no inter-auricular septum.

The ductus arteriosus is usually closed, but Seiler found it permeable in an individual twenty-three years of age. When this canal is absent it is replaced by a remarkable development of the bronchial arteries.

The separation of the aorta and pulmonary artery is always complete, and the auricles are always separated from the ventricles by a septum, even when the ventricles are not completely developed. The pulmonary artery always takes its origin from the right ventricle, while the aorta starts either from the left ventricle, from both ventricles, or sometimes even from the right ventricle.

Symptoms and Diagnosis.

The disturbances caused by this anomaly begin so much earlier the more marked the lesion. Thus, in cases of atresia or complete obliteration the circulatory and respiratory disturbances begin at birth.

When the stenosis is less marked the symptoms appear at a later period. In Landouzy's case they did not begin until the fifth year. We have seen that they may begin at a much later period, when the stenosis is not accompanied by perforation of the septum.

As a rule, the first symptom which strikes the observer is cyanosis or bluish and livid discoloration of the integument. This form of cyanosis persists, unlike that observed in asphyxia of the new-born, which disappears when respiratory efforts are made. It is especially marked upon the lips, cheeks, nose, ears, hands and feet, and the genitals, and increases after movement, particularly after crying. The respirations are often short, and may be so interfered with as to lead to suffocation. The skin is cold and covered with perspiration. Under these circumstances the child may remain asphyxiated and die in collapse.

The patient may survive if the lesion is less advanced and he is not carried off by an acute affection.

Those who survive are usually apathetic; they avoid movements which increase their asphyxia, and remain voluntarily in bed. Morgagni saw a young girl who had remained in bed until she was sixteen years of age.

The bluish color of the face is accompanied by a sort of puffiness, with prominence of the eyes, and often injection of the conjunctivæ. The mucous membranes are also cyanotic, the gums are livid and bleed readily, the tongue is large and cold. The veins are turgescent, particularly in the upper parts of the body. The temperature of the body is lowered several degrees, and the children often complain of a sensation of cold. However, the temperature rises as high when they are affected by febrile diseases as it does in other children (Schroetter).

Does the cyanosis depend solely on the stenosis of the pulmonary artery? In my opinion it does not, for in the cases of pulmonary stenosis under my observation cyanosis was not present. Nor does it depend on the mixture of venous and arterial blood, since cases have been observed in which a communication existed between the two sides of the heart without the occurrence of cyanosis. At the present time the opinion of Louis[1] and Cruveilhier is more generally accepted, viz., that the cyanosis is due, as in all cardiac diseases, to deficient oxygenation on account of stasis, and consequently that it is favored by every obstacle to circulation, the passive obstructions on the one hand and the feebleness of contraction of the heart on the other hand.

Palpation often discloses a purring thrill.

Auscultation almost always reveals a peculiar blowing murmur. It is situated in the middle of the cardiac region, and is scarcely heard over the orifice of the pulmonary artery. It does not occupy the site of any of the valvular murmurs, but is a substernal median murmur at the level of the insertion of the third costal cartilages. According to Roger it is propagated equally in all directions; Féréol has found that it is prolonged more particularly in the direction of the apex of the heart.

The murmur begins and ends with systole, covering the first sound and the period of lesser silence; it is loud, rough, and very vibrating.

The patient complains almost constantly of dyspnœa, and suffers very often from palpitation with a tendency to lipothymia and syncope. He is very much subject to bronchitis, which is apt to be attended with asphyxia.

The pulse is variable, sometimes normal, sometimes rapid. Clubbed fingers are often observed in these cases. The development of the children is usually below the average.

Apart from bronchitis, the patients are predisposed to other diseases, viz., enlargement of the liver and spleen, albuminuria, and dropsy. But the most important complication is tuberculosis. Lebert[2] especially has

---

[1] Louis: Arch. Gén. de Méd., 1823.
[2] Lebert: Ueber d. Einfluss d. Stenose des Conus arteriosus, des Ostium pulmonale u. d. Pulmonal Arterie auf Entstehung von Tuberculose, Berl klin Wschr., 22, 1867.

insisted upon the frequent coincidence of stenosis of the pulmonary artery and tuberculosis. He shows that this complication has been observed much more frequently since attention has been called to its occurrence. From 1815 to 1844 it was observed 5 times in 52 cases ; from 1845 to 1864, 11 times in 38 cases. My own investigations corroborate the frequency of secondary tuberculosis in this disease.

It must be admitted that pulmonary stenosis, whether congenital or acquired, predisposes strongly to phthisis. This fact is so much the more remarkable since phthisis is rare as a complication of aortic stenosis in particular, and of organic affections of the heart in general, and since Rokitansky thought there was an antagonism between phthisis and cyanosis.

The phthisis which complicates pulmonary stenosis is slow in its course ; it is a cheesy pneumonia with successive exacerbations. Now, the latter disease, as it generally occurs, is much more frequent on the right side than the left, or, at least, the right lung is involved more seriously. On the other hand, both lungs were affected in the sixteen cases reported by me of phthisis secondary to stenosis of the pulmonary artery.

I am confirmed in the opinion that both lungs are equally exposed, by the fact that in Villigk's case, in which the stenosis affected only the right branch of the pulmonary artery, both lungs were affected by phthisis, together with the larynx and intestines.

The manner in which phthisis is produced in such cases has not been ascertained, though numerous hypotheses have been proposed.

ENDOCARDITIS OF THE RIGHT HEART—INSUFFICIENCY OF THE TRICUSPID VALVE.

Very few cases of this lesion have been observed ; Ebstein [1] and Rokitansky [2] have reported examples, and a third has been observed by Prof. Peter.

OBSERVATION LXXIV. *Fœtal Endocarditis, Diagnosed before Birth by the Recognition of a Blowing Murmur with the First Sound ; Death of the Child during Delivery ; Autopsy ; Endocarditis of the Right Heart causing Tricuspid Insufficiency ; Enormous Cardiac Hypertrophy, affecting Chiefly the Right Cavities.*—J——, seventeen years of age, entered the maternity wards at the end of January, 1880 ; has never had rheumatism. She became pregnant in May, 1879. Auscultation showed the double pulsations of the fœtal heart a little to the right of the median line, and four finger breadths below the umbilicus. But instead of the regular tic-tac, a loud, rough, blowing murmur was heard, followed almost immediately by a dry click like the normal second sound ; then followed an appreciable silence. This was repeated about 130 times a minute. The sounds could be followed along an oblique line leading from the umbilicus to the left iliac fossa ; then they

---

[1] Ebstein: Verhandl. d. Schw. Gesellsch. f. vaterlaend. Cultur, February 16, 1866.

[2] Rokitansky: Die Defecte der Scheidewande des Herzens, Pathol.-Anat. Abhandlung, Vienna, 1875.

disappeared with the fœtal pulsations themselves. Upon passing a little to the right of this region, toward the median line, merely the very loud blowing murmur was heard; this disappeared a few centimetres further on.

I thought at first that I had to deal with the umbilical souffle of Naegele, which is sometimes produced by compression of the funis between the fœtus and uterine walls. But this was excluded by the limitation of the murmur to the region of the cardiac pulsations of the fœtus, the impossibility of detecting the latter independently of the bruit, and the absence of any uterine contractions capable of producing compression of the funis. I then made a diagnosis of endocarditis, or rather of a lesion of an orifice.

Labor progressed normally, and the patient was delivered of a well-formed, still-born boy, who had evidently died during the last stage of labor.

At the autopsy, it was found that all the fœtal organs were healthy with the exception of the heart, which appeared to fill almost the entire thorax. It is almost spherical in shape, and is especially remarkable by the predominance of the right half over the left. The valves of the left heart and pulmonary artery are healthy, but the free edge of the tricuspid valve is thickened and covered with small vegetations.

Its upper surface is roughened. The shortened and thickened chordæ tendineæ keep the valve applied to the ventricular walls; in a word, there is considerable tricuspid insufficiency. The right ventricle is dilated, and its walls much thicker than those of the left. The following table shows the measurements of the heart compared with those of a child of the same weight, who died at the age of five weeks.

	Still-born child (Endocarditis of right heart).	Child of 5 weeks.
Weight of heart.........................	25 grms.	16 grms.
Vertical diameter	45 mm.	32 mm.
Transverse diameter of base of ventricles.....	40 "	25 "
Thickness of walls of right ventricle	6 "	2 "
Thickness of walls of left ventricle	4 "	4 "

This hypertrophy must have been produced very rapidly, as the endocarditis evidently had developed after the closure of the septum. (Observation reported by Dr. H. Barth, Interne to the Hospital.)

STENOSIS AND OBLITERATION OF THE AORTIC ORIFICE.

Corvisart and Meckel have observed slight malformations of the aorta, but these are very rare. They are found in cases in which the pulmonary artery is narrowed and the ventricular septum incomplete. The lesion often is prolonged far into the aorta. The blood then passes through the persistent aortic arches to enter the arteries of the body. If the septum is complete, the foramen ovale is open, and the blood which arrives from the pulmonary veins passes into the right ventricle and thence into the lungs, and, on the other hand, passes through the ductus arteriosus into the aorta.

This lesion is always the result of endocarditis with myocarditis. It produces such disturbance of circulation that extra-uterine life is very short. However, Corvisart and Meckel have seen such patients who survived to the age of twelve, fourteen, and even thirty years.

Stenosis of the Mitral Orifice.

Stenosis and obliteration of this orifice are extremely rare; in these cases the foramen ovale remains widely open. The interventricular septum is almost entirely absent, or is, at least, very incomplete.

Transposition of the Arterial Trunks.

In certain cases the aorta arises from the right ventricle, and the pulmonary artery from the left ventricle. In these cases the foramen ovale remains open, the interventricular septum is incomplete, and the two kinds of blood are mingled together. The pulmonary and general circulation are not separate, and the blood arriving from the veins in part passes directly into the arteries without going through the lungs. In the same way the blood which has come from the lungs partly returns through them without having passed into the general circulation.

Nevertheless, life is possible under such conditions, and Peacock observed a patient who lived to the age of thirty-two years.

Transposition of the Venous Trunks.

In these cases the persistence of the foramen ovale permits the admixture of arterial and venous blood and partly repairs the anomaly of development. Individuals have been known to live for a long time under such conditions.

TREATMENT.

CHAPTER XLII.

THE CHIEF DRUGS WHICH ACT UPON THE CARDIAC MOVEMENTS.

MODERATORS OF THE HEART.

DIGITALIS.

PATHOGENETIC EFFECT.—1. *Effect upon the Digestive Tract.*—In therapeutic doses, the powder of the leaves often causes anorexia and a feeling of weight in the stomach and, in somewhat larger quantity, nausea and vomiting. In toxic doses emesis is the rule; in small doses, constipation is produced; in large doses, diarrhœa.

This effect upon the digestive tract is less marked with the infusion, still less with the tincture. It is probably due to a direct irritating action upon the mucous membrane, as these symptoms are produced especially when the preparation has been introduced by the mouth. However, intolerance also occurs when digitalis is introduced into the subcutaneous tissue or into the veins, thus tending to show that the drug is eliminated by the stomach.

2. *Effect upon the Circulatory Organs. Modification of the Frequency of the Pulse.*—When digitalis is given in moderate doses and progressively to a young and healthy individual, the frequency of the pulse is always reduced, but this result is not observed until the second or third day of its administration. The pulse may thus be reduced to 60, 50, or even 40, per minute, without interfering with the individual's health. If the administration of the drug is discontinued, either suddenly or gradually, the diminution in the rapidity of the pulse is found to continue two days later; then it gradually returns to the normal. It must be added that the rhythm of the pulse remains intact and there are no intermissions. *When given in toxic doses, digitalis changes the rhythm and produces intermissions.* In such cases the pulse is not alone slow, but irregular and intermittent, and this intermittence is to a certain extent periodical (once in about four or even three beats).

Some observers state that the diminution in the frequency of the pulse is preceded by acceleration. I have sometimes observed this phenomenon when the drug was given in therapeutic doses, but I have asked myself whether it was not due to the excitement always produced in patients by a careful examination. I have never observed it in cases of poisoning.

3. *Effect upon the Tension of the Pulse.*—If digitalis is given in small or very moderate doses to a healthy subject, the pulse appears to retain its force and fulness. But if larger doses, sufficient to produce diminution in the rapidity of the pulse, are given to patients suffering from febrile affections, diminution in the strength of the pulse is observed. This phenomenon is still more marked if the dose administered has produced toxic effects. The pulse then becomes feeble and irregular, and sphygmographic traces show that the tension is much diminished, particularly on the second, third, and fourth days of the administration of the drug.

4. *Effect upon the Heart.*—If some digitaline is dropped upon a frog's heart, the ventricle is seen to contract immediately at the point touched. But as the contact of digitaline is very irritating, the solution of digitaline should be injected subcutaneously, as far as possible from the heart. The heart is then arrested in systole, before voluntary motion, sensation, and the other functions have been affected. Under these conditions it is found that the contraction of the frog's heart begins at the apex, whereas, in the normal condition, it progresses from the base to the apex. King has observed the same phenomenon upon digitalized rabbits, and he has remarked that the contraction of the apex precedes that of the base by three to four hundredths of a second. After appearing at the apex, the muscular contraction shows itself in isolated places, which soon unite to form a sort of peristaltic movement. When once contracted the ventricle remains a certain length of time in this condition. At the next systole of the auricles it does not budge, and only moves on the occurrence of the second auricular systole. Thus the ventricle has only one systole to two of the auricle. A little later, the ventricular contraction does not respond to the auricular systole, but the ventricle remains contracted while the auricles are distended.

But the question arises, Does digitalis act primarily on the heart or on the cardiac nerves? Traube, basing his views upon numerous experiments on dogs, thinks that digitalis acts primarily upon the nerve-centres, first upon the medulla oblongata, then upon the pneumogastric nerves. A second theory places the primary action of digitalis in the peripheral extremities of the cardiac nerves and the intracardiac ganglia. But according to the most plausible theory, digitalis acts directly on the cardiac muscles. This view is upheld not alone by experiments on animals, but also by the fact that digitalis produces an alteration of cardiac rhythm as well as a diminution in the frequency of the pulse, and it is well known that the rhythm is one of the attributes of the myocardium.

5. *Effect on the Vaso-Motor Nerves.*—Legroux reports that after the administration of 1 ctgr. of digitaline to a rabbit, the central artery of the ear became filiform, and he concludes that digitalis acts first upon the sympathetic. Vulpian opposes this theory, and has shown that the drug has no influence on the lymph heart of the frog.

6. *Effect upon the Spinal Cord and Nerves.*—In moderate doses, digitalis produces merely ringing in the ears, slight giddiness, yawning, and insomnia. In toxic doses, it produces an atrocious headache, vertigo, delirium, and an acute pain along the spine. Reflex action does not appear to be influenced even by toxic doses.

7. *Effect upon Striated Muscles.*—Animals that are poisoned by digitalis suffer from weakness, prostration, sometimes from spasmodic tremor. Vulpian and Gourvat have shown that frogs, in whom 1 to 3 milligr. of digitaline have been injected, lose muscular contractility directly, and that this cannot be attributed to arrest of the heart or an alteration of the motor nerves. This is proven by the fact that frogs whose heart has been arrested, or the motor nerves paralyzed by curare, maintain the muscles intact for a very long time, while the same frogs, if poisoned with digitalis, promptly lose their muscular contractility.

8. *Effect upon Temperature.*—The temperature is lowered as the peripheral circulation is reduced. In a case of poisoning, I found that the temperature fell below 36°, while the pulse was scarcely perceptible, slow and intermittent.

9. *Effect upon Respiration.*—When the dose of digitalis is moderate, respiration is very slightly affected, being merely slightly slowed. When toxic doses are given, the respirations are accelerated, particularly in mammals, toward the end.

10. *Effect upon Nutrition.*—After digitalis has been administered for a short time, the appetite disappears, and vomiting and diarrhœa occur. These phenomena, joined to the contraction of the vessels, gives the patient the appearance of an individual who is suffering from cholera.

11. *Effect upon the Urinary Secretion.*—The antidropsical effect of digitalis upon patients suffering from dropsy will be discussed at a later period. But in healthy individuals, to whom toxic doses of digitalis have been administered, considerable diminution and even suppression of urine are often observed.

12. *Effect on the Visual Apparatus.*—This influence becomes appreciable only after the administration of toxic doses. According to Stannius and Hervieux, dilatation of the pupil is one of the first signs of intolerance. At the same time there is disturbance of vision which may advance to complete blindness. The eyes are injected and prominent, the pupils remain dilated and immovable.

13. *Effect on the Genital Organs.*—Digitalis produces contraction of the uterine capillaries and, apart from its slowness of action, is as powerful an

hemostatic as ergot in arresting hemorrhages of the menopause and those due to fibroid tumors.

VERATRINE.

PATHOGENETIC EFFECTS.—Veratrine is a drastic purgative and also acts upon the nervous system and the muscles, particularly the cardiac muscle.

When administered to dogs, after having produced its drastic action, the animals fall into a condition of prostration, marked by notable slowness of circulation and diminution of sensibility. At a later period tetanic convulsions develop and the animal dies of asphyxia.

The effects of veratrine in the human species vary greatly according to the dose. In some, 3 to 4 milligr. are sufficient to produce intolerance ; at other times I have given 20, and even 30, milligr. without effect.

Intolerance is shown by a feeling of heat in the stomach, soon followed by a burning sensation which is often accompanied by nausea and vomiting ; then colic occurs, and finally diarrhœa, which is often bloody.

Formication in the limbs soon develops, the pulse becomes infrequent and irregular, the temperature is lowered and the pupils dilated, and the patient finally sinks into collapse.

Bezold found, upon taking traces of the muscular contractions in frogs to whom veratrine had been administered, that the beginning of the contraction is normal, but that the second period, during which the muscle returns to its former condition, is exceedingly long. This condition is not one of tetanus, nor is it the result of nervous action. For, if the frog is previously curarized and the muscle thus separated physiologically from the nervous system, the same effect upon the muscular fibre is produced. In larger doses, the muscle is paralyzed.

When veratrine is applied directly to the heart, it produces the same effects as upon the other striated muscles, but death of the heart does not occur until two or three hours after life has ceased in the rest of the organism.

The following consideration is important from a therapeutical point of view. Can we hope to obtain a condition of increased contractility of the heart without the production of distressing symptoms? We are hindered usually by the intolerance of the digestive canal, as the patient suffers from colic, diarrhœa, and even vomiting. Great hopes were based on the employment of veratrine in acute articular rheumatism, in which it was expected that the pains and fever would be relieved thereby ; but this plan of treatment was soon abandoned.

BROMIDE OF POTASSIUM.

PATHOGENETIC EFFECTS.—The experiments of Clarke and Amory have shown that this drug is absorbed by all the mucous membranes, but not by the integument. It is eliminated with great rapidity, and has been found in the urine ten minutes after ingestion. The greater part is eliminated in twelve hours, but the remainder is excreted more slowly, and it is not rare to find it in the urine on the third day after administration.

In large doses the drug produces an irritant effect upon the gastric mucous membrane.

Soon after administration it produces a diminution of sensibility, particularly of the reflex sensibility of the pharynx. This anæsthesia sometimes extends to the nasal mucous membrane and the conjunctiva.

In moderate doses, it slows the circulation, and in certain cases suffering from fatty degeneration of the myocardium, it regulates the rhythm. The slowness of the pulse, which is the result of a probable action of the bromide of potassium upon the cardiac muscle, is aided by the fact that the drug contracts the small vessels and thus increases the tension of the blood in the vessels.

This contraction of the small vessels and local anæmia accounts for the cerebral phenomena produced by large doses, and which consist of vertigo, titubation, somnolence, and cerebral incapacity. Attention, memory, and intelligence are manifestly affected, and constitute a peculiar condition known as bromic drunkenness. Finally, the bromide impairs nutrition and diminishes the quantity of urea in the urine.

The elimination of the drug from the skin accounts in a certain measure for bromic acne. Finally, the drug diminishes virility or extinguishes it entirely.

Experiments made upon animals show that the bromide of potassium diminishes very greatly the reflex actions, but the mechanism of its action is determined with difficulty.

In fact, when bromide of potassium, either solid or in solution, is brought in contact with a striated muscle, the latter loses its irritability in a few minutes. The irritability of the mixed nerves and the spinal is destroyed in the same way. But when the drug is carried to the nervous elements through the circulation, the nervous functions are abolished successively.

The centripetal functions are first affected, then the reflex functions of the spinal cord, finally the motor functions. But after reflex power is entirely lost, the animal may still perform certain voluntary movements. In these cases the muscles are affected later than the nerves.

When the drug is carried to the heart through the circulation, the action of the organ is slowed upon the occurrence of insensibility, muscular resolution, and arrest of respiration. The contractions of the heart diminish in frequency, may fall to eight per minute during the anæsthetic sleep into which the animal falls, and increase in rapidity upon awakening. If the intoxication is pushed to a fatal termination, the heart is arrested in systole as after the administration of digitalis, with this difference that the rigidity ceases soon after death, and later, the heart is found distended with blood.

The observations which I have made upon patients suffering from fatty degeneration of the heart, in which the bromide has re-established the cardiac rhythm, show that the myocardium is affected directly, since the rhythm does not depend upon the cardiac nerves.

The drug lowers the temperature in a manner proportionate to the reduction of the peripheral circulation, on account of the contraction of the small vessels.

In man, the action upon respiration is scarcely perceptible, but the diminution of bronchial sensibility and the expiratory reflexes is readily observed.

No experiments have been made upon animals concerning the prolonged use of the bromides; in man, this has been done only in cases of epilepsy, hystero-epilepsy, and hysteria. The term bromism has been applied to a chronic condition similar to that produced by a slight acute poisoning, with the addition of certain phenomena on the part of respiration (a sort of chronic catarrh with spasmodic cough), and disordered nutrition leading to emaciation and anæmia.

In 1874 Falck advocated the theory that a part of the action of the bromide of potassium is due to the potassium. He based his theory on the fact that the potash salts are much more toxic than the soda salts. It must also be remarked that the soda salts, even in toxic doses, have no effect on the heart, temperature, nerve-centres, muscles, or the peripheral nerves. The potash salts, on the contrary, have a toxic effect upon the heart, nerves, and muscles.

But Falck's theory is not admissible for the following reasons: in the first place, bromide of potassium contains 67 parts of bromine and only 33 parts of potassium. Secondly, if bromide of potassium is replaced by bromide of sodium, sensibly the same effects are produced, while no similar results ensue from the administration of some other salt of potash. It has been claimed also that the bromine acts upon innervation, while the potassium acts upon the heart; this is likewise erroneous, as neither the acetate, sulphate, nor chlorate of potash is a cardiac depressor.

CARDIAC TONICS.

CONVALLARIA MAIALIS.

A chemical analysis of this plant has been made several times. In 1830 Valz extracted two principles: a glycoside derived from the aqueous extract of the flowers and dried roots, which he called convallaramine, and another principle, convallarine, obtained from the alcoholic extract. The latter has very little effect.

Stanislas Martin, who made an analysis in 1865, found that the infusion of the flowers is very acid. He extracted an alkaloid which he called maialine (is it the equivalent of convallaramine?) and an acid called maialic acid.

An analysis had been made recently by Hardy, the chief of the analytical laboratory of the Academy, but he could not completely isolate the convallaramine of Valz.

PATHOGENETIC EFFECTS.—According to G. Sée, one drop of the extract placed upon the heart of a frog acts like digitaline. At the end of one or two minutes the ventricle is arrested in systole and the auricles in diastole, though the animal retains its reflex and voluntary movements. Injected under the skin, the extract of the flowers acts in the same manner, though somewhat more slowly.

Bogayavlenski[1] injected an aqueous solution into the lymphatics of a frog. In a little while he observed retardation with increased vigor of the cardiac pulsations. He also noted irregularity of the ventricular diastole from irregular relaxation of the muscular tissue.

G. Sée describes the action of convallaria upon the dog as follows:

First period: *a*, retardation of the cardiac movements; *b*, increase of arterial pressure; *c*, the respiratory movements become deeper and less frequent.

Second period: extreme irregularity of rhythm, intermissions of the heart followed by rapid systoles. The respiration becoming fuller and slower appears at times to be arrested in a movement of profound inspiration. Emesis occurs during this period.

Third period: the pressure of the blood increases, the very frequent pulse becomes imperceptible, the amplitude of the respiratory movements increases. The heart's action is first arrested, then respiration ceases.

The therapeutic effects of the drug will be considered later.

---
[1] Médicin practicien, p. 360. 1882.

CHAPTER XLIII.

SPECIAL TREATMENT.

PERICARDITIS.

TREATMENT OF PSEUDO-MEMBRANOUS PERICARDITIS.

THE first condition is that the patient should be kept absolutely quiet, in the position which is least distressing to him. The head should not be raised very high.

The senses also should not be disturbed; no noise, light, conversation, etc., and care should be taken that the patient be not subjected to any excitement.

The patient may take acidulated drinks in small quantities frequently repeated. Liquid nourishment should alone be given.

1. ANTIPHLOGISTICS.—*A.* Local abstraction of blood by means of cups is still employed at times in the treatment of this disease. I think that this is a mischievous practice, for the cups, unless they relieve the pain, have no effect on the local lesion and cause wounds which may interfere at a later period with the application of blisters.

B. Mercurials. We possess remedies which are more vigorous and more effective than mercurials, and the use of the latter has been generally abandoned at the present time.

2. REVULSIVES.—*A. Blisters.* Corvisart was the first to insist upon the use of large and repeated blisters, and I agree entirely in his opinion. There is perhaps no instance in therapeutics in which such a prompt action of blisters is observed if care be taken to make them very large. Their rapid action may be noted by the application of the stethoscope to the raw surface.

B. The effect of other revulsives, which are advised when the disease is prolonged, is doubtful. Bouillaud mentions the actual cautery, moxa, tartar emetic ointment, and croton oil. Perhaps preference should be given to the actual cautery, which has such a decided action upon chronic phlegmasiæ of the lungs and pleura.

By the side of these local measures must be placed sedatives of the circulation.

A. The first to be employed, if the affection is painful, is the ice-bag, proposed by Gendrin. "It immediately diminishes the local pains, calms the tumultuous pulsations of the heart and the extreme anxiety of the patient. As a rule, also, its refrigerant action diminishes the febrile movement in a little while and lowers the frequency of the pulse."

The ice-bag should not be applied too long, in order not to depress the patient. The pulse and temperature should be watched and the bag removed as soon as they return to the normal. In some patients the effect is obtained in less than an hour; in others, two or three hours are required. After this if the reaction is moderate and develops slowly, it is not necessary to make a fresh application; if this does become necessary, its effects should be graduated by making less prolonged applications or gradually elevating the temperature of the mixture. This measure is not contra-indicated by complication with pneumonia or pleurisy.

B. Digitalis. The great frequency of the pulse which is often observed in pericarditis shows that the pains and the sensibility of the pericardium produce a reflex action which passes through the pneumogastrics and is reflected to the accelerator nerves or the sympathetic. This sensibility may be moderated by the use of digitalis. As a rule, small doses will suffice. In Germany it is given in large doses.

I believe that small doses of digitalis may be useful, but that it is better to employ the ice-bag, or wine or opium internally, but I do not think that the large doses recommended by Friedreich are justifiable. Bauer also states that digitalis in large doses is a sovereign remedy, but that it should be discontinued as soon as the pulse is reduced in frequency or becomes irregular.

C. Veratrine has been proposed by Friedreich to replace digitalis. This drug possesses the double effect of relieving the pains and reducing the pulse and temperature. It should be given in pills or granules, and, if well tolerated, the dose may be increased to thirty, even fifty milligrammes.

D. Hydrocyanic acid has been proposed by Stokes, but timidly, as this is a dangerous remedy. Five to ten drops of the medicinal solution may be given without fear.

The pain may be relieved by the administration of opium internally, or subcutaneous injections of morphine. Insomnia may be combated by hydrate of chloral in doses of one to four grammes.

The depression of the vital energies should be combated by wine, brandy, champagne, punch, sulphate of quinine, musk; if chilly sensations are marked, by subcutaneous injections of ether or camphor.

TREATMENT OF SEROUS PERICARDITIS.

When the diagnosis of pericarditis with effusion has been made, very little can be expected from antiphlogistics or revulsives.

A. Permanent application of blisters has been employed, but generally proves insufficient.

B. Diuretics (squills, nitrate of potash, acetate of potash, etc.) and hydragogue cathartics (jalap, scammony, colocynth, elaterium, etc.) have also been employed.

C. Puncture or Paracentesis. The first puncture of the pericardium was made by Schuh in the service of Professor Skoda, and since that time it has been performed a number of times.

The following rules will serve as a guide in such cases: The opening should always be made to the left of the sternum. In operations made on the right side, along the sternum, the patients have died suddenly. In one case the trocar, which was introduced near the sternum in the third intercostal space, passed directly into the right auricle. In another case the trocar remained engaged in the adhesions in front of the right auricle and produced arrest of the heart's action. The point on the left side selected by almost all operators is the fifth, more rarely the fourth or sixth, intercostal space, according as the pericardial region is more or less enlarged. The site chosen is more or less remote from the sternum, according as we wish to avoid passing through the pleura or not. Jobert, in order to avoid the pleura, makes the puncture 3 ctm. from the median line, *i.e.*, 1½ ctm. from the edge of the sternum. Trousseau operated in the same way, taking care to avoid the internal mammary artery, which passes about 1 ctm. from the edge of the sternum. But if we believe with Roger that there is little risk in perforating the pleura, we may operate farther from the median line. In fact, the more we approach the apex the less the danger of wounding the heart.

Aran punctured at 5 ctm. from the median line, Frémy and Heger at 6 ctm., Chairon at 7 ctm., and Rendu at 8 ctm. in the sixth intercostal space. I believe with Rendu that the puncture should be made as far as possible from the median line, *i.e.*, at the apex, or, if possible, outside of the apex.

Having chosen the site for the operation, how should this be performed? Trousseau and others first made an incision layer by layer; then, having reached the pericardium, they satisfied themselves that the membrane was raised by fluid before opening it.

To make a positive diagnosis, and to determine whether the site chosen for the operation is suitable, I cannot recommend too strongly a preliminary explorative puncture with the hypodermic needle.

At the present time, thanks to the small diameter of the needles (2 mm.) used in performing aspiration, the puncture is effected with the greatest ease.

The sharpness of the needles and their readiness of entrance enables us to advance slowly, and to arrest and even withdraw them, if the slightest alarming symptom should be produced. Roger also states that experiments on animals would lead us to think that if the needle touched the heart the puncture would only be superficial, and would not be so serious as might be feared.

When the canula is introduced, if aspiration is not performed the fluid flows at first in an interrupted spurt, then in drops. This proves that the fluid is escaping from the pericardium. When the puncture is made in hydrothorax the jet is at first continuous, then it becomes jerky, finally intermittent. In the former the jets correspond to the pulse, in the latter to the respiratory movements.

It must not be forgotten that adhesions form almost constantly in pericarditis, and the trocar should be left in place for an hour or even an hour and a half in order to give the fluid time to run out.

This is preferable to aspiration, which gives rise to a sort of cardiac adhesion after the greater part of the effusion has escaped.

The quantity of fluid which may be withdrawn varies from 300 grms. (Trousseau) to 950 (Rendu), and even 1,500 grms. (Hindenlang). The fluid is serous and slightly sanguinolent, and coagulates with great rapidity. Pus has been removed in two cases, one reported by Aran and the other by Reiz and Levison.

These writers injected a fluid of the following composition : water, 50 grms.; tincture of iodine, 15 grms. ; iodide of potassium, 1 grm.

As a rule, the patient experience an almost immediate relief, but this is by no means permanent. The fluid is often reproduced, perhaps more abundantly than before.

It must be confessed that few of the patients operated upon recover, so that paracentesis of the pericardium merely constitutes a last resort when the effusion threatens to suffocate the patient.

Treatment of Endocarditis.

The endocarditis may appear at the beginning and constitute the sole disease, or it may be merely the first manifestation of an acute rheumatism which, in a few days, will attack the joints. At other times, endocarditis appears at a later period as one of the symptomatic affections of rheumatism, gonorrhœa, scarlatina, variola, rubeola, erysipelas, diphtheria, nephritis.

Many have maintained that, in such cases, treatment with sulphate of quinia, nitrate of potash, tartrate of soda, will prevent the development of endocarditis. Stricker has claimed similar results from the use of salicylate of soda. Unfortunately, however, none of these claims can be substantiated.

After the endocarditis has developed it should first be treated locally, the best means being the application of a blister as large as the præcordial region. Bouillaud applied daily to the raw surface 40 to 60 ctgrms. of powdered leaves of digitalis, the irritating action of which maintained the vesicant action.

As in pericarditis, the general treatment includes at first rest and the prevention of excitement, such as light, sound, conversation, etc., then cool, acidulated drinks.

The administration of salicylate of soda and sulphate of quinine may be continued if it appears to act favorably upon the rheumatism and if the cardiac agitation is not great, if the pulse is not too frequent and is regular.

Tincture of digitalis may be administered in doses of 20 drops or 1 grm., repeated at first twice a day, and gradually increasing the dose. But it must not be forgotten that the drug requires at least twenty-four hours before its effect becomes perceptible; we should therefore wait two days before increasing the dose. When its effect has been secured we may diminish or entirely suspend it, until after the lapse of a few days its effect begins to subside.

The other cardiac sedatives, such as veratrine, bromide of potassium, chloral, are far from rendering the same services. Convallaria has appeared to me to regulate the action of the heart in such cases and to stimulate the patient, but my experience dates back only six months—an insufficient period.

After sedation has been obtained by any of these means we should hasten to employ tonics, particularly the ferruginous preparations, which are cardiac sedatives. The perchloride, citrate, and pyrophosphate of iron may be employed.

TREATMENT OF ULCERATIVE OR INFECTIOUS ENDOCARDITIS.

This disease terminates almost always in death. The following remedies have been employed in its treatment:

1. Externally, the ice-bag.
2. Internally, antiseptics.

A. Salicylate of soda, in doses of 6 to 12 grms.
B. Benzoate of soda, in doses of 4 to 10 grms.
C. Sulphate of quinia, in doses of 2 to 4 grms.
D. Tincture of digitalis, in doses of 2 grms., or the infusion, 50 to 75 ctgrms.
E. Camphor, in doses of 50 ctgrms. to 1 grm., in an emulsion made with the yolk of an egg.
F. Musk.
G. Carbonate of ammonia.
H. Corrosive sublimate, in the form of Van Swieten's solution.[1]

---

[1] Bichloride of mercury, 1 grm.; alcohol, 100 grms.; water, 900 grms.

CHAPTER XLIV.

HYGIENE IN HEART DISEASE.

WHEN a patient is convalescent from an acute disease which has left over a cardiac lesion, particularly of the valves, he will do well during repose if the other organs of circulation are in a good condition and perform their function in a satisfactory manner.

But if the equilibrium is disturbed by excitement, or if the organism must perform some work which demands a certain effort, the insufficiency of the heart is speedily manifested by dyspnœa and palpitations.

In order not to overwork this organ, we must select for the patient those conditions in which the heart requires the least expenditure of force.

In the same manner, it is evident that in the treatment of the accidents which are secondary to diseases of the heart the treatment consists less in strengthening the heart than in removing the fresh obstacles which it must overcome.

We will now consider the daily hygiene during the period of tolerance.

The first advice which should be given to the patient is that he rise early. The first rule of hygiene is the avoidance of the obesity which is one of the chief annoyances of patients suffering from cardiac disease.

A. Toilet ; Hydrotherapy.—The patient should accustom himself gradually on rising to the contact of cold water, its temperature being made lower every day. The reflex impression made upon the heart by cold water ceases very soon, as has been proven by Fleury, Bouillaud, Auburtin, Hirtz, and others.

The patient may without fear employ cold washings at a temperature of 12 to 16° C. If we desire to employ hydrotherapy in its entirety we should begin with vigorous frictions for a few seconds with a wet cloth. At the end of a few days we may begin with a douche of a few seconds' duration, or with the alternating douche (warm and cold) if it is followed by slight reaction. It has been found in all hydropathic establishments that the impression produced by cold water is greatly diminished if the cold douche has been preceded by a warm one (30 to 32°).

But tolerance is established so quickly that at the end of a week the patients endure a cold douche, lasting one to one and a half minute, upon the præcordial region.

Age does not constitute a hindrance, and Fleury mentions cases of patients sixty years old who employed hydrotherapy successfully under these circumstances. It must be remembered, however, that though the douche may be cold, the air of the room must be warm. If the room cannot be heated, reaction should be established by warm foot-baths.

B. Baths.—Baths may be classified as cold (0 to 25°), tepid (25 to 30°), and warm (30 to 40°). The cold bath does not agree with cardiac patients, as it produces congestion of the internal organs and causes retardation of the circulation and interference with respiration. At the moment of reaction the pulse becomes somewhat fuller and more frequent.

Cold baths, therefore, should not be recommended unless the water has a temperature of 25°, the air is very warm and calm, and the bath is very short. Reaction will be facilitated by a warm room, foot-bath, warm clothes, etc.

Tepid or warm baths may be permitted under the condition that they do not last more than fifteen minutes and are not repeated oftener than once a week.

C. Hot-air Bath.—This bath is tolerated comparatively well by old people and patients affected with heart disease, provided they do not suffer from aneurism. The patient may take the bath if the temperature of the air does not rise above 55° C. Perspiration may be facilitated by the ingestion of cold water; the bath may end with rubbing, and, if necessary, with a shower-bath lasting a few seconds. The cold plunge and massage should not be employed. Not that massage is not good for these patients, but it should be performed at another period, in order not to induce too much fatigue.

D. Vapor Bath.—A vapor bath, in which the patients are suffocated by the entrance into the bronchi of hot air saturated with vapor, is not to be thought of. But they may be encased in a vapor bath if the head is free and they are enabled to breathe the dry air of the chamber.

E. Compressed-air Baths.—These baths, which are so useful in emphysema and catarrh, even if there is beginning dilatation of the right heart, are absolutely interdicted in patients suffering from cardiac disease.

F. Clothing.—Two points should be considered. In the first place, the patient should avoid constriction either around the neck, waist, or limbs. Females should avoid compression of the chest or liver by corsets. This effect of the clothing is so great that A. Meyers [1] has called attention to the frequency of cardiac affections in the English army, and regards as an important cause the constriction of certain parts of the body by the equipment.

G. Regimen.—The diet should be of such a character as to avoid obes-

---

[1] A. Meyers: On the Etiology and Prevalence of Heart Disease among Soldiers. 1870. Quoted by Beaumetz: Clinique thérapeutique, t. i., p. 17.

ity. The patient should drink as little water as possible. He should take nothing in the morning, or perhaps a little black coffee with some toasted bread or a biscuit. At other meals he should abstain from feculent or farinaceous substances, as these require a large amount of water for their digestion.

The patient should not take more than 200 grms. of bread daily. He may eat all kinds of meat, game, fish, and shell-fish; also fried or baked potatoes, and beans. He should abstain as much as possible from fatty articles, butter, and oil. Nuts, dried fruits, and cheese are allowed for dessert. About a bottle of wine may be taken daily; brandy must be taken in great moderation. He should not drink coffee unless it is weak and does not produce palpitations. Few sweetmeats are permitted. After dinner, a little tea, coca, or guarana. The patient may also take a little cognac or aromatic liqueur such as curaçao, chartreuse, etc.

In order to obviate the constipation which may arise from such a diet, a pill may be taken at night, containing podophyllin, or aloes, scammony, etc.

H. Tobacco.—This should be used in great moderation, as it is capable of producing angina pectoris.

I. Exercise.—The patient should take moderate exercise—for example, walking on a level surface. Billiards forms a very good method of exercise, if the air of the room is not impregnated too strongly with tobacco.

Manual exercises, like piano-playing, may be continued, but all forms of exercise which require effort, such as running, jumping, dancing, swimming, fencing, and particularly riding, must be strictly interdicted.

J. Social Life.—Patients suffering from heart disease should be spared all forms of excitement. The sexual relations should be very moderate; more than one disease of the heart has been revealed by an attack of suffocation occurring after coitus (Oulmont).

Many professions cannot be practised by these patients. They cannot be soldiers or sailors, still less physicians, surgeons, or even accoucheurs. They can scarcely endure the excitements of business, as in the stock exchange, etc.

K. Passive Exercise.—Carriage-riding is not injurious, if there is not too much jolting. Massage, especially when it consists of friction of the limbs in the direction of the venous circulation, furnishes the patients with a sort of accessory venous heart, and does them the greatest good.

L. Temperature.—Patients suffering from heart disease dread the cold. It is well to advise them to pass the winter in a mild climate, like that of Cannes, Nice, Menton, Hyères, Algiers, Madeira, particularly as the salt air is beneficial to them.

M. Altitude.—Diminution of barometric pressure is tolerated very well by the patients. They may live with benefit in moderately elevated localities—for example, 300 to 600 metres—but only under the condition that they have level walks, which render it unnecessary to ascend or descend in order to obtain exercise.

N. Whey Cure.—Whey is difficult to digest, particularly at the outset, and it must be mixed, usually, with mineral waters (Geisshuebel, Johannisbrunnen, Tatmausdorf), or with sodic chloride waters, charged with carbonic acid (Nauheim, Schwalheim, Œyenhausen). This cure is composed of drinks and baths.

It is a remarkable fact that a whey bath causes a rapid diminution in the rapidity of the pulse. This plan of treatment is also laxative in its action and relieves the congestions which are secondary to heart disease.

O. Grape Cure.—This fulfils almost the same indications as the whey-cure. Under its influence the circulation becomes more active at first, the pulse acquires greater amplitude and force, the face assumes a better color, the skin is more active, and the patient has a feeling of increasing strength. Then the secretions increase, the urine becomes more abundant, the patients have free evacuations from the bowels, and sometimes slight diarrhœa.

The quantity of grapes consumed daily varies from $1\frac{1}{2}$ to 4 kilos. A small quantity is generally taken at the beginning, perhaps $\frac{1}{2}$ to 1 kilo. When a greater amount is taken the daily ration is divided into three parts, each of which is taken about half an hour before meals. It must be remembered that the grapes taken early in the morning have a greater diuretic and laxative effect. In some parts of Germany the juice extracted from the grape by means of a press is administered.

Sometimes the diet consists exclusively of grapes; at other times, other articles of food are allowed. In such cases we begin with vegetable food, and then allow white meat; at a later period dark meat is allowed. Toward the close of the treatment wine, coffee, and tea are also allowed.

The duration of the cure varies from three to six weeks. It is contra-indicated by pregnancy.

CHAPTER XLV.

TREATMENT OF THE CLOSE OF THE PERIOD OF TOLERANCE.

AT the end of a longer or shorter period, but which, under certain circumstances, may last for a number of years, the heart no longer is able to fulfil its functions, either because it is overworked and is beginning to degenerate, or because disturbances have arisen in the organs through which it must circulate. The treatment presents two indications: The first consists in sustaining the heart and relieving the palpitations and dyspnœa; the second, in causing the disappearance of the secondary complications which create new obstacles to the circulation.

TREATMENT OF PALPITATIONS—CARDIAC MODERATORS.

The first means to be employed is rest in the horizontal position, the head being a little elevated. Cool drinks should then be given and solid food in small quantities. At the end of a few days the cardiac rhythm is improved; we may then judge more readily of the cardiac affection.

The drugs which regulate the circulation are called sedatives.

A. DIGITALIS.

Administration by the Mouth.—Powdered digitalis leaves are an unreliable preparation; they may be given in the form of pills, but are poorly tolerated by the stomach on account of their irritating action.

To relieve the palpitation of this period, in which the heart begins to lose its force and rhythm, the tincture of digitalis should be given in doses of 25 drops, once or twice a day, and then gradually increasing, if the effect is not produced, up to 100 or even 150 drops at a dose. But before digitalis is administered we should endeavor to relieve any congestions or dropsies which may be present, and should delay the use of digitalis until somewhat later.

If the cardiac affection is less advanced and none of these secondary disturbances have been produced, digitalis may be given from the start in small doses.

It will be found that one, two, or even three days elapse before its effect upon the pulse is manifested. On the other hand, when this effect has been obtained, it is prolonged for several days after the administration of the drug has been discontinued. This is due to the fact that digitalis is eliminated more slowly than it is absorbed, and that it accumulates, therefore, in the organism. This occurs so much more readily if the kidneys are diseased and interfere with its elimination.

Accordingly, when a notable reduction in the number of pulsations has been effected, the use of the drug should be discontinued, or at least diminished.

Digitaline may also be employed. The preparations used are those of Homolle & Quévenne, Merck, and Nativelle. All three preparations are active, but that of Nativelle, to judge from my experiments, is the most constant and regular in its action.

Administration through the Integument.—I have frequently rubbed in the tincture of digitalis in the cardiac region, and have so often seen the patients relieved and the rapidity of the pulse appreciably diminished in consequence that I do not doubt the possibility of absorption in this way. But it must be remembered that there is no comparison between the effect obtained by this means and that secured by the internal administration of the tincture.

Brown [1] and Reynolds [2] have obtained a diuretic action and a marked diminution of the rapidity of the pulse, at the end of an hour, by the application to the belly either of flaxseed poultices to which the tincture has been added, or of poultices made with the fresh leaves and boiling water.

Tourangin and Dujardin-Beaumetz have also observed a noticeable effect in young subjects.

Subcutaneous Administration.—In my experiments on dogs to which an alcoholic solution of digitaline has been administered subcutaneously, this has constantly given rise to local irritation, and so often to diffuse phlegmons that I have been careful not to make such injections in man. This has been done, however, by Otto and Wilkowski, and has resulted in terrible accidents. Gerber and Eulenburg have employed it, but have not obtained favorable results.

Gubler has succeeded in making this injection more tolerable by resorting to the following formula : alcohol, 50 grms. ; digitaline of Homolle & Quévenne, 0.10 grm. ; dissolve and add water 50 grms. A gramme of this solution contains one milligramme of digitaline. Gubler made the subcutaneous injections in the region of the back, and these gave rise to no bad results, although they were repeated as many as sixteen times in one patient.

Contra-indications.—It has been said that digitalis should not be given to those suffering from fatty heart, since under such conditions it no

---

[1] Brown: Med. Times and Gaz., 1868.   [2] Reynolds: The Lancet, 1869.

longer acts well and exposes the patient to the danger of poisoning. I believe that this a mistake. When digitalis no longer acts it is not because there are no more healthy heart-fibres, but because the obstructions to circulation in the limbs and viscera are such that the remaining fibres have no power to overcome them, even with all their energy. In such cases we must remove the infiltrations by punctures in dependent parts and replace the digitalis by opium.

It has also been said that digitalis is dangerous at the beginning of aortic insufficiency. It would be more correct to say that it is useless, as the pulse is then regular and sufficient. But digitalis again becomes useful at a later period, when the hypertrophied heart begins to give out.

B. BROMIDE OF POTASSIUM.

Gubler was the first to employ bromide of potassium in order to relieve palpitations of organic origin, giving it in doses of 2 to 4 grms. His example has been followed by G. Sée, Dujardin-Beaumetz, and myself.

C. CHLORAL.

The depressing action of chloral on the heart is perfectly well known. It is known that an animal poisoned with this drug dies from paralysis of the heart, the ventricles and auricles being distended by black blood. When it has entered the blood in a certain quantity it alters the blood-globules.

By this action on the blood chloral may stimulate the medulla oblongata or the central origin of the spinal accessory nerves; it may also act on the cardiac ganglia and the myocardium itself. Finally, it may paralyze the peripheral vessels and thus produce notable weakness of the movements of the heart.

Accordingly, chloral acts upon the heart by depressing it, and this action may advance to the production of paralysis of the organ.

It follows that chloral is a valuable agent in producing sleep in those patients who are deprived of it on account of the violence of the palpitations, but that this drug becomes more dangerous than digitalis when the degenerated heart has reached a condition of asystole.

CHAPTER XLVI.

TREATMENT OF CARDIAC DISTRESS AND ASYSTOLE.

As in the preceding chapter, we must here take into consideration the dyspnœa which is due properly to the cardiac affection, and not mistake it for the dyspnœa produced by the lesions which are the cause or consequence of the heart disease. Thus, the dyspnœa due to secondary pulmonary congestion should be treated at the beginning by dry cups, bleeding, and tannin; the dyspnœa due to pulmonary œdema calls for vesicants and diuretics; that due to hydrothorax requires paracentesis; uræmic dyspnœa may be treated by the milk diet and bromide of potassium.

The latter variety is particularly apt to be mistaken for cardiac dyspnœa strictly speaking. Cardiac dyspnœa is, in the beginning, only marked during effort. At a later period it becomes marked during left decubitus when hypertrophy occurs, and during right decubitus when tricuspid insufficiency from dilatation has supervened; finally, it passes into orthopnœa, and the patient is obliged to assume a sitting posture, especially after notable pulmonary œdema has developed.

Uræmic dyspnœa, on the contrary, develops suddenly in paroxysms of distress which, up to a certain point, resemble asthma. These paroxysms occur particularly at the period when the patient is about to fall asleep and ceases to contribute voluntarily to respiration. The patient is then seized with dyspnœa and nightmare, which awaken him with a start.

I will here repeat my previous advice that, before treating the cardiac disturbance directly, we should combat the abnormal obstructions to the general circulation.

From a therapeutic point of view, we must include in one chapter the phenomena of asystole, *i.e.*, dyspnœa, lipothymia, and syncope. We now no longer desire to depress the violence of the cardiac impulses, but, on the contrary, to stimulate them. This indication is fulfilled by the class of stimulants which may be administered by the mouth, the respiratory passages, and the subcutaneous connective tissue.

Those administered by the mouth include wine, alcohol, coffee, ether, and the diffusible stimulants.

WINE.

The richer a wine is in alcohol, the better adapted it is to our present purpose. It is also important that it contain as little sugar as possible; in this respect, white wines are better than the red. Finally, warm wine is more stimulating than cold wine.

The following table gives the proportion of alcohol in the different wines to 100 parts by volume :

Marsala	24.09	Roussillon	18.13
Lissa	23.41	White Hermitage	17.26
Port	23.99	Madeira Malmsey	16.48
Madeira	22.27	Bordeaux	15.10
Sherry	19.87	Sauterne	14.22
Teneriffe	19.79	Burgundy	14.57
Lachryma Christi	19.70	Rhine	12.08
White Constance	19.75	Tokay	9.88
Red Constance	18.92	Moselle	8.00
Muscatel	18.25		

ALCOHOL.

Alcohol taken as a drink should mark 36 to 50° at the most. In this grade of density it produces, at the moment of deglutition, a sensation of heat in the throat, pharynx, and even in the stomach. When cognac is old and of good quality, its topical action on the mouth and pharynx is very little marked, and the sensation of heat does not develop in the stomach.

When alcohol is ingested in too great a degree of density, particularly if large quantities are taken, it coagulates the gastric mucus, destroys the pepsine, arrests digestion, and the indigestion leads to vomiting.

In moderate quantities it facilitates digestion and stimulates the solar plexus, one of the centres of action of the sympathetic nerve, *i.e.*, of the accelerator nerves of the heart.

Alcohol is absorbed by the stomach, but if the quantity is considerable it passes into the intestines and there is absorbed in great part by the veins ; perhaps to a slight extent by the lymphatics.

After it has passed into the blood its action is very little known. All that can be said is that when alcohol is absorbed in large quantities, the arterialization of the blood-globules upon contact with oxygen is diminished (Brouardel, Jouffroy).

In moderate doses it stimulates the circulation and respiration, making the inspirations fuller and more frequent ; at the same time the quantity of carbonic acid exhaled is diminished. It also increases the evolution of animal heat.

Alcohol stimulates the peripheral circulation and the muscular energy. It also acts in this manner if taken in sufficiently large quantities to produce the first stage of intoxication; but if given in still larger quantities it depresses all the functions.

ETHER.

Ether can only be taken internally in the form of capsules or pearls. Its action under such circumstances in the stomach can readily be determined by placing a capsule in water at the temperature of 40°. At the end of a few seconds under the influence of this temperature, a little vapor of ether forms within the capsule and breaks that portion of the capsule which is thinnest or most softened by the water. A filiform jet of liquid is produced, but is converted suddenly into vapor and causes a sort of explosion.

When this enormous production of vapor occurs in the stomach it is manifested by a sensation of heat and tension, and also by a few eructations of vapor of ether.

This agent is an excitant of the circulation and calorification; it is at the same time an excitant of respiration. From the promptitude with which it relieves patients suffering from asystole, it may be supposed that this is due to the rapidity of its diffusion. The slight duration of its action may be explained by its ready elimination through the respiratory passages.

CONVALLARIA MAIALIS.

It is evident, from what I have said about the pathogenetic action of this drug, that it is a true heart tonic, particularly of the myocardium.

It must be employed in large doses. I am in the habit of prescribing the tincture in doses of a teaspoonful taken in half a glass of sugar-and-water. This dose may be repeated twice a day in cases of cardiac ataxia or arhythmia.

STIMULANTS ADMINISTERED BY INHALATION.

ETHER.

When the vapor of ether is inhaled in small quantities to avoid producing anæsthesia, it is found that one of the first effects is to render the peripheral circulation more active. The brain is stimulated and, at the same time, the reflex acts are diminished. It is this effect which we endeavor to obtain in cardiac asystole by the inhalation of ether. This measure is useful, but its effect is short and wears off at the end of a certain period.

CHLOROFORM.

Vergely has employed inhalations of chloroform in order to relieve cardiac dyspnœa accompanied by violent palpitations. He concludes from his experiments : 1, that heart disease is not a contra-indication to the use of anæsthetics ; 2, that chloroform acts as a sedative in these affections ; 3, that it should be administered with caution. This writer has found chloroform serviceable in relieving attacks of angina pectoris.

NITRITE OF AMYL.

PATHOGENETIC ACTION.—When the vapor of nitrite of amyl passes into the nasal fossæ it causes a sensation of coolness, but this ceases at the throat and, unlike chloroform, does not pass into the bronchial ramifications. In the pharynx and larynx the sensation is rather that of tickling ; it often excites cough. Almost at the same time a feeling of fulness is felt in the head and even more in the face, as if wine had been taken in excess. The face is soon injected, the eyes shining. The turgescence and redness of the face increase rapidly, and soon exceed that produced by emotional excitement. Not alone the superficial capillaries, but all the vessels of the head—arteries, capillaries, and veins—are in a condition of manifest tension. The carotids pulsate with vigor ; the pulse is accelerated and often exceeds 100 per minute.

It is a striking fact that the phenomena produced are limited to the head. At the end of four or five minutes the stimulation diminishes, and in about ten minutes more all is over. No vertigo is experienced during the entire period, unless the inhalation of the drug is continued too long.

I have frequently employed this drug in advanced cardiac disease, when the patient's suffered from lipothymia, and particularly in attacks of angina pectoris. Unfortunately the action of nitrite of amyl is fleeting, and it can only be used from time to time to counteract attacks of weakness or syncope. Under these conditions it is a valuable auxiliary.

STIMULANTS ADMINISTERED BY THE SUBCUTANEOUS CONNECTIVE TISSUE.

MORPHINE.

These were first employed by Levy to relieve attacks of cardiac distress. Huchard has insisted particularly upon their efficacy in relieving the cerebral anæmia secondary to aortic affections and to asystole. According to him, morphia congests the brain, thus relieving the cerebral anæmia, and consequently the dyspnœa. In 1877 Gubler drew a parallel between the indications for subcutaneous injections of morphine and digitalis. He comes to the conclusion that digitalis is useful at the close of the period of tolerance, and that, at a later period, recourse should be had to morphine.

In fact, hypodermic injections of morphine, given during the period of cardiac cachexia, relieve the patients so much that they imperatively demand them.

From a theoretical point of view it may be assumed that morphine constitutes an excellent vaso-dilator which, on account of the dilatation of the peripheral vessels, causes diminution of the intravascular tension and of the number of cardiac pulsations. However this may be, hypodermic injections of morphine render the greatest service in cases of cardiac dyspnœa and asystole.

ETHER.

Hypodermic injections of ether, recommended recently by Verneuil, may be used in the same manner as morphine. These injections will stimulate calorification and the circulation, but they do not possess the same virtues as injections of morphine.

CHAPTER XLVII.

TREATMENT OF HYPERTROPHY.

It must not be forgotten that hypertrophy is always secondary. Consequently, when it is the result of disease of some other organ than the heart, this affection should be treated at the same time.

We refer here to hypertrophy caused by a lesion of the heart itself, either of the valves or muscular tissue. It has been frequently said that the heart hypertrophies in order to establish a sort of compensation, and this process has even been called providential. This view would be correct if the hypertrophy remained stationary; but experience has shown that the excess of work imposed upon the heart finally deteriorates its fibres, which become changed either by fatty degeneration or by the process of irritation of the connective tissue, which develops excessively and finally strangulates the muscular fibres.

If it is true that this degeneration occurs especially when the other organs of circulation lose their elasticity and contractility on account of age, it must also be added that the arthritic, alcoholic, or syphilitic diathesis often hastens this form of senility of the organ, and that, finally, the heart may rapidly undergo change if the cardiac obstacle is considerable. We should therefore endeavor to relieve the latter, in order that the degeneration may occur as late as possible.

Iodide of potassium, which is administered in a routine manner in all hypertrophies, may be of some service also in cardiac hypertrophy.

To my great astonishment, I have seen blowing murmurs, due to organic disease of the heart, diminish and finally disappear for a long time, in individuals leading an equable and regular life, under the influence of hygiene and treatment. But this is an exceptional result, and one upon which we cannot rely.

A certain distinction must be made between the different forms of cardiac hypertrophy.

Hypertrophy secondary to mitral lesions, if these are not very extensive, offers the patient the best chance for prolonged life. As is well known, they predispose to stasis of blood, and consequently to congestion and dropsy. In these cases cardiac moderators exercise a great influence and may be used for a long time.

When, on the contrary, the hypertrophy follows a lesion of the orifice or beginning of the aorta, the anæmia, which is so often present, indicates that recourse should be had to ferruginous preparations. As these preparations, in large doses, act as sedatives to the circulation, we should rely upon them and not upon digitalis, which is poorly tolerated by these patients.

The soluble preparations are the best in such cases, viz., the tartrate of iron and potash, of which 30 to 60 ctgrms. and even more may be given daily; the perchloride of iron in doses of 50 to 60 drops a day in two or three doses; the protochloride of iron; the pyrophosphate of iron; finally, the carbonate and iodide of iron.

Patients suffering from atheroma of the origin of the aorta are also relieved by iodide of potassium in small doses (30 to 50 ctgrms.), unless the lesion is of a syphilitic origin, when the dose should be increased to two grammes.

When the hypertrophy is secondary to stenosis of the pulmonary artery phthisis is the complication to be dreaded. Reliance must be placed on arsenic in small doses, the phosphates, cod-liver oil, weak alkaline sulphates, as in the treatment of torpid phthisis.

Care should also be taken to treat the patient's diathesis (rheumatism, gout, syphilis), and to remove toxic causes, such as tobacco, alcohol, etc.

TREATMENT OF THE SYMPTOMS DUE TO HYPERTROPHY.

PULMONARY SYMPTOMS.

The pulmonary accidents which follow hypertrophy are produced by the stasis of blood, and include congestion, œdema, apoplexy, and pleural effusions.

The treatment is carried out according to the ordinary methods of revulsion. Internal congestions are combated by external congestions, produced by dry cups in large numbers and applied daily. These cups cause no pain to the patient and produce immediate relief. They should give rise to a persistent violet ecchymosis, but need not be applied more than eight or ten minutes. They may be renewed frequently, but their application should be discontinued if the patient is suffering from anasarca, and if the skin or subcutaneous cellular tissue is infiltrated. In such cases the cups give rise to superficial wounds, which quickly become inflamed and are painful upon contact with the clothes.

Wet cups are not as useful as dry cups. If there is a tendency to asphyxia and venous stasis, the patient will be relieved much more effectually by venesection of 250 to 300 grms., but it will become necessary to repeat them, and their good effect diminishes each time.

So long as the patient can walk around we may use to advantage baths of compressed air. The compression should be increased slowly for fifteen or twenty minutes, and the pressure should not exceed one atmosphere and a half. The sitting should last an hour, and the pressure should be removed as gradually as it was added.

The patients should also be informed that, upon coming out of the apparatus, they are apt to suffer from cold and insufficient reaction. They should wait some time, therefore, before leaving the establishment, and should be warmly covered.

When the congestion passes into œdema or apoplexy it becomes necessary to apply blisters, though the condition of the kidneys may demand that they be employed with caution. The blisters should be very large, and it is best to cover them with oiled paper, not alone because its action will not be so severe, but particularly because, on withdrawing the plaster, fragments will not remain attached to the skin.

The plaster should be removed as soon as the epidermis becomes detached; when this occurs the plaster becomes wrinkled and is readily removed. In this way we will almost always avoid cantharidism.

We should also follow Cullen's recommendation, that the patients drink a great deal of water during the period of vesication, in order that the cantharidine should not be present in the urine in too large quantities at one time. Finally, there is a last precaution which cannot be recommended too strongly.

A patient suffering from cardiac dyspnœa, and upon whom a blister is applied, almost always lies in a half-sitting position in the bed, or at least the head is very much elevated. The force of gravity, accordingly, always tends to make him slide down in bed. Hence a groove is made in which the patient slides, while the pieces of dressing adherent to the linen and the pillows descend to a less extent; consequently the epidermis is torn and the derma, which is laid bare, rubs against the dry linen. The wound, which is thus in contact with more or less rough substances, becomes red and painful, and is a source of distress to the patient. In order to obviate this annoyance, the following slight modification will prove sufficient. A slightly cerated paper is applied to the skin over the wound and the surrounding healthy integument. Upon this is applied another piece of paper which is thickly cerated on the side next to the patient's body, and then the ordinary bandage. It follows that when the patient slides down in bed this will occur most readily between the two pieces of paper on account of the abundance of cerate. The first dressing thus remains applied to the body of the patient and the wound is not laid bare.

This may be a matter of slight importance with regard to the evolution and issue of the disease, but it will protect the poor patient from fresh suffering, and he will be correspondingly grateful to the physician.

Under such circumstances blisters relieve considerably and promptly, but their effect soon diminishes, and after having repeated them a number of times with decreasing success they must be abandoned.

The discussion of external measures of treatment will conclude with a few words on thoracentesis in hydrothorax.

When the fluid is scanty, and particularly when it is present on both sides, it is improper to perform thoracentesis on account of the immediate reproduction of the fluid. But when the effusion, although double, is much more abundant on one side than the other, and examination shows that one side contains more than one and a half litre, it will be useful to perform puncture with the ordinary trocar or with the capillary trocar, with or without the aspirator.

In addition to these external measures, we may employ all the stimuli mentioned in the preceding chapter, provided that they have a more or less favorable effect on the dyspnœa.

Two other remedies deserve mention, viz., lobelia inflata and oxygen.

Tincture of lobelia inflata has been recommended in America and Germany for all kinds of dyspnœa, even for that due to heart disease.

At the close of the period of cardiac cachexia, when the lungs are œdematous and effusion has occurred into the pleuræ or even into the pericardium, oxygen constitutes a valuable agent for diminishing the dyspnœa of the patients. This agent is so much more appropriate because at this period of the disease the kidneys are implicated and the dyspnœa is often uræmic in character. The oxygen not alone aids the respiratory acts, but also diminishes the quantity of albumen in the urine.

POSITION OF THE PATIENTS.

I have stated on several occasions that when the heart is hypertrophied the patient cannot rest on the left side, because the enlarged heart pulls upon its attachments and compresses the left lung, while in right decubitus the organ rests upon its base and does not drag upon its ligaments. At a later period, when dilatation of the right heart and tricuspid insufficiency have developed, the patient cannot lie upon the right side and must remain seated. The dropsy, however, soon extends to the lungs and pleura, the patient can no longer lean backward, and consequently rests on his pillows. In order to sleep he allows the head to fall upon the chest.

A last resort consists in making a wooden slab which rests on the bed and, its upper surface being inclined toward the patient, serves as a desk, upon which is placed a pillow. The patient folds his arms and sleeps bending forward, the head resting on his hands. A similar arrangement may be used if the patient is seated in a chair. This will enable him to obtain a little rest during his last days.

CHAPTER XLVIII.

TREATMENT OF SECONDARY AFFECTIONS OF THE DIGESTIVE ORGANS.

THE symptoms considered in this chapter are due to the tension produced in the hepatic and portal veins, which give rise to congestion of the liver, intestinal venous stasis, and finally ascites.

These symptoms are treated by means of purgatives. At the beginning the abdominal symptoms resulting from heart disease should be treated by saline purgatives, such as Rochelle and Epsom salts, carbonate of magnesia, Glauber's salts, Pullna, Friedrichshall waters, etc.

When the liver is congested, the saline purgatives which had been first employed in relieving œdema of the limbs and intestinal venous stasis should be abandoned on account of a certain degree of antagonism between the intestinal and cholagogue action of the purgatives.

Sulphate of soda or colocynth may serve in the transition from one form of purgative to the other; then we may resort to jalap and scammony; finally to the new purgatives recommended by Rutherford, particularly evonymin, iridin, hydrastin, etc. These will be combined profitably with the benzoate and salicylate of soda.

These purgatives may still be employed to advantage in relieving œdema of the lower limbs.

Finally, in proportion as the tension increases the ascites will become more pronounced and puncture must be performed. But this operation is merely palliative and is not indicated until the effusion, which has become considerable, pushes up the diaphragm and increases the dyspnœa. When the abdomen has been punctured it is not surprising to find it refill almost entirely at the end of two or three days. This is due to the œdema of the adjacent parts which, on account of the removal of pressure in the abdominal cavity, finds its way into the latter.

TREATMENT OF SECONDARY RENAL AFFECTIONS AND DROPSY.

MILK DIET.

Milk diet is the first measure which should be prescribed in such cases. Experience teaches that three litres of milk are sufficient for a patient while resting; if he is walking about or working the quantity should be increased by one litre. I am in the habit of giving 100 to 200 grammes every hour or two. In order to correct the insipid and somewhat pasty taste of the milk, we may add slightly alkaline and somewhat effervescent waters, such as Selters. Some patients ask for the addition of coffee or cognac ; Serré adds a raw onion. All these substances are permissible if given in small quantities.

When an exclusively milk diet is no longer given the midday meal may consist of eggs, fish, and white meat. We may then allow a light supper, diminishing the quantity of milk given during the day, but continuing the amount given at night as long as possible.

The milk diet possesses numerous advantages. It does not require mastication or salivation. A very small quantity of gastric juice will digest it, as the principal constituents, with the exception of the caseine, are already prepared to enter the chyle ducts.

Under a milk diet the pulse generally diminishes in frequency and increases in amplitude, the urine increases in quantity, and the serous effusions diminish.

Potain presented the indications for a milk diet in a communication made at a meeting of the Association for the Advancement of Science, held at Rheims in 1880.

It is not suitable in nervous affections of the heart, in nervous or hypochondriacal palpitations, or even in Basedow's disease. In cases of hypertrophy with valvular lesion, fatty degeneration of the myocardium, and asystole, milk may serve as an adjuvant. The administration of milk is best adapted to those cases in which cardiac hypertrophy is secondary to Bright's disease. The milk then diminishes the dropsy and thus relieves the cardiac functions, and although it does not diminish to any considerable extent the amount of albumen excreted in the urine, it relieves the patient in a very remarkable manner.

This is also true of cardiac hypertrophy secondary to gastro-hepatic affections. Here the cause of the cardiac disease is relieved, and the affection itself is therefore improved.

DIGITALIS.

Digitalis is undoubtedly the best of all diuretics, the infusion constituting the best preparation. The tincture is a less active diuretic than the infusion ; this also holds good of the various forms of digitaline.

It must be remembered that the diuretic action of this drug is slowly produced, and does not appear until the fourth day of its administration. On account of its slow elimination, however, its diuretic action continues for four or five days.

If the diuretic action is marked in a patient suffering from dropsy, particularly anasarca, and if the kidneys are not very much affected, the diuresis readily increases to 4 or 5 litres a day, and the dropsy rapidly diminishes. On the following days its effect diminishes in proportion to the dropsy. It would be a serious mistake to continue the administration of the digitalis in the same dose until the dropsy disappears completely, for in such cases the patient remains digitalized and is restored with difficulty. I make it a rule to continue the use of digitalis so long as the pulse does not descend below 60 per minute.

Many physicians have such a dread of the cumulative effects of the drug that they discontinue its administration as soon as diuresis is established.

SQUILLS.

The pathogenetic action of this drug when administered internally is poorly known. According to Huseman and Koenig, it acts upon frogs in the same manner as digitalis.

In man squills has a diuretic effect in cases of dropsy, and is at the same time expectorant; but if the doses are too large it irritates the stomach and gives rise to vomiting.

Other diuretics possess very little efficacy.

MASSAGE.

Massage was recommended by Galen for the relief of serous infiltrations. It consists in such cases in making friction in the direction of the venous circulation. As Gendrin has said, this constitutes a sort of accessory venous heart. I have employed it successfully very often in order to relieve œdema of the lower limbs (*vide* the ninth edition of the Traité de Thérapeutique, by Trousseau, Pidoux, and Constantin Paul, t. II., pp. 122 and 129).

PUNCTURES, SCARIFICATIONS.

When the dropsy is considerable and the kidneys are diseased we can no longer hope to cause an excretion of the fluid through the kidneys, and it may then be removed directly through solutions of continuity in the skin.

1. *Position of the Patient.*—In order that the punctures may be effective, the patient should be seated so that the fluid may flow from its own weight. It is only in certain cases that we may permit the patient to remain in a recumbent position.

2. *Site of Election of the Punctures.*—Those parts should be selected in which the integument presents the least resistance to the pressure of the fluid.

The site of election in the leg is the inner surface, 3 or 4 ctm. from the tibia and at the level of the lower end of the calf. The punctures may also be made in the upper part of the leg, at the level of the ham-strings.

In the thighs the punctures are made in the most dependent parts of the inner surface, either in the upper or lower third. Upon the abdomen the punctures are made in the suprapubic region, and in the fold formed by the skin of the hypogastric region.

3. *The Operative Procedure* consists generally in making punctures with a somewhat large needle, which is first held in a flame and then dipped in oil. It is introduced to the depth of 2 or 3 ctm. until it comes in contact with the aponeurosis, and is retained in situ for a minute. Dr. Southey has proposed a small trocar provided with a silver canula. This canula, 21 mm. in length, is provided at the end with a little enlargement measuring 3 mm. in order to arrest it externally. Affixed to the enlargement is a rubber tube of variable length which conducts the fluid into a vessel. This apparatus is useful if the patient lies on his back and is kept perfectly quiet.

After the punctures have been made the fluid escapes drop by drop in a continuous manner, and several litres may be discharged in a day. The discharge is arrested at the end of a few days, and sometimes in twenty-four hours, because the serum coagulates and closes the wound.

As a rule the patient feels relieved and demands fresh punctures. Sometimes the skin becomes red and erysipelatous, and a wandering erysipelas is produced which causes little fever, but is a fresh source of exhaustion.

CHAPTER XLIX.

TREATMENT OF THE SECONDARY AFFECTIONS OF THE NERVOUS SYSTEM AND OF EXOPHTHALMIC GOITRE.

As I have shown in a previous chapter, the secondary affections of the nervous system are almost always the result of deficient energy of the heart, which sends an insufficient supply of blood to the brain and spinal cord. In these cases the indication consists in the administration of the stimulants which have been previously described.

Care must also be taken to ascertain whether the cerebral anæmia is not due to the depressing influence of digitalis or other agents, in which event their employment must be discontinued.

When the cerebral symptoms are due to thrombosis, embolism, or apoplexy, the treatment consists:

1. In placing the patient in a semi-recumbent position.
2. In making cold applications to the head.
3. In producing intestinal derivation by the administration of drastic purgatives; aloes pills are very useful in such cases.
4. In the administration of cerebral sedatives, such as hydrate of chloral and bromide of potassium.

Treatment of Exophthalmic Goitre.

Noël Gueneau de Mussy has obtained prompt and unexpected success from the use of iodine in this affection, and he is positive that, when given in small doses, this remedy acts very effectively in relieving the palpitations and nervous agitation. Three to six drops are given three times a day in a little rice-water.

Bromide of potassium and digitalis have also been employed, but they have very little good effect.

Duboisine was employed by Dujardin-Beaumetz in two cases of exophthalmic goitre, a quarter to a half of a milligramme being administered hypodermically. The condition of the patients was notably improved; the pulsations in the tumor diminished and the palpitations disappeared almost entirely. But at the end of a week there were symptoms of intoxication and temporary delirium, so that it became necessary to discontinue the use of the drug for a time.

According to Béni-Bardé,[1] the use of cold water constitutes the best method of treatment of Basedow's disease. The method employed is the use of the movable douche; it should be general, cold, short, and somewhat forcible, particularly at the beginning of treatment. If it is poorly tolerated the tepid douche is given, or simply rubbings with water. Little by little the vigor of the douche is increased and the temperature is lowered. Complete recovery has been secured by treatment of four to eight months' duration, or even longer.

TREATMENT OF AORTIC ANEURISMS.

Aneurism of the aorta is not necessarily a fatal malady, although it proves so in the large majority of cases. Hodgson reports three observations of spontaneous recovery with autopsy, the patients having died of another disease. Thorens[2] has recently presented to the Anatomical Society a cured case of aortic aneurism.

1. *The Method of Albertini and Valsalva.*—Valsalva obtained a successful result in a case by the following plan: The patient remained in bed for forty days in the most complete possible physical and mental rest. In the beginning two venesections were performed, and then the amount of nourishment administered was diminished as much as possible. After a while the diet consisted of half a pound of pap in the morning and half as much in the evening, and no drink except water. Finally, only 125 grms. of solid food and 250 grms. of water were given daily. Purgatives were administered from time to time.

The patient emaciated in consequence, and grew so feeble that he could scarcely raise his hand from the bed. Then the amount of food was slowly increased until the patient acquired sufficient energy to rise.

Under this treatment the pulsations of the tumor disappeared completely. When the patient rose they reappeared, but they ceased in a little while and did not return again.

Lancisi, Guattani, Hodgson, Corvisart, Laennec, Sabatier, and Pelletan improved their patients by the adoption of this plan of treatment, but they did not effect real recovery. This plan has been abandoned at the present time.

2. *Dry Diet.*—In 1819 Kirby adopted a method diametrically opposed to that of Valsalva, and gave the patients substantial food and alcoholic stimulants. He was soon followed by Proudfoot and Beatty, then by Stokes, who made it a rule to give a generous diet to patients suffering from aneurism. There is in fact a means of reconciling these two apparently contradictory theories. The treating by venesection tends to diminish the amount of fluid which must be moved by the heart, and a similar

---

[1] Béni-Bardé: Traité d'Hydrothérapie, p. 846. 1874.
[2] Thorens: Bulletin de la Soc. Anat., p. 637. 1873.

result follows the diminished ingestion of water. It is therefore sufficient to regulate the fluids and solids ingested by the patient.

The bread eaten should be stale and very thoroughly baked; the amount of fruit and vegetables ingested should be very much diminished, or they should be discontinued entirely. The diet should consist of broth, eggs, meat, fish, dried fruit; the patient may take a small quantity of weak tea, 250 grms. at each meal, or an equal quantity of light wine, a little coffee or brandy. This diet gives rise to constipation, which may be relieved from time to time by the use of hydragogue purgatives, such as scammony, jalap, aloes.

3. *Iodide of Potassium.*—The first observations on the treatment of aortic aneurisms by iodide of potassium were made in 1862 by Dr. Chukerbutty, of Calcutta. He relieved his patients considerably by giving 0.60 daily. Iodide of potassium has since been prescribed in numerous cases, but the dose given is much larger; Dreschfeld has even given 6 grms. a day.

The success of this drug in such cases has given rise to the suspicion that the cases may have been syphilitic in their origin, but it proves beneficial even in the absence of this etiological factor. In fact, iodide of potassium is the best remedy which can be administered in aortic aneurism. The dose is usually from one half to two grammes daily, and it may even be raised in exceptional cases to 6 grms.

4. The local application of ice to the site of the aneurism is painful and almost always badly tolerated by the patient; this plan of treatment should be entirely discarded.

5. In case of pain or dyspnœa, local abstraction of blood by means of leeches or wet cups almost always produces great relief.

6. One of our most precious means of treatment is the subcutaneous injection of morphine for the relief of pain in this disease; the injection should not be made in the neighborhood of the tumor.

7. *Compression of the Arteries.*—This plan of treatment, which renders such good services in diseases of the arteries of the limbs, cannot be employed in thoracic aneurism except when rupture is imminent. In this event gentle compression may be resorted to in order to support the walls and prevent rupture. In a case of this kind Pelletan applied with great success a plate of lead covered with a piece of flannel. Broca relieved two patients by the application of several layers of collodion.

In a case of aortic aneurism which had perforated the sternum, Tillaux made slight pressure upon the tumor, when the patient suddenly became pale, the eyes glassy and expressionless, and the left arm and then the right were paralyzed. At the end of half an hour consciousness returned but the patient was aphasic. At the end of three hours the patient was able to get up and urinate voluntarily; the aphasia disappeared on the seventh day.[1]

---

[1] Tillaux: Bulletin Gén. de Thérapeutique, t. lxxxv., p. 231. 1873.

8. *Acupuncture.*—In 1878 I made the first series of punctures in this affection. The application was not painful; the needles, to the number of five, introduced at a distance of 1 ctm. from one another, bent like reeds in the current of blood. The needles were made of gold, and when withdrawn, at the end of a quarter of an hour, were unchanged and presented no trace of the coagulation of fibrin.

The following was the method of operation: I buried in the tumor a certain number of needles as fine as a hair at the distance of 1 ctm. from one another. In order to do this it was sufficient to take a conductor and to give with the finger a slight push to the end of the needle opposite to the point. This slight push made it pass through the skin, and then, by grasping it near the integument, the needle was gradually shoved onward. The passage through the thoracic walls is not painful, but there is pain in passing through the walls of the sac when the latter are inflamed. As a rule, I leave the needles in situ for fifteen minutes. Healt kept large needles for four days in a subclavian aneurism; at the end of that time the tumor had solidified and the pulsations had disappeared.

In this operation the sac undergoes inflammation as the result of puncture, and thickens at the point which has been pierced by the needles. In fact, at each new sitting I experienced greater resistance in traversing the sac, and finally introduction became impossible. This method therefore gives rise to an obstacle to external rupture of the sac.

9. *Electropuncture.*—Broca has shown that electropuncture has no other effect beyond that the traumatism due to the needles slightly inflames the sac and causes it to thicken. It follows that the electricity adds nothing to acupuncture.

10. *Palliative Operations.*—Tracheotomy is only justifiable if there is spasm of the larynx, since in cases of compression the obstruction is situated too low. The laryngoscope enables us to determine the difference between these conditions.

Stokes has proposed, when the aneurism passes under the clavicle, to divide the sterno-clavicular ligaments. This may give the patient a little longer lease of life, but probably for a very short time.

INDEX.

ABORTION in heart disease, 244
Abscesses of myocardium, 100
Accelerator nerves, 38
Acupuncture in the treatment of aneurism, 330
Acute endocarditis, 85
   myocarditis, 100
Adhesions, pericardial, 71
Afferent cardiac nerves, 31
Albertini's treatment of aneurism, 328
Anæmia, 47
   cardiac murmurs of, 47
   cardiac murmurs of, cause of, 53
   in articular rheumatism, 54
Anæmic murmurs, modifications of, 51
   murmurs, differential diagnosis from extra-cardiac murmurs, 52
   murmurs, differential diagnosis from endocarditis, 58
Anatomy of aorta, 134
   of heart, 5
   of pericardium, 1
Aneurism of aorta, 268
   of coronary artery, 280
   of ductus arteriosus, 285
   of pulmonary artery, 183, 280
Aneurismal bronchocele, 260
Angina pectoris, 207
Ankylosis of the heart, 71
Ante-mortem heart-clots, 266
Aorta, anatomy of, 134
   aneurismal dilatation of, 137
   dimensions of, 120
   lesion of descending portion of, 151
Aortic insufficiency, 118
   insufficiency, pathological anatomy of, 118
   insufficiency, symptoms and diagnosis of, 121
   insufficiency, traces of pulse, 122
   insufficiency, murmur, 125
   insufficiency, course, 126
   insufficiency, retardation of pulse, 126
   orifice, congenital stenosis and obliteration, 293
   stenosis, 129
   stenosis, pathological anatomy of, 129
   stenosis, symptoms of, 129
   stenosis and insufficiency, 132

Aortitis, acute, 135
   chronic, 136
Apoplexy, cerebral, 258
   retinal, 258
Arterio-capillary fibrosis, 235
Asystolic bruit, 225
Ataxia, relations to cardiac affections, 258
Atheroma of aorta, 136
   of pulmonary artery, 182
Atrophic sclerotic myocarditis, 193
Auscultation, 23

BASEDOW's disease, 260
   clinical history of, 260
   treatment of, 327
Baths in the treatment of cardiac diseases, 308
Beraud's ligament, 2
Bi-auricular stethoscope, 24
Bi-mammillary line, 8
Blisters, method of application, 321
Bright's disease as a cause of hypertrophy, 233
Bromide of potassium, 299, 313
Bronchocele, aneurismal, 260
Brown induration of the lungs, 249
Bruit de la caille, 107
   de moulin, 71
   de rappel, 107

CALCULI of the heart, 263
Cancer of the heart, 264
Cardiac distress and asystole, treatment of, 314
   revolution, duration of, 28
   dilatation, 223
   dilatation, temporary, 224
   ganglia, 34
   plexus, 32
   reflexes, 39
   rhythm, 188
   topography, 1
Cardiodynia, 209
Cardioliths, 76
Catadicrotism, 163
Centrifugal nerves of heart, origin of, 38
Cephalalgia, 256

Cerebral apoplexy, 257
 softening, 257
 emboli, 258
Cervical blowing murmur, 49
Cervico-pericardial pseudo-aponeurosis, 2
Cheyne-Stokes respiration, the, 249
 respiration, symptoms of, 249
 respiration, pathogeny of, 250
Chloral, its use in heart disease, 313
Chlorosis, 225
Chronic endocarditis, 91
 myocarditis, 193
Closure of the valves, physiology of, 95
Compressed-air baths in heart disease, 308
Compression of arteries in aneurism, 329
Concentric hypertrophy, 214
Conus pulmonalis, 156
Convallaria maialis, 301, 316
Coronary artery, aneurism of, 280
Corrigan's disease, 118
Costo-pericardial ligament, 2
Cysts of the heart, 204

DIAPHRAGM, its position during inspiration, 3
Digitalis, 295, 303, 311, 324
Dilatation of the heart, 223
 of the heart, temporary, 224
Displacements of the heart, 16
 of the heart, downward, 17
 of the heart, to the left, 18
 of the heart, to the right, 19
 of the heart, upward, 20
 of the heart, backward, 20
Dry diet in treatment of aneurism, 328
Duboisine, 327
Ductus arteriosus, persistence of, 283
 arteriosus, aneurism of, 285
Duplication of mitral murmurs, 106
Dyspnœa on exertion, 201
 of nervous asthma, 201
 of emphysema, 202
 of obesity, 202
 of chloro-anæmia, 202
 of hysteria, 202
 of flatulent dyspepsia, 203

ELECTROPUNCTURE, 330
Emboli, 79
 cerebral, 258
Emphysema, condition of heart in, 17
Endarteritis of pulmonary artery, 182
Endocardium, 77
 anatomy of, 77
 post-mortem imbibition, 78
Endocarditis, 77
 pathological anatomy of, 78
 diagnosis of, 80
 differential diagnosis of, 83
 acute or subacute, 85
 ulcerative, 87
 chronic, 91
 treatment of, 305

Epigastric pulsation in pericardial adhesions, 72
Exceutric hypertrophy, 214
Exophthalmic goitre, 260
Extracardiac murmurs, 52

False aneurisms of aorta, 137
 aneurisms of heart, 196
Fibromata of heart, 263
Fœtal heart, 281
 heart, diseases of, 282
Foramen ovale, isolated persistence of, 282
Foreign bodies in pericardium, 76
Friction murmur of pericarditis, 63
 murmur of pericarditis, characteristics of, 65
 murmur of pericarditis, differential diagnosis of, 66

GALLOPING bruit, 107
Ganglia of the heart, action of, 34
Gastro-hepatic diseases, effect on the heart of, 241
Goitre, exophthalmic, 260
Grape cure, 310
Gummata of myocardium, 194

HÆMOPERICARDIUM, 75
Hæmophilia, 244
Hæmopytsis, 247
Hallucinations, 256
Heart, relations of, 4
 anatomy of, 5
 clinical measurement of, 8
 clinical measurement of, new method, 10
 displacement of, 16
 inversion of, 16
 displacement downward, 17
 displacement to the left, 18
 displacement to the right, 19
 displacement upward, 20
 displacement backward, 20
 torsion, 20
 examination by inspection, 22
 examination by palpation, 22
 examination by percussion, 22
 examination by auscultation, 23
 normal sounds, 24
 normal trace, 26
 muscular tissue, 30
 nerves, 30
 sensory nerves, 39
 painful affections, 207
 dilatation, 223
 influence of aortic lesions on, 225
 influence of pregnancy on, 228
 acute partial aneurisms of, 196
 chronic partial aneurisms of, 196
 hypertrophy of, 210
 hypertrophy of, pathology of, 215
 hypertrophy of, symptoms of, 217

INDEX.

Heart, hypertrophy of, causes of, 222
  hypertrophy of, treatment of, 319
  weight of, 210
  capacity of, 211
  thickness of walls of, 213
Hepatic cardiopathies, 241
  congestion, 252
  venous pulse, 162
  venous pulse, cause of, 164
Hodgson's disease, 137
  symptoms and diagnosis of, 140
  murmur, 142
  compression of adjacent organs, 147
  course, 148
  etiology of, 148
Hot-air baths, 308
Hunchbacks, heart of, 238
Hydatids of myocardium, 265
  of pericardium, 76
Hydropericardium, 74
Hydrotherapy, 307
Hydrothorax, 248
Hygiene of patients suffering from heart disease, 307
Hypertrophy and dilatation secondary to renal lesions, 233
  and dilatation secondary to pulmonary affections, 237
  and dilatation secondary to gastro-hepatic disorders, 241
  and dilatation, effects on menstruation, gestation, and delivery, 244
  and dilatation, effects on the respiratory passages, 246
  and dilatation, effects on the liver, 251
  and dilatation, effects on the gastro-intestinal tract, 252
  and dilatation, extracardiac causes, 225
Hypertrophic sclerotic myocarditis, 194
Hysterical dyspnœa, 202

INFARCTIONS of the kidneys, 254
Infectious endocarditis, 87
Inferior cardiac nerve, 31
Infundibulum of pulmonary artery, 156
Inhibitory nerves of heart, 35
Insomnia, 256
Inspection of heart, 22
Insufficiency of aortic valve, 118
  of pulmonary valve, 180
Intercostal muscles, action of, 3
Intermittence of the pulse, true, 204
  of the pulse, false, 205
Intestinal congestion, 252
Intracardiac nervous system, action of, 34
Inversion of the heart, 16

JAUNDICE as a cause of mitral blowing murmurs, 241
Jugular reflux, 160
  venous pulse, 160

KIDNEY diseases, effect on the heart, 233
Kidneys, congestion of, 254
  infarctions of, 254

LARYNGEAL spasm in aortic aneurism, 276
Latent pericarditis, 62
Left ventricle, anatomy of, 91
Lipomata of the heart, 264
Lipothymia, 45
Liver, congestion of, 252
Locomotor ataxia, 258
Ludwig and Cyon, nerve of, 39

MAL de théâtre, 45
Maniacal excitement in heart disease, 256
Massage, 325
Measurement of heart, 8
  of heart, new method of, 10
Medical pulse, 43
Menorrhagia in heart disease, 244
Middle cardiac nerve, 32
Milk diet, 323
Milk spots of pericardium, 60
Mitral disease, diagnosis of, 99, 115
  disease, pulse in, 110
  disease, cause of, 116
  disease, duration of, 116
  disease, prognosis of, 117
  disease, termination of, 117
  insufficiency, murmur of, 113
  insufficiency, thrill of, 99
  murmurs, topography of, 100
  murmurs, duplication of, 106
  orifice, congenital stenosis of, 294
  stenosis, murmur of, 113
  stenosis, thrill of, 99
  valve, lesions of, 91
  valve, anatomy of, 93
  valve, physiology of, 95
  valve, pathological anatomy of, 93
Monoauricular stethoscope, 23
Morphine in heart disease, 317
Muscular tissue of heart, 30
Myocardium, diseases of, 187
  diseases of, physiology of, 187
  diseases of, abscesses of, 190
Myocarditis, acute, 189
  interstitial, 190
  chronic, 193
Myomata, 264

NERVES of heart, 30
  of heart, afferent branches of, 31
  of heart, efferent branches of, 33
  of heart, physiology of, 34
  of heart, terminations of, 34
Nerve of Ludwig and Cyon, 39
Neuralgia of the heart, 207
Nightmare in heart disease, 256
Nitrite of amyl, 317
Nutmeg liver, 252

OBLITERATION of superior vena cava, 167
Œdema of the lungs, 248

PALPITATION, 42
  treatment of, 311
Palpation of heart, 22
Paracentesis of pericardium, 304
Paradox bruit, 81, 102
Parasites of the heart, 265
Parenchymatous nephritis, 255
Partial aneurism of heart, 196
Percussion of heart, 22
Pericardium, anatomy of, 1, 59
  suspensory ligament of, 2
  tumors of, 76
  foreign bodies, 76
  puncture of, 304
Pericardial adhesions, 71
  adhesions, diagnosis of, 72
Pericarditis, 60
  symptoms of, 61
  physical signs of, 62
  complications of, 68
  differential diagnosis of, 70
  sudden death in, 70
  etiology of, 71
Phthisis in heart disease, 249
  in aortic aneurism, 276
Plexus, cardiac, 32
Pneumogastric nerve, 31
  nerve, action of, 35
Pneumopericardium, 75
Post-mortem heart-clots, 266
Pregnancy, influence on the heart, 228
Premature labor in heart disease, 244
Presystolic murmur, 104
Pulmonary artery, diseases of, 172
  artery, narrowing of orifice of, 172
  artery, narrowing of trunk and branches of, 179
  artery, endarteritis of, 182
  artery, atheroma of, 182
  artery, aneurism of, 183, 280
  artery, thrombosis and embolism of, 184
  artery, stenosis of infundibulum with incomplete development of ventricular septum and communication between both ventricles, 288
  artery, congenital stenosis with persistent communication between both ventricles, 288
    anatomical lesions of, 289
    symptoms and diagnosis of, 290
  artery, congenital stenosis with closure of foramen ovale, 285
  artery, congenital stenosis with persistence of foramen ovale and obliteration of ductus arteriosus, 286
  congestion and hemorrhage, 246
  œdema, 248
  diseases, influence on the heart, 237
  infundibulum, 156

Pulmonary valve, insufficiency of, 180
  valve, stenosis of, 172
Pulse, 29
  variations of, 41
Punctures in dropsy, 325
Purring thrill of true aneurism of the aorta, 140
Purulent endocarditis, 87

RENAL congestion, 254
Retinal apoplexy, 258
  hyperæmia, 258
  pulsations, 258
Regimen in heart disease, 309
Resonator of stethoscope, 24
Rhythm of the heart, 188
Right heart, affections of, 154
  heart, anatomy of, 155
  heart, insufficiency and stenosis of tricuspid orifice, 158
  heart, congenital endocarditis of, 292

SCARIFICATIONS in dropsy, 325
Sclerotic endocarditis, 91
Second cardiac nerve, 31
Septic endocarditis, 87
Squills, 325
Stenosis of pulmonary orifice, 172
  of pulmonary orifice, pathology of, 173
  of pulmonary orifice, diagnosis of, 175
  of trunk and branches of pulmonary artery, 179
Sterno-pericardial ligament, 2
Stethoscope, 23
Stimulants administered by inhalation, 316
  by the subcutaneous tissue, 317
Sudden death in pericarditis, 70
Superior cardiac nerve, 31
  vena cava, obliteration of, 167
Suspensory ligament of the pericardium, 2
Sympathetic cardiac nerves, 32
Syncope, 45, 209
Syphilitic myocarditis, 194
  myocarditis, vascular lesions of, 149

TEMPORARY dilatation of heart, 224
  hemiplegia, 257
Theatre sickness, 45
Third cardiac nerve, 31
Thoracic aorta, false aneurism of, 268
  diagnosis of, 269
  effects on adjacent organs, 273
  signs of, 278
Thrombi of heart, ante-mortem, 266
  of heart, post-mortem, 266
Thrombosis of superior vena cava, 167
  and embolism of pulmonary artery, 184

Torsion of heart, 30
Tolerance, period of, 200
Tracheotomy in aortic aneurism, 330
Tracheo-bronchial murmur of Hodgson's disease, 147
Transposition of arterial trunks, 294
  of venous trunks, 294
Trichinæ of myocardium, 265
Tricuspid valve, anatomy of, 155
  valve, compressor muscle of, 156
  valve, physiology of, 157
  insufficiency and stenosis, 157
  insufficiency and stenosis, symptoms of, 158
  insufficiency and stenosis, murmur of, 158
  stenosis, 166
  congenital insufficiency, 292
True aneurism of aorta, 137

Tubercles of heart, 265
Tumors of pericardium, 76

ULCERATIVE endocarditis, 87, 306
  endocarditis, morbid anatomy of, 88
  endocarditis, pathogeny of, 88
  endocarditis, diagnosis of, 89

VALSALVA's treatment of aneurism, 328
Valvular aneurisms, 196
Vapor baths, 308
Vegetations, 78
Veratrine, 298, 303
Visible pulse, 121
Vocal cords, paralysis of, in thoracic aneurism, 274

WHEY cure, 310

www.ingramcontent.com/pod-product-compliance
Lightning Source LLC
Chambersburg PA
CBHW031851220426
43663CB00006B/581